Advances in Experimental Medicine and Biology

Advances in Experimental Medicine and Biology presents multidisciplinary and dynamic findings in the broad fields of experimental medicine and biology. The wide variety in topics it presents offers readers multiple perspectives on a variety of disciplines including neuroscience, microbiology, immunology, biochemistry, biomedical engineering and cancer research. Advances in Experimental Medicine and Biology has been publishing exceptional works in the field for over 30 years and is indexed in Medline, Scopus, EMBASE, BIOSIS, Biological Abstracts, CSA, Biological Sciences and Living Resources (ASFA-1), and Biological Sciences. The series also provides scientists with up to date information on emerging topics and techniques. 2015 Impact Factor: 1.953

More information about this series at http://www.springer.com/series/5584

Junjie Xiao • Sanda Cretoiu
Editors

Exosomes in Cardiovascular Diseases

Biomarkers, Pathological and Therapeutic Effects

Springer

Editors
Junjie Xiao
Shanghai University
Shanghai
China

Sanda Cretoiu
Victor Babeş National Institute of Patho
Bucharest
Romania

ISSN 0065-2598 ISSN 2214-8019 (electronic)
Advances in Experimental Medicine and Biology
ISBN 978-981-10-4396-3 ISBN 978-981-10-4397-0 (eBook)
DOI 10.1007/978-981-10-4397-0

Library of Congress Control Number: 2017952039

Printed on acid-free paper

This Springer imprint is published by Springer Nature
The registered company is Springer Nature Singapore Pte Ltd.
The registered company address is: 152 Beach Road, #21-01/04 Gateway East, Singapore 189721, Singapore

Contents

Part I
Overview

Chapter 1
The Multifaceted Functions of Exosomes in Health and Disease: An Overview

Claudia Arenaccio and Maurizio Federico

1.1 Introduction

Cytoplasm of eukaryotic cells contains several compartments, including *trans*-Golgi network, mitochondria, peroxisomes, endoplasmic reticulum, having different functions. Transport of macromolecules among these dynamic structures is mediated by vesicles moving in a densely populated microenvironment [1, 2]. In some instances, part of these vesicles are released into the extracellular milieu. Extracellular vesicles (EVs) are part of mechanism of intercellular communication, a function of vital importance for multicellular organisms. For decades, intercellular communication has been thought to be solely regulated by cell-to-cell contact and release of soluble molecules into the extracellular space. These molecules transmit the signal through their uptake or binding to specific receptors on target cells. However, the discovery of vesicular structures released into the extracellular space containing a multitude of factors including signaling molecules, proteins and nucleic acids, has opened a new frontier in the study of signal transduction, thereby adding a new level of complexity to our understanding of cell-to-cell communication.

Body fluids (e.g., blood, urine, saliva, amniotic fluid, bronchoalveolar lavage fluid, synovial fluid, breast milk) contain various types of membrane-enclosed vesicles [3] recognizing different pathways of biogenesis. These vesicles possess different biophysical features and functions in health, e.g., protein clearance [4], immune regulation [5], cell signaling [6–8], as well as in disease, such as in infections [9–12] and cancer [13, 14]. Originally, EVs were thought to be garbage bags through which cells eject their waste. Today, it is widely accepted that EVs are key components of the intercellular communication network.

C. Arenaccio (✉) • M. Federico
National Center for Global Health, Istituto Superiore di Sanità,
Viale Regina Elena 299, 00161 Rome, Italy
e-mail: claudia.arenaccio@iss.it

J. Xiao, S. Cretoiu (eds.), *Exosomes in Cardiovascular Diseases*, Advances in Experimental Medicine and Biology 998, DOI 10.1007/978-981-10-4397-0_1

All EV subtypes are limited by a lipid bilayer membrane surrounding a specific cargo of molecules, and having different sizes and buoyant densities. The variety of vesicles released from cells as well as the methods used to isolate them led to some confusion in their nomenclature. Current research mainly considers two types of EVs according to their biogenesis, i.e., ectosomes and exosomes. The term ectosomes indicates vesicles of 150–1000 nm in diameter directly budding from plasma membrane, whereas exosomes refer to vesicles of 30–150 nm in diameter generated intracellularly by inward invagination of endosome membranes leading to formation of intraluminal vesicles (ILVs). ILVs became part of multivesicular bodies (MVBs) which are released in the extracellular space upon fusion with plasma membrane [15]. The term exosomes was coined in 1981 by Trams and coll. who described the release from various normal and neoplastic cell lines of EVs with an average diameter of 500–1000 nm accompanied by a vesicle sub-population having a diameter of ~40 nm [16]. Some years later, it had been reported that reticulocytes actively secrete vesicles of 50–100 nm in diameter through a process mediated by fusion events of multivesicular endosomes with the plasma membrane [4].

Exosomes contain DNA, RNA, proteins, lipids, and metabolites of producing cells, and are released into the extracellular space under both physiological and pathological conditions. In recent years, the effects of exosomes are being studied in several pathological conditions, such as neurodegenerative, viral, cancer, and cardiovascular diseases. Their presence in many biological fluids prompted many research groups to investigate their possible use as disease biomarkers and tools for the development of new therapies.

In this introductive chapter, an overview about biogenesis, structure, and functions of exosomes in both physiological and pathological conditions is provided. In addition, some clues about current and future utilizations of exosomes in both diagnostic and therapy are summarized.

1.2 Biogenesis of Exosomes

Cell vesiculation can be induced by multiple stimuli, including cell differentiation, activation, senescence, hypoxia, transformation, and viral infections. Among the different types of EVs, exosomes are the best characterized. They have a buoyant density of 1.10–1.14 g/mL, and display either a round spherical shape (Fig. 1.1), or a cup-like morphology depending on the transmission electron microscopy technique used [17]. Exosomes are the only known secreted cellular vesicles originating from internal membranes. They are essentially ILVs generated by inward budding of endosomal MVBs and targeted to plasma membrane [18].

The processes leading to generation of ILVs in MVBs and their fusion with plasma membrane are not completely known. Two independent pathways have been proposed (Fig. 1.2). The first one involves the endosomal sorting complex required for transport (ESCRT). This multi-molecular machinery comprises ESCRT0, ESCRTI, ESCRTII and ESCRTIII, and is recruited to the endosomal membranes

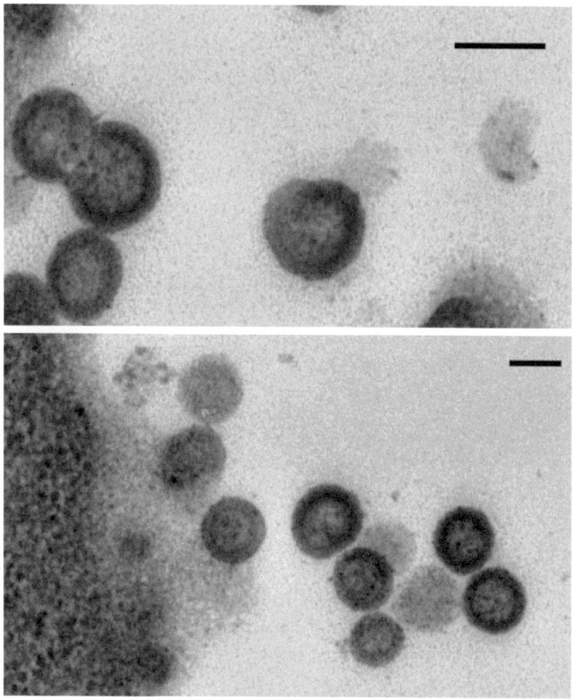

Fig. 1.1 Exosomes as detected by transmission electron microscopy upon negative staining. Bar: 0.1 μm

where ILVs are generated. In detail, ESCRT0, ESCRTI and ESCRTII recognize ubiquitinated proteins, whereas ESCRTI and ESCRTII induce, together with additional factors, the invagination of the late endosomal membrane [7, 19]. Afterwards, ESCRTIII binds ESCRTII thereby leading to the deubiquitination of cargo proteins, the promotion of vesicle abscission and, ultimately, the generation of ILVs [20].

Recently published evidences describe the existence of an ESCRT-independent pathway based on the specific lipid composition of the endosomal membranes. This hypothesis stemmed from the evidence that MVB can be formed in cells depleted of the four ESCRT components [21]. Membranes of endosomal compartments include lipid rafts comprising high quantities of sphingolipids, which are substrates for the neutral sphingomyelinase 2 (nSMase2) [22]. This enzyme converts sphingolipids to ceramide, whose accumulation induces microdomain coalescence thereby triggering ILV budding. As a matter of fact, ILV formation and exosome release are reduced when nSMase2 is inhibited [23].

Once ILVs are released into MVBs, they are either forwarded to degradation through the lysosomal pathway, or transferred to the cell periphery for the secretory pathway. Both processes are regulated by RabGTPases. While Rab7 mediates the ILV degradation through the fusion of MVBs with lysosomes, several other Rab proteins (i.e., Rab27a, Rab27b, and Rab11) are responsible, together with

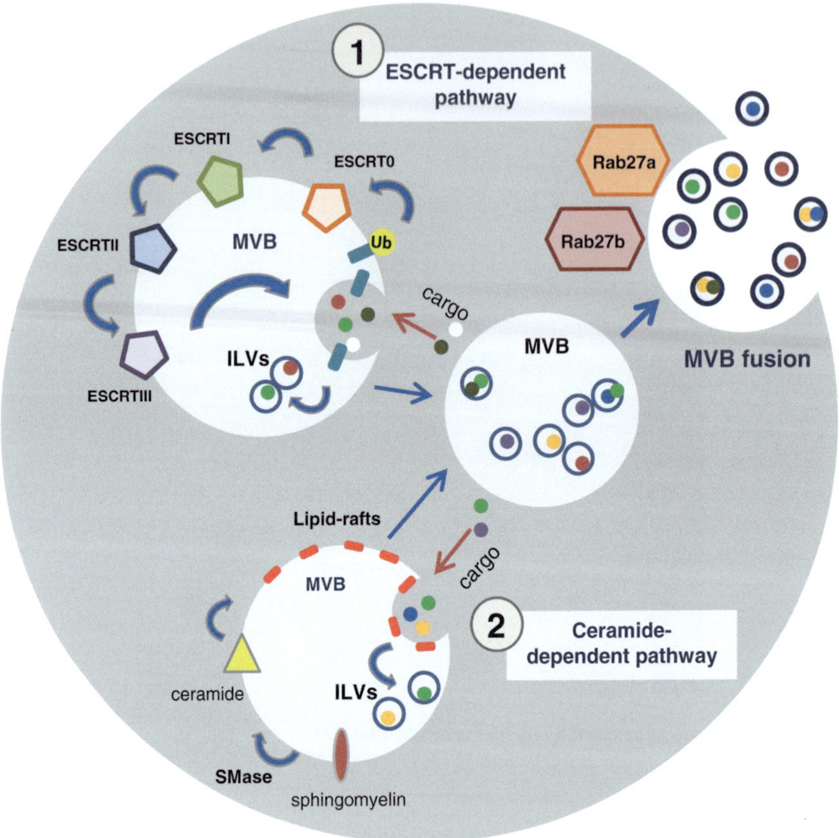

Fig. 1.2 Biogenesis and secretion of exosomes**.** Exosome biogenesis is mediated by ESCRT (*1*) and/or ceramide (*2*)-dependent pathways. In ESCRT depend pathway, sequential recruitment of ESCRT0, ESCRT I, ESCRT II to the endosomal membrane induces membrane curvature, as well as recruitment of ubiquitinated (Ub) proteins for sorting into the vesicles. Binding of ESCRTII to ESCRTIII leads to deubiquitination of cargo proteins, promotion of vesicle abscission, and thereby generation of ILVs. In ceramide dependent pathway, nSMase2 converts sphingolipids to ceramide whose accumulation leads to ILV budding. After ILV formation, MVBs fuse with plasma membrane. ILVs released into extracellular space are referred to as exosomes

tetraspanins, for intracellular MVB trafficking and secretion [22, 24]. In detail, Rab27b induces the mobilization of MVBs to the actin-rich cortex beneath the plasma membrane to which MVBs contact and fuse as consequence of the action of Rab27a. In cells defective for Rab27a functions, the fusion of MVBs with plasma membrane is induced by Rab11 in response to increased cytosolic calcium [25].

Endosome-like domains rich in exosomal proteins, lipids, and carbohydrates have been found within the plasma membrane of certain cell types [26]. These domains are supposed to be involved in either trafficking of cargo from plasma membrane back to MVBs, or in vesicle formation and budding from the plasma membrane [20]. For instance, it was shown that vesicles with the typical size of exosomes bud from the plasma membrane of both lymphocytes [27] and muscle cells [28, 29].

1.3 Structure and Composition of Exosomes

In recent years, many research groups have focused their efforts on the identification of the content of EVs and exosomes. These works led to the development of two constantly updated databases, i.e., Vesiclepedia (http://microvesicles.org), a compendium where the characteristics of all EVs are summarized [30], and ExoCarta (http://www.exocarta.org), a manually updated list of proteins, RNAs, and lipids identified in exosomes [31, 32].

Exosomes are formed by a lipid bilayer membrane enclosing a small organelle-free cytosol containing a heterogeneous array of macromolecules defined luminal cargo [33, 34]. It includes proteins, RNA, DNA, and lipid-derivatives, such as ceramide, cholesterol, phosphatidylserine and sphingolipids. Similarly to plasma membrane, the composition of lipid bilayer of these vesicles includes lipid rafts, i.e., detergent-resistant microdomains enriched in specific proteins such as flotillins and caveolins [35, 36]. At the same time, exosome membrane comprises components not present in plasma membrane of the exosome-producer cells and vice versa. For instance, exosome membranes do not contain lysobisphosphatidic acid (LBPA) [37] which, on the contrary, has been isolated from both plasma membrane and ILVs [38]. Starting from this evidence, it was hypothesized that LBPA has an a role exclusively in the formation of MVBs targeted to lysosomes [39].

Recent studies based on mass spectrometry highlighted two key aspects regarding the protein contents of exosomes. First, some exosome proteins are cell type-specific, while others are invariable part of exosomes independently from the cell of origin. Second, the exosome protein composition does not necessarily reflect the proteome of the parental cell. Typical proteins found in exosomes include those involved in MVB formation (e.g., Alix, TSG101), membrane transport and fusion (e.g., annexins, flotillins, GTPases), adhesion (e.g., integrins), tetraspanins (e.g., CD9, CD63, CD81, CD82), and antigen presentation (MHC class I and II molecules). Heat shock proteins (e.g., HSP70, HSP90) and lipid-related proteins [17, 40] were also found in exosomes. Some proteins are preferentially uploaded in exosomes, but it is still unclear how proteins are targeted specifically to exosomes. More studies are needed to unravel possible mechanisms of exosome sorting/incorporation, hence addressing the question of selectivity versus randomness. In particular, current research aimed at improving the methods of vesicle isolation, protein purification and detection will allow to identify the vesicle proteome more precisely [41].

Exosomes contain both short and long RNAs [42]. When transferred to target cells, mRNAs are translated into proteins [43, 44], and microRNAs (miRs) can silence target genes [45]. These findings have given way to study the role of exosome-delivered extracellular RNA in different biological processes, such as immune response, cancer, viral infections, formation of immunological synapse, and angiogenesis. Besides mRNAs and miRs, other RNA species have been found within exosomes, such as viral RNAs, Y-RNAs, fragments of tRNAs, small nuclear RNA, small nucleolar RNA, piwi-interacting RNAs, and long non-coding RNAs [46–48]. However, mechanisms controlling the specific loading of RNA species into exosomes are only partly known. Recently, it has been identified a short nucleotide motif regulating the sorting of RNA into exosomes through binding with the heterogeneous nuclear ribonucleoprotein (hnRNP)-A2B1, i.e., a ubiquitously expressed RNA-binding protein [49]. Interestingly, an additional short nucleotide sequence has been identified as binding motif for the hnRNP–Q-mediated delivery of miRs into exosomes released by hepatocytes [50].

Exosomes also incorporate genomic DNA through unknown mechanism. Likely, this process is mediated by the release of DNA fragments in cytoplasm during mitosis after breaking of nuclear membrane. Genomic DNA has been found in a panel of tumor cell lines of nervous and gut origin [51]. They can contain oncogenes as well as transposable elements of the genomic DNA [52, 53]. However, the function of the DNA incorporated into exosomes is still unclear, and further studies are needed to understand its possible role in physiological and pathological processes.

1.4 Interaction of Exosomes with Bystander Cells

Experimental evidences indicate that exosomes can transfer their contents into the cytoplasm of target cells. Since exosomes have been isolated from many biological fluids [34], it is likely that these vesicles can reach very distant recipient cells while protecting their cargo from enzymatic degradation during transit into the extracellular environment [54–56]. Exosome contents can be delivered through fusion of exosome lipid membrane with either plasma or endosomal membrane, in the latter case upon endocytosis. After release of luminal cargo inside the recipient cells, exosome macromolecules can induce pre- and/or post-translational alterations of gene expression [57].

Given the emerging role of exosomes in both physiological and pathological conditions, as well as their therapeutic potential, understanding the molecular processes by which they are taken up by recipient cells is relevant. Exosome uptake has been monitored mainly using both flow cytometry and confocal microscopy. These techniques allowed to analyze the dynamic localization of exosomes through the labeling with fluorescent lipid membrane dyes. Examples of such dyes include PKH67 [58], PKH26 [59], rhodamine B [60], DiI [61] and DiD [62]. The use of GFP-tagged exosomal proteins also (e.g., GFP-CD63) allowed direct vesicle visualization, confirming their rapid incorporation into recipient cells [58, 63]. The treatment of target cells with either acidic buffers [63] or trypsin [64] allowed to discriminate between internalized and surface-bound fluorescent vesicles.

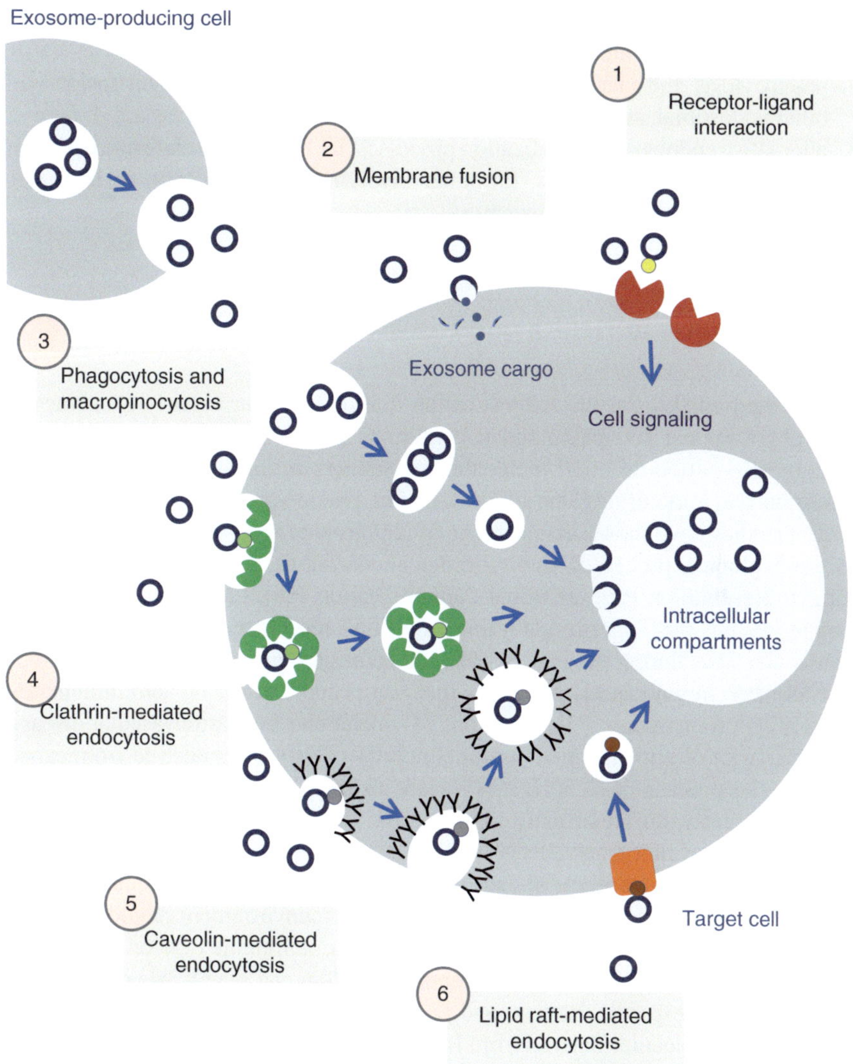

Fig. 1.3 Pathways involved in exosome uptake by target cells. (*1*) Binding of exosome membrane proteins with cellular receptor inducing intracellular signaling cascades. (*2*) Direct fusion of exosome lipid bilayer with cell plasma membrane, delivering lumen cargo in the cytosol. (*3*) Phagocytosis and macropinocytosis of exosomes. (*4*) Clathrin-mediated endocytosis. (*5*) Caveolin-mediated endocytosis. (*6*) Lipid-raft mediated endocytosis dependent on specific ligand-receptor interaction. Exosomes after endocytosis blend with the intracellular endosomal membranes changing gene expression/phenotype

Concerning the mechanisms underlying exosome internalization in target cells, four processes have been proposed (Fig. 1.3):

1. a direct interaction of exosome lipids and/or *trans*-membrane proteins with receptors on plasma membrane of the target cell, thereby inducing intracellular signaling cascades;

2. fusion events of exosome membrane with plasma membrane, delivering luminal cargo directly into the cytosol;
3. phagocytosis and macropinocytosis of exosomes, with subsequent fusion with other endosomal structures;
4. alternative endocytic internalization processes, including both clathrin-dependent and -independent pathways, the latter involving either caveolin or lipid rafts upon binding with specific receptors.

1.5 Exosomes in Health and Disease

During the past decade, the interest in the role of EVs, particularly exosomes, in both physiological and pathological conditions significantly increased. They are gaining recognition as multi-molecular messengers acting in both autocrine and paracrine ways modifying the activity and/or phenotype of recipient cells [34]. Recent studies have shown a wide range of pleiotropic functions of these vesicles in several biological processes. In physiological conditions, exosomes are involved in antigen presentation [65], neuronal communication [66], blood coagulation [67], wound healing [68], sperm maturation [69], and regulation of immune response against the fetus during pregnancy [70]. Exosomes are proposed to have an important role also in pathogenic processes including cancer [14, 71], autoimmune diseases [72], inflammation [73], infection [74], metabolic and cardiovascular diseases [75, 76]. A lot of knowledge about the functions of exosomes derived from cancer studies. The process of tumorigenesis leads to increased exosome secretion [77] and, consistently, an abnormally high number of circulating exosomes has been found in blood of cancer patients [78]. Recent studies have shown the importance of exosomes in multiple aspects of cancer biology and disease progression, including transformation [79], tumor growth [80], tumor microenvironment remodeling [14], invasion [81], cell migration [82], metastasis [83] and immune evasion [34].

Exosomes secreted by cells of the immune system can play a role of mediators of the immune response. Raposo and coll. demonstrated that exosomes carrying MHC class II molecules derived from B-lymphocytes induce antigen specific $CD4^+$ cell responses [65]. Many researchers have also investigated the effects of exosomes released by dendritic cells (DCs) on cell activation [84]. It was found that exosomes from DCs carrying peptides MHC-I and II, as well as other costimulatory factors such as CD80 and CD86, can induce $CD8^+$ and $CD4^+$ T-lymphocyte activation [5].

In recent years, the relevance of exosomes in viral infections has been strongly highlighted. Exosomes generated by virus-infected cells can incorporate viral proteins and fragments of viral RNA which can be functional in target cells. In particular, it has been shown that: (1) exosomes from HTLV-1 infected cells decrease apoptosis and cell migration [85]; (2) exosomes from HCV-infected cells contain non-enveloped virions which can enter target cells [10]; (3) exosomes from HHV-6-infected cells contain mature virions, an event which was proven to be relevant for viral spread [86]; (4) exosomes from HSV-1 infected cells produce exosomes incorporating viral RNA [87] (5) and exosomes from EBV-infected cells increase cytokine expression in target cells, consequently inducing inflammation events [88].

In the last few years, the potential functions of exosomes during HIV-1 pathogenesis began to emerge. Exosomes from HIV-1 infected cells contain viral proteins such as Nef [89] and viral RNAs [45, 57, 90]. Published evidence demonstrated that HIV-1 infected cells secrete large amount of nanovesicles in a Nef-dependent manner [91]. Exosomes containing HIV-1 Nef protein have multiple pathogenic effects such as induction of T-cell apoptosis [89] and down-modulation of cell surface molecules (i.e., MHC-I and CD4) for immune evasion [92]. The expression of HIV-1 Nef induces the release of exosomes incorporating active ADAM17 [93], i.e., a multi-domain, transmembrane, Zn^{2+}-dependent proteinase whose most studied function is processing pro-TNFα to its active form [94]. Resting $CD4^{+}$ T lymphocytes targeted by exosome-associated ADAM17 are induced to release TNFα. It initiates events leading to activation of quiescent human primary $CD4^{+}$ T lymphocytes which thereby become competent for HIV-1 expression and replication [95, 96]. A similar mechanism was found involved in the reactivation of HIV-1 latently infecting $CD4^{+}$ T lymphocytes [97].

Exosomes are also involved in a wide range of non-infective human diseases, like obesity and metabolic syndromes. Obesity and other associated metabolic disorders induce increased secretion of vesicles [75] incorporating specific RNAs and proteins as observed in both rodents and humans [98].

Many studies demonstrated the role of exosomes in neuronal protection, regeneration and development, as well as synaptic plasticity. Accordingly, exosomes have been found released by neurons [66], microglia [99], astrocytes [100], oligodendrocytes [101], and neural stem cells [102]. Furthermore, exosomes have the capacity to cross the blood barrier brain (BBB) making them excellent candidates for therapeutic interventions aimed at regenerating damaged CNS districts [103].

In the last years, results suggesting a key role of EVs in cardiovascular diseases is emerging. They are the primary cause of death worldwide, being both myocardial infarction and coronary artery diseases the most common cardiovascular disorders [104]. The progression of these disorders is very complex, involving a variety of pathological processes not fully characterized yet. Heart is composed by many cell types constituting myocardium, endocardium and epicardium communicating each other to control the organ homeostasis. EVs, notably exosomes, have a considerable importance in the transmission of intracardiac signals [105]. Usually, cardiac muscle cells are not considered as typical secretory cells, but it has been demonstrated that cardiomyocytes, endothelial cells, and fibroblasts release exosomes at least in an inducible manner. Heart exosomes have been isolated for the first time from adult rat cardiac myocytes [106]. These vesicles contain both sarcomeric and mitochondrial proteins as well as HSP-60, relevantly contributing to cardiomyocyte apoptosis [107, 108]. The content of exosomes depends on the secretion stimulus. Hypoxia is a potent stimulator of exosome release. In this condition, cardiac exosomes are enriched with both angiogenic and pro-survival factors [106]. After myocardial infarction (MI), exosomes released by damaged cardiac muscle cells contain angiogenic, mitogenic, anti-apoptosis and grow factors inducing cardiac repair. In particular, exosomal miRs have been found fundamental in healing the infarcted myocardium [109]. Exosomes play a key role also in injury protection and regeneration of cardiovascular system. Exosomes released from human cardiac progenitor cells (CPCs) after MI incorporate cardioprotective miRs, like miR146a-3p, as well as anti-apoptotic, anti-fibrotic and pro-angiogenic factors, meanwhile enhancing cardiac differentiation [110, 111].

The relevance of cardiovascular exosomes in both intramyocardial communication and cardiac repair will be detailed along the chapters of this book.

1.6 Exosomes as Biomarkers and Drug Delivery Tools

Exosomes are considered excellent diagnostic biomarkers in view of their ability to alter their cargo according to different cell stimuli. In cancer and other disorders, they can be useful to monitor disease progression as well as evaluate therapy responses. In fact, cargo of exosomes released from cancer cells can vary with the development of the disease. For example, in melanoma patients the proteome of circulating exosomes can be correlated with different clinical tumor stages. In a similar way, a distinctive set of miRs uploaded in exosomes marks the evolution of ovarian cancer [112]. In addition, seven miRNAs species derived from circulating exosomes have been identified as biomarkers of colorectal cancer [113]. Moreover, Skog and coll. demonstrated that exosomes derived from human glioblastoma cells contain proteins and miRs similar to those incorporated in exosomes detectable in the patients' sera [114].

Exosome miRs can also be used as markers of additional pathologies. High levels of exosomes associating miR-1 and miR-133a [115] as well as p53-responsive miRs (miR-192, miR-194, miR-34a) have been detected in sera from patients with acute MI [116]. The expression of miR-126 and miR-199a predicted the occurrence of cardiovascular events in patients with stable coronary artery disease [117].

Since the discovery of EVs and exosomes as messengers of biological information, their potential use as drug delivery vehicles has gained considerable scientific interest. The ability of these vesicles to overcome natural barriers, their intrinsic cell targeting properties and stability in the circulation make them excellent drug delivery vehicles. Exosomes have distinct features, like high biocompatibility, safety, and nano-sized diameters which allow efficient drug loading capacity and overcome many of the limitations of cell-based therapeutics. On this subject, Sun and coll. shown that exosomes can deliver the anti-inflammatory agent curcumin which, in this form, was found more stable than free curcumin [118].

The majority of the studies carried out using exosomes as therapeutic agents was based on their capacity to modulate immune responses with the purpose to develop cell-free cancer vaccines. For instance, DC exosomes carrying melanoma-associated antigen (MAGE)-A3 peptides were used for vaccination of patients bearing MAGE-A3^{+} advanced melanomas. The vaccination of fifteen melanoma patients led to an objective response in one patient, a minor response in another one, and disease stabilizations associated with tumor regression in two additional patients. An enhanced effector function of natural killer (NK) cells has been highlighted in eight patients [119].

Our group developed a strategy to engineer exosomes with full-length proteins which were proven effective in inducing specific, unrestricted cytotoxic T cell (CTL) immunity when injected in mice. For example, the inoculation of exosomes engineered with Human Papilloma Virus (HPV)-E7 protein induced production of HPV-E7 specific CTLs, blocked the growth of syngeneic tumor cells inoculated after immunization, and controlled the development of tumor cells inoculated before the exosome challenge [120].

Exosomes can be also engineered to incorporate mRNA and small interfering RNAs (siRNA), and were proven to be active in strategies of RNAi-based therapies. For instance, exosomes engineered with siRNA have shown to generate clear therapeutic effects in a mouse model of Parkinson's disease [121].

In sum, exosomes have been proven to have great potentialities as disease biomarkers as well as delivery tools of therapeutic/immunogenic molecules. Most recent findings pave the way for a wide use of exosomes and other EVs in both diagnosis and therapy.

References

1. Novick P, Field C, Schekman R (1980) Identification of 23 complementation groups required for post-translational events in the yeast secretory pathway. Cell 21(1):205–215
2. Balch WE, Dunphy WG, Braell WA, Rothman JE (1984) Reconstitution of the transport of protein between successive compartments of the Golgi measured by the coupled incorporation of N-acetylglucosamine. Cell 39(2 Pt 1):405–416
3. Raposo G, Stoorvogel W (2013) Extracellular vesicles: exosomes, microvesicles, and friends. J Cell Biol 200(4):373
4. Johnstone RM, Adam M, Hammond JR, Orr L, Turbide C (1987) Vesicle formation during reticulocyte maturation. Association of plasma membrane activities with released vesicles (exosomes). J Biol Chem 262(19):9412–9420
5. Bobrie A, Colombo M, Raposo G, Thery C (2011) Exosome secretion: molecular mechanisms and roles in immune responses. Traffic 12(12):1659–1668
6. Eder C (2009) Mechanisms of interleukin-1beta release. Immunobiology 214(7):543–553
7. Li J, Liu K, Liu Y, Xu Y, Zhang F, Yang H, Liu J, Pan T, Chen J, Wu M, Zhou X, Yuan Z (2013) Exosomes mediate the cell-to-cell transmission of IFN-alpha-induced antiviral activity. Nat Immunol 14(8):793–803
8. Cossetti C, Iraci N, Mercer TR, Leonardi T, Alpi E, Drago D, Alfaro-Cervello C, Saini HK, Davis MP, Schaeffer J, Vega B, Stefanini M, Zhao C, Muller W, Garcia-Verdugo JM, Mathivanan S, Bachi A, Enright AJ, Mattick JS, Pluchino S (2014) Extracellular vesicles from neural stem cells transfer IFN-gamma via Ifngr1 to activate Stat1 signaling in target cells. Mol Cell 56(2):193–204
9. Wiley RD, Gummuluru S (2006) Immature dendritic cell-derived exosomes can mediate HIV-1 trans infection. Proc Natl Acad Sci U S A 103(3):738
10. Bukong TN, Momen-Heravi F, Kodys K, Bala S, Szabo G (2014) Exosomes from hepatitis C infected patients transmit HCV infection and contain replication competent viral RNA in complex with Ago2-miR122-HSP90. PLoS Pathog 10(10):e1004424

11. Longatti A, Boyd B, Chisari FV (2015) Virion-independent transfer of replication-competent hepatitis C virus RNA between permissive cells. J Virol 89(5):2956–2961
12. Fevrier B, Vilette D, Archer F, Loew D, Faigle W, Vidal M, Laude H, Raposo G (2004) Cells release prions in association with exosomes. Proc Natl Acad Sci U S A 101(26):9683–9688
13. Al-Nedawi K, Meehan B, Rak J (2009) Microvesicles: messengers and mediators of tumor progression. Cell Cycle 8(13):2014–2018
14. Peinado H, Alečković M, Lavotshkin S, Matei I, Costa-Silva B, Moreno-Bueno G, Hergueta-Redondo M, Williams C, García-Santos G, Ghajar C (2012) Melanoma exosomes educate bone marrow progenitor cells toward a pro-metastatic phenotype through MET. Nat Med 18(6):883
15. Colombo M, Raposo G, Thery C (2014) Biogenesis, secretion, and intercellular interactions of exosomes and other extracellular vesicles. Annu Rev Cell Dev Biol 30:255–289
16. Trams EG, Lauter CJ, Salem SN Jr, Heine U (1981) Exfoliation of membrane ecto-enzymes in the form of micro-vesicles. Biochim Biophys Acta 645(1):63–70
17. Conde-Vancells J, Rodriguez-Suarez E, Embade N, Gil D, Matthiesen R, Valle M, Elortza F, Lu SC, Mato JM, Falcon-Perez JM (2008) Characterization and comprehensive proteome profiling of exosomes secreted by hepatocytes. J Proteome Res 7(12):5157–5166
18. Stoorvogel W, Kleijmeer MJ, Geuze HJ, Raposo G (2002) The biogenesis and functions of exosomes. Traffic 3(5):321–330
19. Rusten TE, Vaccari T, Stenmark H (2011) Shaping development with ESCRTs. Nat Cell Biol 14(1):38–45
20. Hurley JH, Hanson PI (2010) Membrane budding and scission by the ESCRT machinery: it's all in the neck. Nat Rev Mol Cell Biol 11(8):556–566
21. Stuffers S, Sem WC, Stenmark H, Brech A (2009) Multivesicular endosome biogenesis in the absence of ESCRTs. Traffic 10(7):925–937
22. Poteryaev D, Datta S, Ackema K, Zerial M, Spang A (2010) Identification of the switch in early-to-late endosome transition. Cell 141(3):497–508
23. Trajkovic K, Hsu C, Chiantia S, Rajendran L, Wenzel D, Wieland F, Schwille P, Brügger B, Simons M (2008) Ceramide triggers budding of exosome vesicles into multivesicular endosomes. Science 319(5867):1244–1247
24. Hsu C, Morohashi Y, Yoshimura S, Manrique-Hoyos N, Jung S, Lauterbach MA, Bakhti M, Gronborg M, Mobius W, Rhee J, Barr FA, Simons M (2010) Regulation of exosome secretion by Rab35 and its GTPase-activating proteins TBC1D10A-C. J Cell Biol 189(2):223–232
25. Ostrowski M, Carmo NB, Krumeich S, Fanget I, Raposo G, Savina A, Moita CF, Schauer K, Hume AN, Freitas RP (2009) Rab27a and Rab27b control different steps of the exosome secretion pathway. Nat Cell Biol 12(1):19–30
26. Fang Y, Wu N, Gan X, Yan W, Morrell JC, Gould SJ (2007) Higher-order oligomerization targets plasma membrane proteins and HIV gag to exosomes. PLoS Biol 5(6):e158
27. Booth AM, Fang Y, Fallon JK, Yang JM, Hildreth JE, Gould SJ (2006) Exosomes and HIV Gag bud from endosome-like domains of the T cell plasma membrane. J Cell Biol 172(6):923–935
28. Romancino DP, Paterniti G, Campos Y, De LA, Di FV, D'Azzo A, Bongiovanni A (2013) Identification and characterization of the nano-sized vesicles released by muscle cells. FEBS Lett 587(9):1379–1384
29. Forterre A, Jalabert A, Berger E, Baudet M, Chikh K, Errazuriz E, De Larichaudy J, Chanon S, Weiss-Gayet M, Hesse AM, Record M, Geloen A, Lefai E, Vidal H, Coute Y, Rome S (2014) Proteomic analysis of C2C12 myoblast and myotube exosome-like vesicles: a new paradigm for myoblast-myotube cross talk? PLoS One 9(1):e84153
30. Kalra H, Simpson RJ, Ji H, Aikawa E, Altevogt P, Askenase P, Bond VC, Borras FE, Breakefield X, Budnik V, Buzas E, Camussi G, Clayton A, Cocucci E, Falcon-Perez JM, Gabrielsson S, Gho YS, Gupta D, Harsha HC, Hendrix A, Hill AF, Inal JM, Jenster G, Kramer-Albers EM, Lim SK, Llorente A, Lotvall J, Marcilla A, Mincheva-Nilsson L, Nazarenko I, Nieuwland R, Nolte-'t Hoen EN, Pandey A, Patel T, Piper MG, Pluchino S, Prasad TS, Rajendran L,

Raposo G, Record M, Reid GE, Sanchez-Madrid F, Schiffelers RM, Siljander P, Stensballe A, Stoorvogel W, Taylor D, Thery C, Valadi H, van Balkom BW, Vazquez J, Vidal M, Wauben MH, Yanez-Mo M, Zoeller M, Mathivanan S (2012) Vesiclepedia: a compendium for extracellular vesicles with continuous community annotation. PLoS Biol 10(12):e1001450
31. Simpson RJ, Kalra H, Mathivanan S (2012) ExoCarta as a resource for exosomal research. J Extracell Vesicles 1:10.3402
32. Mathivanan S, Fahner CJ, Reid GE, Simpson RJ (2012) ExoCarta 2012: database of exosomal proteins, RNA and lipids. Nucleic Acids Res 40:1241–1244
33. Mathivanan S, Ji H, Simpson RJ (2010) Exosomes: extracellular organelles important in intercellular communication. J Proteome 73(10):1907–1920
34. Théry C, Ostrowski M, Segura E (2009) Membrane vesicles as conveyors of immune responses. Nat Rev Immunol 9(8):581
35. Logozzi M, De Milito A, Lugini L, Borghi M, Calabro L, Spada M, Perdicchio M, Marino ML, Federici C, Iessi E, Brambilla D, Venturi G, Lozupone F, Santinami M, Huber V, Maio M, Rivoltini L, Fais S (2009) High levels of exosomes expressing CD63 and caveolin-1 in plasma of melanoma patients. PLoS One 4(4):e5219
36. Saunderson SC, Dunn AC, Crocker PR, Mclellan AD (2014) CD169 mediates the capture of exosomes in spleen and lymph node. Blood 123(2):208–216
37. Laulagnier K, Motta C, Hamdi S, Roy S, Fauvelle F, Pageaux JF, Kobayashi T, Salles JP, Perret B, Bonnerot C, Record M (2004) Mast cell- and dendritic cell-derived exosomes display a specific lipid composition and an unusual membrane organization. Biochem J 380(Pt 1):161–171
38. Matsuo H, Chevallier J, Mayran N, Le Blanc I, Ferguson C, Faure J, Blanc NS, Matile S, Dubochet J, Sadoul R, Parton RG, Vilbois F, Gruenberg J (2004) Role of LBPA and Alix in multivesicular liposome formation and endosome organization. Science 303(5657):531–534
39. Brouwers JF, Aalberts M, Jansen JW, van Niel G, Wauben MH, Stout TA, Helms JB, Stoorvogel W (2013) Distinct lipid compositions of two types of human prostasomes. Proteomics 13(10–11):1660–1666
40. Subra C, Grand D, Laulagnier K, Stella A, Lambeau G, Paillasse M, De MP, Monsarrat B, Perret B, Silvente-Poirot S (2010) Exosomes account for vesicle-mediated transcellular transport of activatable phospholipases and prostaglandins. J Lipid Res 51(8):2105
41. Witwer KW, Buzás EI, Bemis LT, Bora A, Lässer C, Lötvall J, Hoen ENNT, Piper MG, Sivaraman S, Skog J (2013) Standardization of sample collection, isolation and analysis methods in extracellular vesicle research. J Extracell Vesicles 2:1–25
42. Lotvall J, Hill AF, Hochberg F, Buzas EI, Di Vizio D, Gardiner C, Gho YS, Kurochkin IV, Mathivanan S, Quesenberry P, Sahoo S, Tahara H, Wauben MH, Witwer KW, Thery C (2014) Minimal experimental requirements for definition of extracellular vesicles and their functions: a position statement from the International Society for Extracellular Vesicles. J Extracell Vesicles 3:26913
43. Mittelbrunn M, Gutierrez-Vazquez C, Villarroya-Beltri C, Gonzalez S, Sanchez-Cabo F, Gonzalez MA, Bernad A, Sanchez-Madrid F (2011) Unidirectional transfer of microRNA-loaded exosomes from T cells to antigen-presenting cells. Nat Commun 2:282
44. Ratajczak J, ., Miekus K, ., Kucia M, ., Zhang J, ., Reca R, ., Dvorak P, ., Ratajczak MZ (2006) Embryonic stem cell-derived microvesicles reprogram hematopoietic progenitors: evidence for horizontal transfer of mRNA and protein delivery. Leukemia 20 (5):847–856.
45. Pegtel DM, Kieff E (2010) Functional delivery of viral miRNAs via exosomes. Proc Natl Acad Sci U S A 107(14):6328–6333
46. Bellingham SA, Coleman BM, Hill AF (2012) Small RNA deep sequencing reveals a distinct miRNA signature released in exosomes from prion-infected neuronal cells. Nucleic Acids Res 40(21):10937–10949
47. Huang X, Yuan T, Tschannen M, Sun Z, Jacob H, Du M, Liang M, Dittmar RL, Liu Y, Liang M, Kohli M, Thibodeau SN, Boardman L, Wang L (2013) Characterization of human plasma-derived exosomal RNAs by deep sequencing. BMC Genomics 14:319

48. Vojtech L, Woo S, Hughes S, Levy C, Ballweber L, Sauteraud RP, Strobl J, Westerberg K, Gottardo R, Tewari M, Hladik F (2014) Exosomes in human semen carry a distinctive repertoire of small non-coding RNAs with potential regulatory functions. Nucleic Acids Res 42(11):7290–7304
49. Villarroya-Beltri C, Gutiérrez-Vázquez C, Sánchez-Cabo F, Pérez-Hernández D, Vázquez J, Martin-Cofreces N, Martinez-Herrera DJ, Pascual-Montano A, Mittelbrunn M, Sánchez-Madrid F (2011) Sumoylated hnRNPA2B1 controls the sorting of miRNAs into exosomes through binding to specific motifs. Nat Commun 4(1):2980
50. Santangelo L, Giurato G, Cicchini C, Montaldo C, Mancone C, Tarallo R, Battistelli C, Alonzi T, Weisz A, Tripodi M (2016) The RNA-binding protein SYNCRIP is a component of the hepatocyte exosomal machinery controlling microRNA sorting. Cell Rep 17(3):799
51. Lee TH, D'Asti E, Magnus N, Al-Nedawi K, Meehan B, Rak J (2011) Microvesicles as mediators of intercellular communication in cancer—the emerging science of cellular 'debris'. Semin Immunopathol 33(5):455–467
52. Thakur BK, Zhang H, Becker A, Matei I, Huang Y, Costasilva B, Zheng Y, Hoshino A, Brazier H, Xiang J (2014) Double-stranded DNA in exosomes: a novel biomarker in cancer detection. Cell Res 24(6):766
53. Minciacchi VR, You S, Spinelli C, Morley S, Zandian M, Aspuria PJ, Cavallini L, Ciardiello C, Reis Sobreiro M, Morello M, Kharmate G, Jang SC, Kim DK, Hosseini-Beheshti E, Tomlinson Guns E, Gleave M, Gho YS, Mathivanan S, Yang W, Freeman MR, Di Vizio D (2015) Large oncosomes contain distinct protein cargo and represent a separate functional class of tumor-derived extracellular vesicles. Oncotarget 6(13):11327–11341
54. Klibi J, Niki T, Riedel A, Pioche-Durieu C, Souquere S, Rubinstein E, Le Moulec S, Guigay J, Hirashima M, Guemira F, Adhikary D, Mautner J, Busson P (2009) Blood diffusion and Th1-suppressive effects of galectin-9-containing exosomes released by Epstein-Barr virus-infected nasopharyngeal carcinoma cells. Blood 113(9):1957–1966
55. Silverman JM, Clos J, De'Oliveira CC, Shirvani O, Fang Y, Wang C, Foster LJ, Reiner NE (2010) An exosome-based secretion pathway is responsible for protein export from Leishmania and communication with macrophages. J Cell Sci 123(Pt 6):842–852
56. Keller S, Ridinger J, Rupp AK, Janssen JW, Altevogt P (2011) Body fluid derived exosomes as a novel template for clinical diagnostics. J Transl Med 9:86
57. Valadi H, Ekström K, Bossios A, Sjöstrand M, Lee JJ, Lötvall JO (2007) Exosome-mediated transfer of mRNAs and microRNAs is a novel mechanism of genetic exchange between cells. Nat Cell Biol 9(9):654–659
58. Svensson KJ, Christianson HC, Wittrup A, Bourseauguilmain E, Lindqvist E, Svensson LM, Mörgelin M, Belting M (2013) Exosome uptake depends on ERK1/2-heat shock protein 27 signaling and lipid raft-mediated endocytosis negatively regulated by caveolin-1. J Biol Chem 288(24):17713
59. Fitzner D, Schnaars M, van Rossum D, Krishnamoorthy G, Dibaj P, Bakhti M, Regen T, Hanisch UK, Simons M (2011) Selective transfer of exosomes from oligodendrocytes to microglia by macropinocytosis. J Cell Sci 124(Pt 3):447–458
60. Montecalvo A, Larregina AT, Shufesky WJ, Stolz DB, Sullivan ML, Karlsson JM, Baty CJ, Gibson GA, Erdos G, Wang Z, Milosevic J, Tkacheva OA, Divito SJ, Jordan R, Lyons-Weiler J, Watkins SC, Morelli AE (2012) Mechanism of transfer of functional microRNAs between mouse dendritic cells via exosomes. Blood 119(3):756–766
61. Tian T, Zhu YL, Hu FH, Wang YY, Huang NP, Xiao ZD (2013) Dynamics of exosome internalization and trafficking. J Cell Physiol 228(7):1487–1495
62. Tian T, Wang Y, Wang H, Zhu Z, Xiao Z (2010) Visualizing of the cellular uptake and intracellular trafficking of exosomes by live-cell microscopy. J Cell Biochem 111(2):488–496
63. Feng D, Zhao WL, Ye YY, Bai XC, Liu RQ, Chang LF, Zhou Q, Sui SF (2010) Cellular internalization of exosomes occurs through phagocytosis. Traffic 11(5):675–687
64. Franzen CA, Simms PE, Van Huis AF, Foreman KE, Kuo PC, Gupta GN (2014) Characterization of uptake and internalization of exosomes by bladder cancer cells. Biomed Res Int 2014:619829

65. Raposo G, Nijman HW, Stoorvogel W, Liejendekker R, Harding CV, Melief CJ, Geuze HJ (1996) B lymphocytes secrete antigen-presenting vesicles. J Exp Med 183(3):1161–1172
66. Faure J, Lachenal G, Court M, Hirrlinger J, Chatellard-Causse C, Blot B, Grange J, Schoehn G, Goldberg Y, Boyer V, Kirchhoff F, Raposo G, Garin J, Sadoul R (2006) Exosomes are released by cultured cortical neurones. Mol Cell Neurosci 31(4):642–648
67. Berckmans RJ, Sturk A, van Tienen LM, Schaap MC, Nieuwland R (2011) Cell-derived vesicles exposing coagulant tissue factor in saliva. Blood 117(11):3172–3180
68. Biro E, Sturk-Maquelin KN, Vogel GM, Meuleman DG, Smit MJ, Hack CE, Sturk A, Nieuwland R (2003) Human cell-derived microparticles promote thrombus formation in vivo in a tissue factor-dependent manner. J Thromb Haemost 1(12):2561–2568
69. Sullivan R, Saez F, Girouard J, Frenette G (2005) Role of exosomes in sperm maturation during the transit along the male reproductive tract. Blood Cells Mol Dis 35(1):1–10
70. Taylor DD, Akyol S, Gercel-Taylor C (2006) Pregnancy-associated exosomes and their modulation of T cell signaling. J Immunol 176(3):1534–1542
71. Hanahan D, Weinberg RA (2011) Hallmarks of cancer: the next generation. Cell 144(5):646–674
72. Buzas EI, Gyorgy B, Nagy G, Falus A, Gay S (2014) Emerging role of extracellular vesicles in inflammatory diseases. Nat Rev Rheumatol 10(6):356–364
73. Silverman JM, Reiner NE (2011) Exosomes and other microvesicles in infection biology: organelles with unanticipated phenotypes. Cell Microbiol 13(13):1–9
74. Angela S, Meyering SS, Ben L, Sergey I, Van HML, Hakami RM, Fatah K (2015) Extracellular vesicles from infected cells: potential for direct pathogenesis. Front Microbiol 6:1132
75. Stepanian A (2013) Microparticle increase in severe obesity: not related to metabolic syndrome and unchanged after massive weight loss. Obesity 21(11):2236–2243
76. Emanueli C, Shearn AI, Angelini GD, Sahoo S (2015) Exosomes and exosomal miRNAs in cardiovascular protection and repair. Vasc Pharmacol 71:24–30
77. Rak J (2010) Microparticles in cancer. Semin Thromb Hemost 36(8):888–906
78. Kim HK, Song KS, Park YS, Kang YH, Lee YJ, Lee KR, Kim HK, Ryu KW, Bae JM, Kim S (2003) Elevated levels of circulating platelet microparticles, VEGF, IL-6 and RANTES in patients with gastric cancer: possible role of a metastasis predictor. Eur J Cancer 39(2):184–191
79. Antonyak MA, Li B, Boroughs LK, Johnson JL, Druso JE, Bryant KL, Holowka DA, Cerione RA (2011) Cancer cell-derived microvesicles induce transformation by transferring tissue transglutaminase and fibronectin to recipient cells. Proc Natl Acad Sci U S A 108(12):4852–4857
80. Taylor DD, Gerceltaylor C (2011) Exosomes/microvesicles: mediators of cancer-associated immunosuppressive microenvironments. Semin Immunopathol 33(5):441–454
81. Higginbotham JN, Demory Beckler M, Gephart JD, Franklin JL, Bogatcheva G, Kremers GJ, Piston DW, Ayers GD, McConnell RE, Tyska MJ, Coffey RJ (2011) Amphiregulin exosomes increase cancer cell invasion. Curr Biol 21(9):779–786
82. Park JE, Tan HS, Datta A, Lai RC, Zhang H, Meng W, Lim SK, Sze SK (2010) Hypoxic tumor cell modulates its microenvironment to enhance angiogenic and metastatic potential by secretion of proteins and exosomes. Mol Cell Proteomics 9(6):1085
83. Hood JL, Pan H, Lanza GM, Wickline SA, Consortium for Translational Research in Advanced Imaging and Nanomedicine (2009) Paracrine induction of endothelium by tumor exosomes. Lab Investig 89(11):1317–1328
84. Sobo-Vujanovic A, Munich S, Vujanovic NL (2014) Dendritic-cell exosomes cross-present Toll-like receptor-ligands and activate bystander dendritic cells. Cell Immunol 289(1–2):119–127
85. Jaworski E, Narayanan A, Van Duyne R, Shabbeer-Meyering S, Iordanskiy S, Saifuddin M, Das R, Afonso PV, Sampey GC, Chung M, Popratiloff A, Shrestha B, Sehgal M, Jain P, Vertes A, Mahieux R, Kashanchi F (2014) Human T-lymphotropic virus type 1-infected cells secrete exosomes that contain Tax protein. J Biol Chem 289(32):22284–22305
86. Mori Y, Koike M, Moriishi E, Kawabata A, Tang H, Oyaizu H, Uchiyama Y, Yamanishi K (2008) Human herpesvirus-6 induces MVB formation, and virus egress occurs by an exosomal release pathway. Traffic 9(10):1728–1742

87. Kalamvoki M, Du T, Roizman B (2014) Cells infected with herpes simplex virus 1 export to uninfected cells exosomes containing STING, viral mRNAs, and microRNAs. Proc Natl Acad Sci U S A 111(46):E4991–E4996
88. Ariza ME, Rivailler P, Glaser R, Chen M, Williams MV (2013) Epstein-Barr virus encoded dUTPase containing exosomes modulate innate and adaptive immune responses in human dendritic cells and peripheral blood mononuclear cells. PLoS One 8(7):e69827
89. Lenassi M, Cagney G, Liao M, Vaupotic T, Bartholomeeusen K, Cheng Y, Krogan NJ, Plemenitas A, Peterlin BM (2010) HIV Nef is secreted in exosomes and triggers apoptosis in bystander CD4+ T cells. Traffic 11(1):110–122
90. Columba Cabezas S, Federico M (2013) Sequences within RNA coding for HIV-1 Gag p17 are efficiently targeted to exosomes. Cell Microbiol 15(3):412–429
91. Muratori C, Cavallin LE, Kratzel K, Tinari A, De Milito A, Fais S, D'Aloja P, Federico M, Vullo V, Fomina A, Mesri EA, Superti F, Baur AS (2009) Massive secretion by T cells is caused by HIV Nef in infected cells and by Nef transfer to bystander cells. Cell Host Microbe 6(3):218–230
92. Gray LR, Gabuzda D, Cowley D, Ellett A, Chiavaroli L, Wesselingh SL, Churchill MJ, Gorry PR (2011) CD4 and MHC class 1 down-modulation activities of nef alleles from brain- and lymphoid tissue-derived primary HIV-1 isolates. J Neurovirol 17(1):82–91
93. Lee JH, Wittki S, Brau T, Dreyer FS, Kratzel K, Dindorf J, Johnston IC, Gross S, Kremmer E, Zeidler R, Schlotzer-Schrehardt U, Lichtenheld M, Saksela K, Harrer T, Schuler G, Federico M, Baur AS (2013) HIV Nef, paxillin, and Pak1/2 regulate activation and secretion of TACE/ADAM10 proteases. Mol Cell 49(4):668–679
94. Gooz M (2010) ADAM-17: the enzyme that does it all. Crit Rev Biochem Mol Biol 45(2):146–169
95. Arenaccio C, Chiozzini C, Columba-Cabezas S, Manfredi F, Affabris E, Baur A, Federico M (2014) Exosomes from human immunodeficiency virus type 1 (HIV-1)-infected cells license quiescent CD4+ T lymphocytes to replicate HIV-1 through a Nef- and ADAM17-dependent mechanism. J Virol 88(19):11529–11539
96. Arenaccio C, Chiozzini C, Columba-Cabezas S, Manfredi F, Federico M (2014) Cell activation and HIV-1 replication in unstimulated CD4+ T lymphocytes ingesting exosomes from cells expressing defective HIV-1. Retrovirology 11:46
97. Arenaccio C, Anticoli S, Manfredi F, Chiozzini C, Olivetta E, Federico M (2015) Latent HIV-1 is activated by exosomes from cells infected with either replication-competent or defective HIV-1. Retrovirology 12:87
98. Ferrante SC, Nadler EP, Pillai DK, Hubal MJ, Wang Z, Wang JM, Gordish-Dressman H, Koeck E, Sevilla S, Wiles AA, Freishtat RJ (2015) Adipocyte-derived exosomal miRNAs: a novel mechanism for obesity-related disease. Pediatr Res 77(3):447–454
99. Potolicchio I, Carven GJ, Xu X, Stipp C, Riese RJ, Stern LJ, Santambrogio L (2005) Proteomic analysis of microglia-derived exosomes: metabolic role of the aminopeptidase CD13 in neuropeptide catabolism. J Immunol 175(4):2237
100. Taylor AR, Robinson MB, Gifondorwa DJ, Tytell M, Milligan CE (2007) Regulation of heat shock protein 70 release in astrocytes: role of signaling kinases. Dev Neurobiol 67(13):1815–1829
101. Kramer-Albers EM, Bretz N, Tenzer S, Winterstein C, Mobius W, Berger H, Nave KA, Schild H, Trotter J (2007) Oligodendrocytes secrete exosomes containing major myelin and stress-protective proteins: trophic support for axons? Proteomics Clin Appl 1(11):1446–1461
102. Marzesco AM, Janich P, Wilsch-Brauninger M, Dubreuil V, Langenfeld K, Corbeil D, Huttner WB (2005) Release of extracellular membrane particles carrying the stem cell marker prominin-1 (CD133) from neural progenitors and other epithelial cells. J Cell Sci 118(Pt 13):2849–2858
103. Lakhal S, Wood MJ (2011) Exosome nanotechnology: an emerging paradigm shift in drug delivery: exploitation of exosome nanovesicles for systemic in vivo delivery of RNAi heralds new horizons for drug delivery across biological barriers. BioEssays 33(10):737–741
104. Roger VL, Go AS, Lloyd-Jones DM, Benjamin EJ, Berry JD, Borden WB, Bravata DM, Dai S, Ford ES, Fox CS, Fullerton HJ, Gillespie C, Hailpern SM, Heit JA, Howard VJ, Kissela

BM, Kittner SJ, Lackland DT, Lichtman JH, Lisabeth LD, Makuc DM, Marcus GM, Marelli A, Matchar DB, Moy CS, Mozaffarian D, Mussolino ME, Nichol G, Paynter NP, Soliman EZ, Sorlie PD, Sotoodehnia N, Turan TN, Virani SS, Wong ND, Woo D, Turner MB, American Heart Association Statistics Committee and Stroke Statistics Subcommittee (2012) Executive summary: heart disease and stroke statistics—2012 update: a report from the American Heart Association. Circulation 125(1):188–197
105. Ailawadi S, Wang X, Gu H, Fan GC (2015) Pathologic function and therapeutic potential of exosomes in cardiovascular disease. BBA-Biomembranes 1852(1):1–11
106. Gupta S, Knowlton AA (2007) HSP60 trafficking in adult cardiac myocytes: role of the exosomal pathway. Am J Physiol Heart Circ Physiol 292(6):H3052–H3056
107. Kim SC, Stice JP, Chen L, Jung JS, Gupta S, Wang Y, Baumgarten G, Trial J, Knowlton AA (2009) Extracellular heat shock protein 60, cardiac myocytes, and apoptosis. Circ Res 105(12):1186–1195
108. Tian J, Guo X, Liu XM, Liu L, Weng QF, Dong SJ, Knowlton AA, Yuan WJ, Lin L (2013) Extracellular HSP60 induces inflammation through activating and up-regulating TLRs in cardiomyocytes. Cardiovasc Res 98(3):391–401
109. Chistiakov DA, Orekhov AN, Bobryshev YV (2016) The role of cardiac fibroblasts in post-myocardial heart tissue repair. Exp Mol Pathol 101(2):231–240
110. Ibrahim AG, Cheng K, Marban E (2014) Exosomes as critical agents of cardiac regeneration triggered by cell therapy. Stem Cell Rep 2(5):606–619
111. Barile L, Lionetti V, Cervio E, Matteucci M, Gherghiceanu M, Popescu LM, Torre T, Siclari F, Moccetti T, Vassalli G (2014) Extracellular vesicles from human cardiac progenitor cells inhibit cardiomyocyte apoptosis and improve cardiac function after myocardial infarction. Cardiovasc Res 103(4):530–541
112. Schorey JS, Bhatnagar S (2008) Exosome function: from tumor immunology to pathogen biology. Traffic 9(6):871
113. Azmi AS, Bao B, Sarkar FH (2013) Exosomes in cancer development, metastasis, and drug resistance: a comprehensive review. Cancer Metastasis Rev 32(3–4):623–642
114. Skog J, Würdinger T, Van RS, Meijer DH, Gainche L, Sena-Esteves M, Curry WT Jr, Carter BS, Krichevsky AM, Breakefield XO (2008) Glioblastoma microvesicles transport RNA and proteins that promote tumour growth and provide diagnostic biomarkers. Nat Cell Biol 10(12):1470–1476
115. Kuwabara Y, Ono K, Horie T, Nishi H, Nagao K, Kinoshita M, Watanabe S, Baba O, Kojima Y, Shizuta S, Imai M, Tamura T, Kita T, Kimura T (2011) Increased microRNA-1 and microRNA-133a levels in serum of patients with cardiovascular disease indicate myocardial damage. Circ Cardiovasc Genet 4(4):446–454
116. Matsumoto S, Sakata Y, Suna S, Nakatani D, Usami M, Hara M, Kitamura T, Hamasaki T, Nanto S, Kawahara Y, Komuro I (2013) Circulating p53-responsive microRNAs are predictive indicators of heart failure after acute myocardial infarction. Circ Res 113(3):322–326
117. Jansen F, Yang X, Proebsting S, Hoelscher M, Przybilla D, Baumann K, Schmitz T, Dolf A, Endl E, Franklin BS, Sinning JM, Vasa-Nicotera M, Nickenig G, Werner N (2014) MicroRNA expression in circulating microvesicles predicts cardiovascular events in patients with coronary artery disease. J Am Heart Assoc 3(6):e001249
118. Sun D, Zhuang X, Xiang X, Liu Y, Zhang S, Liu C, Barnes S, Grizzle W, Miller D, Zhang HG (2010) A novel nanoparticle drug delivery system: the anti-inflammatory activity of curcumin is enhanced when encapsulated in exosomes. Mol Ther 18(9):1606–1614
119. Morse MA, Garst J, Osada T, Khan S, Hobeika A, Clay TM, Valente N, Shreeniwas R, Sutton MA, Delcayre A, Hsu DH, Le Pecq JB, Lyerly HK (2005) A phase I study of dexosome immunotherapy in patients with advanced non-small cell lung cancer. J Transl Med 3(1):9
120. Di Bonito P, Ridolfi B, Columba-Cabezas S, Giovannelli A, Chiozzini C, Manfredi F, Anticoli S, Arenaccio C, Federico M (2015) HPV-E7 delivered by engineered exosomes elicits a protective CD8(+) T cell-mediated immune response. Viruses 7(3):1079–1099
121. Wu X, Zheng T, Zhang B (2016) Exosomes in Parkinson's disease. Neurosci Bull 7:501. doi:10.1007/s12264-016-0092-z

Part II
Basic Aspects of Exosomes

Chapter 2
Exosomes: Nanocarriers of Biological Messages

Alice Conigliaro, Simona Fontana, Stefania Raimondo, and Riccardo Alessandro

2.1 Introduction

Trafficking of biological materials across cellular membranes is part of normal cell homeostasis and the cellular release of molecules through extracellular membrane vesicles (EVs) is used by cells for a number of physiological functions including cell-cell communication, cellular differentiation, immunity and inflammation [1].

As such, EVs are a heterogeneous population of vesicles, which possess different biophysical properties, dimension and have different biogenesis routes. The release of EVs is a process extremely conserved across eukaryotes and prokaryotes as well as in plants [2]. In bacteria, low eukaryotes and in plants, EVs are involved in the host-pathogen interaction and in the release of compounds such as virulence factors or toxins in the microenvironment [2]. These vesicles have been named during these

---

Alice Conigliaro, Simona Fontana and Stefania Raimondo contributed equally to this work.

A. Conigliaro
Dipartimento di Biotecnologie Cellulari ed Ematologia, Sapienza University of Rome, Rome 00185, Italy

Dipartimento di Biopatologia e Biotecnologie Mediche, University of Palermo, Palermo 90133, Italy

S. Fontana • S. Raimondo
Dipartimento di Biopatologia e Biotecnologie Mediche, University of Palermo, Palermo 90133, Italy

R. Alessandro (✉)
Dipartimento di Biopatologia e Biotecnologie Mediche, University of Palermo, Palermo 90133, Italy

Institute of Biomedicine and Molecular Immunology (IBIM), National Research Council, Palermo, Italy
e-mail: riccardo.alessandro@unipa.it

J. Xiao, S. Cretoiu (eds.), *Exosomes in Cardiovascular Diseases*, Advances in Experimental Medicine and Biology 998, DOI 10.1007/978-981-10-4397-0_2

last 15 years with a multitude of terms including ectosomes, shedding vesicles, microparticles, microvesicles and exosomes [3]. Contrary to other types of extracellular vesicles, exosomes have a smaller diameter, between 30 and 100 nm, an endocytic origin and are released into the extracellular compartment when the multivesicular bodies (MVB) fuse with the plasma membrane [4]. The secretion of exosomes occurs in a constitutive manner although cellular stress or receptor activation may modulate their secretion [5]. The mechanisms of assembly and sorting of exosomes are not well defined, but several molecules have been shown to regulate this process, such as RAB11, RAB27, RAB35 and syndecan-syntenin-ALIX [6, 7]. Moreover the ESCRT (endosomal sorting complex required for transport) member TSG101 (tumor susceptibility gene 101), and the tetraspanin CD63, which is enriched in specific plasma membrane domains involved in microvesicle budding, have been both described as involved in exosome formation [8]. The exosome molecular content is strictly related to the type and functional state of the producing cell and its characterization can be affected by the purification methods; the currently available techniques are based on ultracentrifugation, ultrafiltration and size-exclusion liquid chromatography, gradient methods, immunoaffinity, microfluidics, and polymeric precipitation [9]. Although all these methods have been successfully used for exosomes preparation they present limitations due to possible cellular contaminations caused by the well-known dynamic trafficking between the endosomal compartment and the plasma membrane, which makes the presence of the so-called exosomal markers often enriched, but not unique for exosomes. Anyhow, their biochemical content consists of not only lipids, metabolites and proteins, but also miRNA and mRNA and more recently, the presence of DNA, such as genomic and mitochondrial, has been also reported (Fig. 2.1). In the following sections, a wider and more insight of these components will be given in order also to better understand exosome function.

2.2 Lipids and Metabolites

Exosome lipid composition reflects that of the parental cell but has also some features that share with vesicles of different origin, thus representing potential EV markers. The lipid composition of exosomes has not been completely clarified but several authors agree that common lipids components of extracellular vesicles are present in the lipid bilayer and consist of sphingomyelins, cholesterol, phosphatidylserine, phosphatidylcholine, phosphatidylethanolamine, and ganglioside GM3 [10–13]. Phosphatidylserine has been found enriched in exosomal membrane compared to the cellular one, while is the opposite for phosphatidylcholine [10, 14]. The specific lipid composition allows the *in vivo* exosomes stability.

In addition, lipid raft-like domains have been found in exosome membranes suggesting a role in vesicle formation and structure [15]. Dubois and colleagues isolated lipid rafts from human prostasomes by gradient ultracentrifugation and analyzed them by mass spectrometry; they found several lipid raft associated proteins, some of

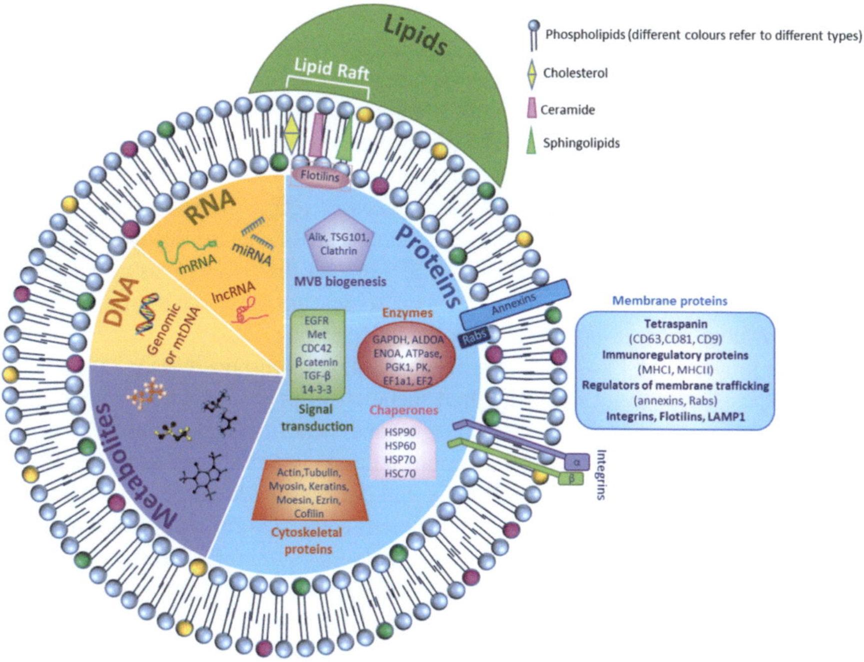

Fig. 2.1 Schematic representation of structure and content of exosomes. Exosomes are enclosed by a membrane phospholipid bilayer enriched in lipid raft-like domains and their cargo includes RNAs (mRNA, lncRNA and miRNA), DNA (both genomic and motochondrial), and several proteins like annexins, tetraspanins, Alix, TSG101, MHC molecules, Integrins, Rab proteins, cytoskeletal proteins, enzymes, and signal transduction proteins, etc. Even if exosomes contain a number of common components, their molecular composition varies depending on specific features of the originating cell

them involved in intraluminal vesicle formation [16]. One of the most extensive molecular lipidomic study was performed by Llorente in 2013 on exosome released by prostate cancer cell lines. The analysis highlighted the exclusive composition of exosome lipids compared with the parental cells; indeed exosomes were highly enriched in glycosphingolipids, sphingomyelin, cholesterol and phosphatidylserine [17]. A detailed lipidomic analysis was performed recently by Haraszti et al. that compared the lipidomic profile of microvesicles and exosomes from three different cell lines: glioblastoma, hepatocellular carcinoma and bone marrow mesenchymal stem cells. Authors found that the enrichment in glycolipids and free fatty acids characterized exosomes, whereas enrichment in ceramides and sphingomyelins characterized microvesicles. They also observed that exosomes from Huh7 and MSC cells were enriched in cardiolipins, while U87 exosomes were enriched in sphingomyelins [18].

Vesicular lipids are essential for exosome biogenesis, release and interaction with target cells, Trajkovic et al. reported that central nervous system cell-derived exosomes are enriched in ceramide, involved in the budding of exosomes into the lumen of multivesicular bodies [12]; indeed, the release of exosomes was reduced by the inhibition of the synthesis of ceramide precursor, neutral sphingomyelin.

In addition to structural function, lipids in exosomes can be considered as bioactive components involved in pathophysiological conditions. One of the first evidence came out in 2002 from Kim and colleagues that showed that tumor-derived extracellular vesicles induced an angiogenic phenotype and correlated these effects with the presence of vesicular sphingomyelin [19]. Some years later, prostaglandins (PGE), physiologically active lipid compounds, have been found into exosomes, displaying biological effects related to inflammatory processes [20]. Authors found that exosomes transported PGE2 to target cells and induced prostaglandin-dependent biological responses [20]. In addition, Xiang et al. showed that PGE2 was contained in murine mammary adenocarcinoma-derived exosomes; tumor exosomes were internalized by myeloid-derived suppressor cells and induced their expansion, thus promoting the tumor growth [21]. Exosomes also transport enzymes responsible for leukotrienes (LT) synthesis, thus influencing the lipid metabolism of recipient cells. Esser et al reported that exosomes from macrophages and dendritic cells contained active enzymes for leukotrienes biosynthesis, contributing to inflammatory responses [22].

In addition to lipids, exosomes contain and deliver a wide range of metabolites, reprogramming the metabolic state of recipient cells [23–25]. Metabolomics is a recent application of "omics" studies in the field of EVs, therefore the related literature is still scarce. The first report of extracellular vesicle metabolic content was provided by Altadill et al., that isolated extracellular vesicles from human plasma and cell cultures. Through a mass spectrometry approach, authors found that, in addition to glycerophospholipids, EVs contain metabolites such as organic acids, cyclic alcohols, steroids, phenols and amino acid conjugates, sugar and sugar conjugates [23]. A very recent study from Royo et al. indicated that EVs are "metabolically active" structures able to affect biological processes [25]. Authors performed a targeted ultra-high performance liquid chromatography–mass spectrometry (UHPLC–MS) metabolomic analysis of serum additionated with extracellular vesicles from hepatocytes treated with different liver damage-inducing drugs. They observed that the metabolic profile of the serum is significantly affected by presence of hepatic EVs [25].

A very challenging aspect, not still fully resolved, to deeply define the exosomes molecular cargo, concerns the effective and selective isolation of these nanovesicles. The plurality of developed protocols determinates qualitative and quantitative variability of the obtained exosomes, which considerably affects the results of downstream analyses and makes difficult to compare, reproduce and interpret data obtained by different research groups.

2.3 Proteins

Recent improvements in proteomic technologies, are allowing to perform a high-level qualitative and quantitative characterization of exosomal proteins, providing new information indispensable for functional characterization and clinical use of exosomes.

Data from multiple proteomics studies have clearly demonstrated that exosome proteins can be sorted into two main groups. The first one is a conserved set of proteins, considered as "exosomal markers", largely associated to exosome biogenesis and irrespective of their cell origin. The second one is formed by proteins defining a unique exosome signature specifically related to the producing cell and that determinate their role in several biological phenomena, such as disease progression, and may represent a source of biomarkers for disease diagnosis, prognosis and response to treatment.

The most common exosome proteins, belonging to the functional classes reported in Table 2.1, are cataloged in the ExoCarta website (http://www.exocarta.org/), a primary resource for high-quality exosomal datasets accessible also from Vesiclepedia (http://www.microvesicles.org) a manually curated compendium that contains molecular data identified in all classes of EVs, including apoptotic bodies, exosomes, large dense core vesicles, microparticles, and shedding microvesicles [26–28].

As more proteome studies are performed, it is becoming even more apparent that beyond the set of conserved proteins, exosomes contain proteins mirroring their cell origin. For example, numerous proteomic studies have demonstrated that tumor-derived exosomes (TDEs) express a discrete set of proteins specifically related to the tumor phenotype and involved in cell proliferation, antigen presentation, signal transduction, migration, invasion and angiogenesis, supporting the hypothesis that exosomes may play a crucial role in modulating tumor progression and preparing the metastatic niche [26]. Recently, it was demonstrated that exosomal integrins (ITGs) direct organ-specific colonization by interacting with receptors on target cells in a tissue-specific fashion, preparing pre-metastatic niche formation. In particular, the authors showed that the exosomal integrins $\alpha6\beta4$ and $\alpha6\beta1$ were associated with

Table 2.1 Most common exosome proteins and related functional classes [26]

Functional Class	Proteins
Membrane adhesion proteins	Integrins
Components of the ESCRT machinery	alix, TSG101, vps-28, vps-4B, ubiquitin ubiquitin-like modifier-activating enzyme,
Membrane transport/trafficking	Annexins, Rab protein family
Cytoskeletal components	Actin, cytokeratins, ezrin, tubulin, myosin
Lysosomal markers	Lysosome membrane protein 2, cathepsin-D, CD63, LAMP-1/2
Antigen presentation proteins	HLA class I and II/peptide complexes
Metabolic enzymes	GAPDH, pyruvate, enolase alpha
Heat shock proteins	Hsc70, Hsp70, Hsp90
Kinases	LYN, MINK1, and MAP4K4
Tetraspanins	CD9, CD81, CD82, tetraspanin-8
Proteases	ADAM10, DPEP1, ST14
Transporters	ATP7A, ATP7B, MRP2, SLC1A4, SLC16A1, CLIC1
Receptors	CD46, CD55, NOTCH1

lung metastasis, while exosomal integrin αvβ5 was linked to liver metastasis. This study also demonstrated that TDEs are sufficient to redirect metastasis of tumor cells that normally lack the capacity to metastasize to a specific organ [29].

Several studies have evidenced that treatment of cancer cells with anti-tumor compounds induces alterations of the basal protein composition of TDEs that reverts their pro-tumor actions. Taverna et al. reported that after treatment with curcumin, a plant-derived compound well known for its anticancer effects, Chronic Myeloid Leukemia cells released exosomes (curcumin/CML-exos) with significant differences in their protein composition in comparison to exosomes released by no-treated cells (CML-exos). In particular, it was reported that curcumin/CML-exos were significantly depleted in pro-angiogenic proteins and enriched in proteins with anti-angiogenic activity and this reflected the loss of CML-exosome's ability to promote the angiogenic phenotype and to alter the endothelial barrier organization [30].

Evidence accumulated in the last years has demonstrated that exosomes derived from mesenchymal stem cells (MSC) are able to mediate several of the functions traditionally associated with canonical secretory proteins such as growth factors of the MSC secretome [31–33]. In a recent proteomic study has been reported that MSC derived exosomes contain a robust profile of angiogenic paracrine effectors, with a potential use for the treatment of ischemic tissue-related diseases. Among them platelet derived growth factor, epidermal growth factor, fibroblast growth factor, and most notably nuclear factor-kappaB (NFkB) signaling pathway proteins were identified [34].

Fen et al. also reported that heat shock induces Sca-1^+ stem cells to release exosomes enriched in the heat shock transcription factor 1 (HSF1) that directs ischemic cardiomyocytes toward a pro-survival phenotype by epigenetic repression of miR-34a [35].

As it was mentioned above, another interesting aspect, correlated with the peculiar protein exosome protein signature, is the possibility to use exosomes as source of biomarkers for several pathologies. Several data published in the last years strongly support the effective clinical impact of exosomes that as multimolecular aggregates also offer the unique opportunity to identify combination of different biomarkers.

Recently, by using mass spectrometry analyses, a cell surface proteoglycan, glypican-1 (GPC1), was found specifically enriched on cancer-cell-derived exosomes. GPC1-positive circulating exosomes were detected in the serum of patients with pancreatic cancer with absolute specificity and sensitivity, discriminating healthy subjects and patients with a benign pancreatic disease from patients with pancreatic cancer [36].

In another interesting paper, the role of exosomal-survivin as a diagnostic and/or prognostic marker in early breast cancer patients was proposed. The authors found that the levels of this protein (and of its splice variant) were significantly higher in all exosome serum samples of women affected by breast cancer compared to controls. Moreover, the variable expression of Survivin-2B levels correlated with cancer stages [37]. Interestingly, in addition to plasma/serum, other biofluids, such as urine, may represent valuable sources of exosomal biomarkers.

The potential use of urinary exosomes was overall reported for the diagnosis and clinical management of urogenital cancers, such as bladder and prostate cancers.

A comparative study of protein profiling by mass spectrometry-based proteomics highlighted the expression of ITGA3 and ITGB1 (proteins involved in migration/invasion processes) on exosomes released by prostate cancer cell lines (LNCaP and PC3T). Afterwards, these proteins were found more abundant in urine exosomes of metastatic patients compared to benign prostate hyperplasia or prostate cancer (PCa), suggesting the potential use of urine exosomes for identification of patients with metastatic PCa in a non-invasive manner [38].

Beside their role in tumor biology, there is increasing evidence that exosomes can play a role in regulating tissue-specific and/or whole-body metabolism via the delivery of different exosomal cargo molecules to adjacent and remote tissues. Although there are still limited data on the clinical applicability of exosomal biomarkers for prediction of metabolic syndromes progression [39], some promising studies have supported the hypothesis that proteins of circulating or urinary exosomes may be potentially associated to obesity and metabolic complications in patients with manifest cardiovascular disease [40] and to renal complications of diabetes [41–43].

Diabetic nephropathy (DN) is a major complication of diabetes mellitus and the most frequent cause of end-stage renal disease. Current markers for DN diagnosis (i.e. creatinine and urinary albumin excretion) have proven limitations and the discovery of new candidate markers for better disease staging, outcome prediction, and monitoring of the response to clinical intervention is needed. Recent proteomic studies of urinary exosomes have provided promising indications for the potential use of these nanovesicles as source of DN biomarkers [41, 42]. For example, a label-free quantitative comparison of DN urinary exosomes vs control group allowed to highlight a panel of three proteins (AMBP, MLL3 and VDAC1) which change in DN and which could be used to develop new diagnostic approaches for monitoring the disease onset and progression [42]. Interestingly, in another study it was underlined that urinary exosomes better reflect protein changes occurred in kidney of diabetic patients. The authors reported that, in agreement with the alterations found in the kidney tissue of diabetic patients, gelatinase and ceruloplasmin were found respectively with decreased activity and increased levels in the urinary exosomes of patients with DN. In contrast, the levels of these two proteins in whole urine were highly variable and did not correlate with levels in the diabetic kidney tissue [43].

Current studies on exosomes are providing promising indications about their effective use in clinical settings and deserve further advance in order to develop new and valid non-invasive diagnostic and prognostic tools in multiple diseases.

2.4 Nucleic Acids

2.4.1 mRNA

The exciting intuition of exosomes as mediators in cell-cell communication drove the efforts of two independent groups that, between 2007 and 2008, proved the horizontal transfer of RNA molecules via exosomes. In these experiments, vesicles obtained from mast cells and glioblastoma cells, were shown to carry and deliver functional messenger RNAs (mRNAs) [44, 45]. The analysis, done by Lotvall's group, identified approximately 1300 different mRNA transcripts in exosomes from mast cells; moreover, the isolated poly adenylate mRNAs were demonstrated to be stable and functional in recipient cells, thus confirming exosomes as mediator of horizontal transfer of genetic information. The formal evidence that mRNAs in exosomes are biologically functional came from subsequent studies, demonstrating that an mRNA, coding for a luciferase reporter gene, could be transferred via exosomes leading, in the recipient cell, to luciferase activity. Interestingly, the enzymatic activity observed in the target cells, was dependent on the amount of exosomes used to treat them [44, 46]. Thereafter, several groups identified mRNAs in exosomes released from different cell types focusing on a better understanding of the mechanism underlying the phenotypic changes induced by these RNAs. Today, exosomes are widely accepted as mediators of cell communication, participating in the maintenance of tissue homeostasis and contributing to modulate cellular microenvironment; however, the majority of scientific publications have described their role into initiation and/or promotion of pathological conditions. Exosome mRNAs, released by cancer cells, have been found to promote tumor growth and/or tumor progression, as well as to drastically transform a normal cell, as recently demonstrated by Gutkin et al. Their experiments, in fact, indicated that hTERT mRNA, transported by tumor exosomes in normal fibroblasts, can be processed and translated, thus changing fibroblast properties such as proliferation rate, senescence and resistance to apoptosis [47].

Deeper analyses of exosomes content, in term of nucleic acids, revealed that the majority of normal and cancer cells load in exosomes different RNAs species with a size distribution between 25 and 700 nucleotides (nt). Small size RNAs (<700 nt) were found in human plasma [48], saliva and breast milk exosomes [49, 50]; moreover, vesicles released by human mesenchymal stem cells [51] and human tracheobronchial epithelial cells [52] were found to contain smaller RNA species (< 500 nt in length). Considering the pleiotropic properties attributed in the last years to small RNAs, great efforts have been focused on the identification of these small exosomal RNAs in order to (1) attribute them biological functions in cell-cell communication and to (2) use them as novel biomarkers.

2.4.2 Non Coding RNA

The human genome project drastically changed the paradigm that most of genetic information encodes for proteins. Recent evidences indicated that only 2% of nuclear DNA is transcribed in messenger RNA while, the majority of the genome of mammals, as well as of other complex organisms, is transcribed into non coding RNAs, most of which are alternatively spliced and/or processed into smaller products. To date, a wide repertoire of biological functions have been identified for non coding RNAs with a predominant role in gene regulation, as predicted 45 years ago by Jacob and Monod [53].

2.4.3 Micro RNAs

Micro RNAs (miRNAs)are a class of noncoding RNAs, long 17–24 nucleotides, well conserved during evolution; these miRNAs produced in the nucleus, once fully processed in the cytoplasm acquire a mature conformation that allow them to regulate gene expression. Mature miRNAs mediate post-transcriptional gene silencing by targeting mRNAs, through a "Watsonian complementarity" binding, thus inducing dsRNA cleavage and translational repression [54]. MiRNAs have been found expressed in all tissues and their aberrant levels have been reported in numerous diseases.

First Valadi et al. identified the presence of a large amount of small RNAs in exosomes thus suggesting and proving that exosomes contain miRNAs [45]. Meanwhile, the first evidence that exosomal miRNA play a functional role in target cells, came from the experiments of Montecalvo et al. that properly demonstrated how miRNAs, once transported by exosomes, repress their specific mRNA targets in dendritic cells [55]. Moreover, a recent publication, demonstrated that exosomes, together with miRNAs and pre-miRNAs, can transport the whole molecular machinery required to induce a miRNA mediated silencing so to "guarantee" their activity once reached the destination [56].

Several studies highlighted that the miRNA content of exosomes did not reflect the miRNA repertoire of the cells of origin and that, some miRNAs, independent by the cellular amount, are selectively exported or retained within the cells [57–59]. These data suggest the ability of cells to specifically load RNA species into the lumen of exosomes.

To date, mechanisms controlling the specific loading of miRNAs in exosomes are still largely unknown even if different pathways and molecules that regulate miRNAs sorting in different cell types and tissues have been described. In silico analysis of over-represented motifs and experiments of directed mutagenesis, allowed the identification of a specific EXO-motif (GGAG) that controls the loading of some miRNAs into exosomes. This conserved motif in fact, can be recognized by the ubiquitous heterogeneous nuclear ribonucleoprotein A2B1 (hnRNPA2B1) that binds miRNAs controlling their loading into exosomes [60]. However, the EXO-motif characterizes only a sub set of miRNAs and it is supposed that other proteins and/or mechanisms are required thus suggesting cellular or tissue specificity. For example, it has been recently demonstrated that the RNA binding protein SYNCRIP is a crucial component of exosomal miRNA sorting machinery in hepatocytes [61].

Overall, it is now largely accepted that miRNA secretion by exosomes is not only a mechanism whereby cells rapidly dispose miRNAs in excess to maintain mRNA homeostasis [62] but also a strategy for the horizontal transfer of RNA. Small-RNAs, once packaged and surrounded by a lipid bilayer are physically protected from enzymatic degradation, and then free to travel in extracellular space including biologic fluids [48]. Once reached the target cells that, in theory, could be everywhere in the body, exosome miRNA can affect cell phenotype resulting in altered cellular function and subsequently promoting pathological conditions. Data obtained from miRNA registry [63] indicated the existence of more than 2000 human miRNA that can influence gene regulation of essential biological pathways such as cellular development [64], proliferation [65], apoptosis [66], cellular signaling and disease progression [67, 68]. In cancer, a variety of processes has been found affected by exosomal miRNA [3, 69] e.g. exosomes released by tumor cells promote angiogenesis by transport of miR9 in endothelial cells [70, 71], exosome-mediated transfer of cancer-secreted miR-105 efficiently destroys tight junctions and the integrity of natural barriers thus favoring metastasis. Also the oncomiR, miR-21, has been found secreted in plasma exosomes from patients affected by different cancer types, and its presence has been positively correlated with tumor progression and aggressiveness [30]. MiRNA exosomes have been found involved also in several pathologies including cardio vascular diseases. MiR-21-3p contained in fibroblast exosomes led to the induction of cardiomyocyte hypertrophy [72] while endothelial exosomes induce in cardiomyocytes an increase of miR-146a levels leading to impaired metabolic activity and contractile function [73]. Moreover, dysregulation of intracellular miRNA expression has been linked to many clinically relevant cardiovascular conditions [74, 75].

Several evidences demonstrated that is possible to use microRNA as biomarkers for diagnosis and therapeutic monitoring of diseases such as cancer, neurodegenerative disorders, heart disease and infection; furthermore these miRNAs are protected inside vesicles and can be easily obtained from biological fluids. As described, exosomes have been found almost in every body fluid including urine, plasma, breast milk, broncho-alveolar lavage (BAL) fluid, saliva, seminal fluid, amniotic liquid, ascites, synovial fluid, breast milk, and cerebrospinal fluid (CSF) [48, 76]. Exosomes provide a consistent source of miRNA for disease biomarker detection and can become the fingerprint of pathology. New technologies are emerging to purify exosomes from biologic fluids in order to develop more accurate and less invasive strategies for diagnosis and prognosis. Today, once isolated the exosomes, the quantity and the quality of miRNAs can be easily estimated by various methods, such as Microarray, Hybridization, Deep-sequencing, qRT-PCR, and microbeads analysis. Shao H et al., recently, described the definition a sensitive and comprehensive microfluidic platform termed immuno-magnetic exosome RNA (iMER) analysis that, integrating immuno-magnetic selection, RNA collection and real-time PCR into a single microfluidic chip format, allows the enrichment of cancer-specific exosomes from blood and subsequent, on-chip analysis, of their RNA contents. The authors, obtaining simultaneously exosome enrichment and RNA isolation, proved that exosomal mRNA profiles could be used as predictor to treatment response in patients with glioblastoma [77].

2.4.4 Long Non Coding RNAs

Even if less abundant than small RNAs, long non coding RNAs (lncRNAs) have been found in exosomes released by different cell types.

To date, several lncRNAs have been associated to carcinogenesis and cancer progression. However, a number of reports have also demonstrated a crucial role for lncRNAs in development and disease including cardiac development and cardiac disorders [78].

LncRNAs are non-protein-coding RNAs, longer then >200 nucleotides. Located in different genomic regions they can be classified as long intergenic non-coding RNA (lincRNA), antisense lncRNA, intronic RNA and circular RNAs. They are transcribed by RNA polymerase II and exist as single-exon or multi-exon transcripts and often undergo 5′ capping and splicing [79]. LncRNAs are now recognized as important regulatory elements and exert this role by different molecular mechanisms; they are actively involved at epigenetic, transcriptional and post-transcriptional level and interact with proteins, RNAs and DNA. LncRNAs can work as scaffolds to bring together different subunits of protein complexes and maintain them in close proximity [80–82] while, intergenic RNA as lincRNA-p21 can control protein translation [83], regulate alternative splicing as for MALAT1 [84] or stabilize mRNA. Very interesting is the role of lncRNA in genetic reprogramming; HOTAIR [85], Braveheart [86] and FENDRR [87] are only few examples of lncRNAs able to recruit epigenetic modifiers at specific loci in the DNA thus regulating gene activity. Moreover, if a single miRNA can silence different mRNAs at the same way, through complementary regions, a single lncRNAs can be able to "sponge" different miRNAs from cell cytoplasm inducing their degradation, e.g. CARL, CHFR, HULC, linc-MD1, H19 [88].

Therefore, considering the multiple roles that a single lncRNA could play inside the cells is possible to appreciate as each single lncRNA, carried by exosomes, might completely modify the phenotype of receiving cells.

It was demonstrated that liver cancer stem cells release and deliver exosomes containing H19 to endothelial cells, thus promoting neo-angiogenesis [89]. Other lncRNAs have been found in exosomes from HCC [90], colon cancer cells [91], glioblastoma cells [92] and have been often recognized as promoters of tumor progression once delivered in target cells. The recent study conducted by Ahadi et al., demonstrated that a comparable abundance of lncRNAs can be found both in healthy and prostate cancer exosomes while significative differences have been discovered in their sequences [93]. They in fact, found an increase of lncRNAs carrying miRNA seed regions in prostate cancer exosomes, thus indicating that the packaging of lncRNA, as it occurs for miRNAs, is a physiological process that is significantly affected by the cellular condition. The investigation of molecular mechanisms driving lncRNA packaging could shed light about their role inside or outside the cells but, to date, few information are known about this process.

Recently collected data indicated exosomal lncRNAs as new prognostic markers in tumour as recently suggested by Zhang et al. [94] that demonstrated the association of exosomal lncRNAs HOTAIR, MALAT1, and MEG3 with cervical cancer. Moreover, urinary exosomes from bladder cancer patients have been found enriched in lncRNA HOTAIR [95].

However, even if several groups published exosomal lncRNA arrays, more stringent analyses are required to formally proof that the intact long sequences, and not a part of them, are packaged and carried by exosomes.

2.4.5 Genomic and Mitochondrial DNA

Circulating DNA (cDNA)is a heterogeneous population of molecules whose origin depends on processes that are still not fully understood. The stability of cDNA suggests that it is protected and encapsulated in specific structures. Indeed, recently it has been demonstrated that cDNA is predominantly released in the extracellular space by exocytosis mediated by extracellular vesicles, including exosomes.

The discovery of cDNA in the blood was described for the first time in 1948 by Mandel and Matais [96] that purified the nucleic acid from human plasma. For several years studies on circulating DNA were focused on autoimmune diseases where is possible to find large amount of cDNA. Some years later, in 1977, elevated levels of cDNA were found in the plasma of cancer patients, correlating with tumor progression [97].

The association between the level and composition of cDNA and morbidity opens to the possibility of considering cDNA as a rich source of biomarkers. Qualitative and quantitative alteration cDNA reflects disease features, thus providing information on disease development, progression and response to therapy. For these reasons cDNA is emerging as a non-invasive tool for patient's stratification and disease monitoring.

Genomic single or double-stranded DNA and mitochondrial DNA has been recently detected in extracellular vesicles from different cell types [98–102]. It is now well established that nucleic acids are selectively packaged in extracellular vesicles [3, 102, 103], therefore the presence of DNA indicates a specialized function in cell to cell communication. Indeed, recent data suggest that EV-associated DNA (EV-DNA) exerts its biological function in physiological and pathological conditions, thus influencing immunomodulation, cancer and cardiovascular diseases.

Accumulating evidences demonstrated that cDNA is contained in microvesicles released by prostate acinar epithelial cells (prostasomes) [101, 104, 105]. Ronquist and colleagues provided the first description in 2009; they purified prostasomes from seminal fluid and from PC-3-cells and found human chromosomal DNA, fragments with a length from over 12 kb and lower, protected from enzymatic attack because encapsulated in vesicles [104]. The same group evaluated the specific role of prostasomal DNA few years later. In 2011 authors provided evidences of the transfer of DNA into sperm from DNA-stained prostasomes, thus demonstrating that the transfer of chromosomal DNA fragments is conceivable from human prostasomes to human sperm [101].

In 2012 Waldenstrom and colleagues reported that also cardiomyocytes transfer genetic information (DNA and RNA) to fibroblasts though the release of microvesicles/exosomes [106].

Recently, Fischer et al. observed high-molecular DNA in association with EV purified from human bone marrow mesenchymal stromal cells (BM-hMSC). They also demonstrated that EV-DNA was transferred to recipient cells and propagated into the host genome, providing the first evidences of horizontal DNA transfer in eukaryotic cells [107].

Circulating DNA fragments contain genetic alterations that reflect those of tumor tissues, including point mutations, rearrangements and amplifications [108]. Several studies highlight the presence of cDNA in tumor-derived extracellular vesicles [98, 109, 110] thus opening to the possibility to consider EV-DNA as marker of disease progression.

In 2011 Balaj and colleague analyzed the nucleic acid contents of glioblastoma and medulloblastoma (which had genomic amplification and high expression levels of the c-Myc oncogene) extracellular vesicles, providing one of the first evidence of the presence of amplified genomic DNA (gDNA), cDNA and retrotransposon in tumor vesicles. In particular authors found elevated levels of c-Myc sequences in microvesicles from medulloblastoma cell lines compared with those found in fibroblasts and tumor cells with diploid c-Myc copy numbers. These data, together with the presence of specific retrotransposon sequences, strongly support the concept that genetic material contained in extracellular vesicles represents useful biomarkers [98].

In 2013 Cai et al. provided evidences of pathophysiological significance of the transfer of EV genomic DNA. Authors first showed substantial amounts of double strand DNA, mostly around 17 kb, in EVs derived from human plasma and supernatants of vascular smooth muscle cells, in addition they demonstrated that EV-DNA is shuttled to target cells, altering their cellular functions. Furthermore, a clear demonstration of the transmission of oncogenic material through extracellular vesicles was provided by the incubation of exosomes, derived from chronic myeloid leukemia cell carrying the BCR/ABL hybrid gene, with neutrophils; authors demonstrated that this incubation led to the transfer of the oncogene to normal cells [111]. One year later the same authors showed that the injection of chronic myeloid leukemia EVs in NOD/SCID mice caused de-novo transcription of BCR/ABL mRNA and protein synthesis, providing *in vivo* evidences of the functional significance of transferred genetic material by EVs [112].

Lyden group, by using two different specific approaches to detect EV-DNA on broad cancer models, found that the majority of exosome DNA is double-stranded, in addition, they found that the amount of EV-DNA is higher in tumor exosomes compared to normal ones. Whole genome sequencing allows a detailed characterization that reveals that the exosome DNA represents the whole genome, including tumor-specific genetic alterations [110].

Rak's group found that H-ras-transformed rat intestinal epithelial cells release more vesicles than the normal counterparts; the vesicles can be internalized by target cells and contain double-stranded genomic DNA, representing the entire genome and including H-ras sequences [109]. Additional studies on the presence of double-stranded genomic DNA in cancer cells comes from Kahlert et al. that found genomic DNA, spanning all chromosomes, in exosomes from pancreatic cancer cells and from serum of patients with pancreatic ductal adenocarcinoma. In addition they observed the presence of DNA with mutated KRAS and p53, confirming the applicability of EV-DNA as diagnostic tool [113].

More recently, Lazaro-Ibanez et al. showed that genomic double-stranded DNA is present in different EV subpopulations (apoptotic bodies, microvesicles and exosomes) from malignant prostate cancer cell lines and from plasma of both PCa patients and healthy donors. Authors found part of both TP53 and PTEN genes, involved in cancer initiation, progression and treatment resistance in the EV subpopulations. Analysis of genomic mutations, by using EV-associated DNA, could therefore represent a source of diagnostic biomarkers [100].

The possibility to use EV-DNA as a source of biomarkers for cancer diagnostics and prognostics is supported by a very recent study from Jin and colleagues that provided evidences of the stability of serum EV and EV DNA under different conditions. In particular they purified EVs from serum stored at 4 °C, room temperature and after repeated freeze-thaw cycles and isolated EV-DNA, observing that the content and function of DNA in EVs was stable [114].

It is now known that mitochondria contain extranuclear DNA (mtDNA) maternally inherited and with a high mutation rate. Mutations in mtDNA are often causes of many mitochondrial diseases. In addition, mtDNA mutations are found to be associated with neurodegenerative diseases, diabetes, cancer and aging. The presence of mtDNA in extracellular vesicles is a more recent report; it is conceivable to speculate that altered mtDNA can be delivered among cells, favoring the diffusion of several pathologies.

Guescini and colleagues provided the first evidence that human glioblastoma cell line and primary astrocyte cells package mtDNA in exosomes. In particular authors showed that mitochondrial DNA was confined inside the exosomes by treating them with DNaseI before DNA purification; in addition they quantified mtDNA by measuring, through Real-time PCR, the expression of specific mitochondrial encoded gene, NADH dehydrogenase subunit 1 (MT-ND1). Authors concluded that, although a marked portion of mtDNA released in the conditioned medium is free, the DNase-resistant mtDNA, enveloped in exosomes, is about 10% of the total [99]. In 2010 the same group reported that also cultured myoblasts-derived microvesicles contained mtDNA, by measuring a region of the mitochondrial D-loop. Authors finally speculated that mtDNA, shuttled among cells, could have physiological relevance in restoring the proper mtDNA or, conversely, favoring the spread of pathological phenotypes, such as skeletal muscle diseases [115].

Zhang and collaborators investigated if the stimulation of mast cell leads to secretion of mitochondrial components that could elicit inflammatory effects. They found that, in response to allergic and neuropeptide trigger, human mast cells secrete mtDNA and that about 25% of this DNA is contained in exosomes. These findings highlight a potential role of EV-associated mtDNA also in inflammatory disorders [116].

2.5 Concluding Remarks

The results accumulated over the last years have shown that exosomes contain RNAs, microRNAs, long non-coding RNAs, DNA, lipids, metabolites and proteins, functioning as intercellular delivery system, even over a long distance. Considering these properties and the development of increasingly innovative technologies for their isolation, characterization and manipulation, exosomes are nowadays considered biomarkers in several conditions as well as vehicles for the delivery of therapeutic cargoes.

Acknowledgment Riccardo Alessandro received a grant (n°18783) from the Associazione Italiana per la Ricerca sul Cancro (AIRC). Stefania Raimondo issupported by a "FIRC" (Fondazione Italiana Ricerca sul Cancro) fellowship.

Conflict of Interests The authors declare no conflict of interest.

References

1. Yanez-Mo M, Siljander PR, Andreu Z, Zavec AB, Borras FE, Buzas EI, Buzas K, Casal E, Cappello F, Carvalho J, Colas E, Cordeiro-da Silva A, Fais S, Falcon-Perez JM, Ghobrial IM, Giebel B, Gimona M, Graner M, Gursel I, Gursel M, Heegaard NH, Hendrix A, Kierulf P, Kokubun K, Kosanovic M, Kralj-Iglic V, Kramer-Albers EM, Laitinen S, Lasser C, Lener T, Ligeti E, Line A, Lipps G, Llorente A, Lotvall J, Mancek-Keber M, Marcilla A, Mittelbrunn M, Nazarenko I, Nolte-'t Hoen EN, Nyman TA, O'Driscoll L, Olivan M, Oliveira C, Pallinger E, Del Portillo HA, Reventos J, Rigau M, Rohde E, Sammar M, Sanchez-Madrid F, Santarem N, Schallmoser K, Ostenfeld MS, Stoorvogel W, Stukelj R, Van der Grein SG, Vasconcelos MH, Wauben MH, De Wever O (2015) Biological properties of extracellular vesicles and their physiological functions. J Extracell Vesicles 4:27066
2. Deatherage BL, Cookson BT (2012) Membrane vesicle release in bacteria, eukaryotes, and archaea: a conserved yet underappreciated aspect of microbial life. Infect Immun 80(6):1948–1957
3. Corrado C, Raimondo S, Chiesi A, Ciccia F, De Leo G, Alessandro R (2013) Exosomes as intercellular signaling organelles involved in health and disease: basic science and clinical applications. Int J Mol Sci 14(3):5338–5366
4. Raposo G, Stoorvogel W (2013) Extracellular vesicles: exosomes, microvesicles, and friends. J Cell Biol 200(4):373–383
5. Colombo M, Raposo G, Thery C (2014) Biogenesis, secretion, and intercellular interactions of exosomes and other extracellular vesicles. Annu Rev Cell Dev Biol 30:255–289
6. Colombo M, Moita C, van Niel G, Kowal J, Vigneron J, Benaroch P, Manel N, Moita LF, Thery C, Raposo G (2013) Analysis of ESCRT functions in exosome biogenesis, composition and secretion highlights the heterogeneity of extracellular vesicles. J Cell Sci 126(Pt 24):5553–5565
7. Ostrowski M, Carmo NB, Krumeich S, Fanget I, Raposo G, Savina A, Moita CF, Schauer K, Hume AN, Freitas RP, Goud B, Benaroch P, Hacohen N, Fukuda M, Desnos C, Seabra MC, Darchen F, Amigorena S, Moita LF, Thery C (2010) Rab27a and Rab27b control different steps of the exosome secretion pathway. Nat Cell Biol 12 (1):19–30; sup pp 11–13
8. Babst M (2011) MVB vesicle formation: ESCRT-dependent, ESCRT-independent and everything in between. Curr Opin Cell Biol 23(4):452–457
9. Zaborowski MP, Balaj L, Breakefield XO, Lai CP (2015) Extracellular vesicles: composition, biological relevance, and methods of study. Bioscience 65(8):783–797
10. Laulagnier K, Motta C, Hamdi S, Roy S, Fauvelle F, Pageaux JF, Kobayashi T, Salles JP, Perret B, Bonnerot C, Record M (2004) Mast cell- and dendritic cell-derived exosomes display a specific lipid composition and an unusual membrane organization. Biochem J 380(Pt 1):161–171
11. Choi DS, Kim DK, Kim YK, Gho YS (2013) Proteomics, transcriptomics and lipidomics of exosomes and ectosomes. Proteomics 13(10–11):1554–1571

12. Trajkovic K, Hsu C, Chiantia S, Rajendran L, Wenzel D, Wieland F, Schwille P, Brugger B, Simons M (2008) Ceramide triggers budding of exosome vesicles into multivesicular endosomes. Science 319(5867):1244–1247
13. Subra C, Laulagnier K, Perret B, Record M (2007) Exosome lipidomics unravels lipid sorting at the level of multivesicular bodies. Biochimie 89(2):205–212
14. Vidal M, Sainte-Marie J, Philippot JR, Bienvenue A (1989) Asymmetric distribution of phospholipids in the membrane of vesicles released during in vitro maturation of guinea pig reticulocytes: evidence precluding a role for "aminophospholipid translocase". J Cell Physiol 140(3):455–462
15. de Gassart A, Geminard C, Fevrier B, Raposo G, Vidal M (2003) Lipid raft-associated protein sorting in exosomes. Blood 102(13):4336–4344
16. Dubois L, Ronquist KK, Ek B, Ronquist G, Larsson A (2015) Proteomic profiling of detergent resistant membranes (lipid rafts) of prostasomes. Mol Cell Proteomics 14(11):3015–3022
17. Llorente A, Skotland T, Sylvanne T, Kauhanen D, Rog T, Orlowski A, Vattulainen I, Ekroos K, Sandvig K (2013) Molecular lipidomics of exosomes released by PC-3 prostate cancer cells. Biochim Biophys Acta 1831(7):1302–1309
18. Haraszti RA, Didiot MC, Sapp E, Leszyk J, Shaffer SA, Rockwell HE, Gao F, Narain NR, DiFiglia M, Kiebish MA, Aronin N, Khvorova A (2016) High-resolution proteomic and lipidomic analysis of exosomes and microvesicles from different cell sources. J Extracell Vesicles 5:32570
19. Kim CW, Lee HM, Lee TH, Kang C, Kleinman HK, Gho YS (2002) Extracellular membrane vesicles from tumor cells promote angiogenesis via sphingomyelin. Cancer Res 62(21):6312–6317
20. Subra C, Grand D, Laulagnier K, Stella A, Lambeau G, Paillasse M, De Medina P, Monsarrat B, Perret B, Silvente-Poirot S, Poirot M, Record M (2010) Exosomes account for vesicle-mediated transcellular transport of activatable phospholipases and prostaglandins. J Lipid Res 51(8):2105–2120
21. Xiang X, Poliakov A, Liu C, Liu Y, Deng ZB, Wang J, Cheng Z, Shah SV, Wang GJ, Zhang L, Grizzle WE, Mobley J, Zhang HG (2009) Induction of myeloid-derived suppressor cells by tumor exosomes. Int J Cancer 124(11):2621–2633
22. Esser J, Gehrmann U, D'Alexandri FL, Hidalgo-Estevez AM, Wheelock CE, Scheynius A, Gabrielsson S, Radmark O (2010) Exosomes from human macrophages and dendritic cells contain enzymes for leukotriene biosynthesis and promote granulocyte migration. J Allergy Clin Immunol 126(5):1032–1040, 1040 e1031–e1034
23. Altadill T, Campoy I, Lanau L, Gill K, Rigau M, Gil-Moreno A, Reventos J, Byers S, Colas E, Cheema AK (2016) Enabling metabolomics based biomarker discovery studies using molecular phenotyping of exosome-like vesicles. PLoS One 11(3):e0151339
24. Palomo L, Casal E, Royo F, Cabrera D, van-Liempd S, Falcon-Perez JM (2014) Considerations for applying metabolomics to the analysis of extracellular vesicles. Front Immunol 5:651
25. Royo F, Palomo L, Mleczko J, Gonzalez E, Alonso C, Martinez I, Perez-Cormenzana M, Castro A, Falcon-Perez JM (2017) Metabolically active extracellular vesicles released from hepatocytes under drug-induced liver-damaging conditions modify serum metabolome and might affect different pathophysiological processes. Eur J Pharm Sci 98:51–57. doi:10.1016/j.ejps.2016.10.020. S0928-0987(16)30449-3 [pii]
26. Fontana S, Saieva L, Taverna S, Alessandro R (2013) Contribution of proteomics to understanding the role of tumor-derived exosomes in cancer progression: state of the art and new perspectives. Proteomics 13(10–11):1581–1594
27. Kalra H, Simpson RJ, Ji H, Aikawa E, Altevogt P, Askenase P, Bond VC, Borras FE, Breakefield X, Budnik V, Buzas E, Camussi G, Clayton A, Cocucci E, Falcon-Perez JM, Gabrielsson S, Gho YS, Gupta D, Harsha HC, Hendrix A, Hill AF, Inal JM, Jenster G, Kramer-Albers EM, Lim SK, Llorente A, Lotvall J, Marcilla A, Mincheva-Nilsson L, Nazarenko I, Nieuwland R, Nolte-'t Hoen EN, Pandey A, Patel T, Piper MG, Pluchino S, Prasad TS, Rajendran L, Raposo G, Record M, Reid GE, Sanchez-Madrid F, Schiffelers RM, Siljander P, Stensballe A, Stoorvogel W, Taylor D, Thery C, Valadi H, van Balkom BW, Vazquez J, Vidal M, Wauben

MH, Yanez-Mo M, Zoeller M, Mathivanan S (2012) Vesiclepedia: a compendium for extracellular vesicles with continuous community annotation. PLoS Biol 10(12):e1001450
28. Simpson RJ, Kalra H, Mathivanan S (2012) ExoCarta as a resource for exosomal research. J Extracell Vesicles 1. doi:10.3402/jev.v1i0.18374
29. Hoshino A, Costa-Silva B, Shen TL, Rodrigues G, Hashimoto A, Tesic Mark M, Molina H, Kohsaka S, Di Giannatale A, Ceder S, Singh S, Williams C, Soplop N, Uryu K, Pharmer L, King T, Bojmar L, Davies AE, Ararso Y, Zhang T, Zhang H, Hernandez J, Weiss JM, Dumont-Cole VD, Kramer K, Wexler LH, Narendran A, Schwartz GK, Healey JH, Sandstrom P, Labori KJ, Kure EH, Grandgenett PM, Hollingsworth MA, de Sousa M, Kaur S, Jain M, Mallya K, Batra SK, Jarnagin WR, Brady MS, Fodstad O, Muller V, Pantel K, Minn AJ, Bissell MJ, Garcia BA, Kang Y, Rajasekhar VK, Ghajar CM, Matei I, Peinado H, Bromberg J, Lyden D (2015) Tumour exosome integrins determine organotropic metastasis. Nature 527(7578):329–335
30. Taverna S, Fontana S, Monteleone F, Pucci M, Saieva L, De Caro V, Cardinale VG, Giallombardo M, Vicario E, Rolfo C, Leo GD, Alessandro R (2016) Curcumin modulates chronic myelogenous leukemia exosomes composition and affects angiogenic phenotype via exosomal miR-21. Oncotarget 7(21):30420–30439
31. Bian S, Zhang L, Duan L, Wang X, Min Y, Yu H (2014) Extracellular vesicles derived from human bone marrow mesenchymal stem cells promote angiogenesis in a rat myocardial infarction model. J Mol Med 92(4):387–397
32. Kordelas L, Rebmann V, Ludwig AK, Radtke S, Ruesing J, Doeppner TR, Epple M, Horn PA, Beelen DW, Giebel B (2014) MSC-derived exosomes: a novel tool to treat therapy-refractory graft-versus-host disease. Leukemia 28(4):970–973
33. Zhang B, Wang M, Gong A, Zhang X, Wu X, Zhu Y, Shi H, Wu L, Zhu W, Qian H, Xu W (2015) HucMSC-exosome mediated-Wnt4 signaling is required for cutaneous wound healing. Stem Cells 33(7):2158–2168
34. Anderson JD, Johansson HJ, Graham CS, Vesterlund M, Pham MT, Bramlett CS, Montgomery EN, Mellema MS, Bardini RL, Contreras Z, Hoon M, Bauer G, Fink KD, Fury B, Hendrix KJ, Chedin F, El-Andaloussi S, Hwang B, Mulligan MS, Lehtio J, Nolta JA (2016) Comprehensive proteomic analysis of mesenchymal stem cell exosomes reveals modulation of angiogenesis via nuclear factor-KappaB signaling. Stem Cells 34(3):601–613
35. Feng Y, Huang W, Meng W, Jegga AG, Wang Y, Cai W, Kim HW, Pasha Z, Wen Z, Rao F, Modi RM, Yu X, Ashraf M (2014) Heat shock improves Sca-1+ stem cell survival and directs ischemic cardiomyocytes toward a prosurvival phenotype via exosomal transfer: a critical role for HSF1/miR-34a/HSP70 pathway. Stem Cells 32(2):462–472
36. Melo SA, Luecke LB, Kahlert C, Fernandez AF, Gammon ST, Kaye J, LeBleu VS, Mittendorf EA, Weitz J, Rahbari N, Reissfelder C, Pilarsky C, Fraga MF, Piwnica-Worms D, Kalluri R (2015) Glypican-1 identifies cancer exosomes and detects early pancreatic cancer. Nature 523(7559):177–182
37. Khan S, Bennit HF, Turay D, Perez M, Mirshahidi S, Yuan Y, Wall NR (2014) Early diagnostic value of survivin and its alternative splice variants in breast cancer. BMC Cancer 14:176
38. Bijnsdorp IV, Geldof AA, Lavaei M, Piersma SR, van Moorselaar RJ, Jimenez CR (2013) Exosomal ITGA3 interferes with non-cancerous prostate cell functions and is increased in urine exosomes of metastatic prostate cancer patients. J Extracell Vesicles 2. doi:10.3402/jev.v2i0.22097
39. Lee MJ, Park DH, Kang JH (2016) Exosomes as the source of biomarkers of metabolic diseases. Ann Pediatr Endocrinol Metab 21(3):119–125
40. Kranendonk ME, de Kleijn DP, Kalkhoven E, Kanhai DA, Uiterwaal CS, van der Graaf Y, Pasterkamp G, Visseren FL (2014) Extracellular vesicle markers in relation to obesity and metabolic complications in patients with manifest cardiovascular disease. Cardiovasc Diabetol 13:37
41. Raimondo F, Corbetta S, Morosi L, Chinello C, Gianazza E, Castoldi G, Di Gioia C, Bombardi C, Stella A, Battaglia C, Bianchi C, Magni F, Pitto M (2013) Urinary exosomes and diabetic nephropathy: a proteomic approach. Mol BioSyst 9(6):1139–1146

42. Zubiri I, Posada-Ayala M, Sanz-Maroto A, Calvo E, Martin-Lorenzo M, Gonzalez-Calero L, de la Cuesta F, Lopez JA, Fernandez-Fernandez B, Ortiz A, Vivanco F, Alvarez-Llamas G (2014) Diabetic nephropathy induces changes in the proteome of human urinary exosomes as revealed by label-free comparative analysis. J Proteome 96:92–102
43. Gudehithlu KP, Garcia-Gomez I, Vernik J, Brecklin C, Kraus M, Cimbaluk DJ, Hart P, Dunea G, Arruda JA, Singh AK (2015) In diabetic kidney disease urinary exosomes better represent kidney specific protein alterations than whole urine. Am J Nephrol 42(6):418–424
44. Skog J, Wurdinger T, van Rijn S, Meijer DH, Gainche L, Sena-Esteves M, Curry WT Jr, Carter BS, Krichevsky AM, Breakefield XO (2008) Glioblastoma microvesicles transport RNA and proteins that promote tumour growth and provide diagnostic biomarkers. Nat Cell Biol 10(12):1470–1476
45. Valadi H, Ekstrom K, Bossios A, Sjostrand M, Lee JJ, Lotvall JO (2007) Exosome-mediated transfer of mRNAs and microRNAs is a novel mechanism of genetic exchange between cells. Nat Cell Biol 9(6):654–659
46. Kogure T, Lin WL, Yan IK, Braconi C, Patel T (2011) Intercellular nanovesicle-mediated microRNA transfer: a mechanism of environmental modulation of hepatocellular cancer cell growth. Hepatology 54(4):1237–1248
47. Gutkin A, Uziel O, Beery E, Nordenberg J, Pinchasi M, Goldvaser H, Henick S, Goldberg M, Lahav M (2016) Tumor cells derived exosomes contain hTERT mRNA and transform nonmalignant fibroblasts into telomerase positive cells. Oncotarget 7(37):59173–59188. doi:10.18632/oncotarget.10384
48. Cheng L, Sharples RA, Scicluna BJ, Hill AF (2014) Exosomes provide a protective and enriched source of miRNA for biomarker profiling compared to intracellular and cell-free blood. J Extracell Vesicles 3. doi:10.3402/jev.v3.23743
49. Lasser C, Alikhani VS, Ekstrom K, Eldh M, Paredes PT, Bossios A, Sjostrand M, Gabrielsson S, Lotvall J, Valadi H (2011) Human saliva, plasma and breast milk exosomes contain RNA: uptake by macrophages. J Transl Med 9:9
50. Palanisamy V, Sharma S, Deshpande A, Zhou H, Gimzewski J, Wong DT (2010) Nanostructural and transcriptomic analyses of human saliva derived exosomes. PLoS One 5(1):e8577
51. Chen TS, Lai RC, Lee MM, Choo AB, Lee CN, Lim SK (2010) Mesenchymal stem cell secretes microparticles enriched in pre-microRNAs. Nucleic Acids Res 38(1):215–224
52. Kesimer M, Scull M, Brighton B, DeMaria G, Burns K, O'Neal W, Pickles RJ, Sheehan JK (2009) Characterization of exosome-like vesicles released from human tracheobronchial ciliated epithelium: a possible role in innate defense. FASEB J 23(6):1858–1868
53. Jacob F, Monod J (1961) Genetic regulatory mechanisms in the synthesis of proteins. J Mol Biol 3:318–356
54. Bartel DP (2004) MicroRNAs: genomics, biogenesis, mechanism, and function. Cell 116(2):281–297
55. Montecalvo A, Larregina AT, Shufesky WJ, Stolz DB, Sullivan ML, Karlsson JM, Baty CJ, Gibson GA, Erdos G, Wang Z, Milosevic J, Tkacheva OA, Divito SJ, Jordan R, Lyons-Weiler J, Watkins SC, Morelli AE (2012) Mechanism of transfer of functional microRNAs between mouse dendritic cells via exosomes. Blood 119(3):756–766
56. Melo SA, Sugimoto H, O'Connell JT, Kato N, Villanueva A, Vidal A, Qiu L, Vitkin E, Perelman LT, Melo CA, Lucci A, Ivan C, Calin GA, Kalluri R (2014) Cancer exosomes perform cell-independent microRNA biogenesis and promote tumorigenesis. Cancer Cell 26(5):707–721
57. Guduric-Fuchs J, O'Connor A, Camp B, O'Neill CL, Medina RJ, Simpson DA (2012) Selective extracellular vesicle-mediated export of an overlapping set of microRNAs from multiple cell types. BMC Genomics 13:357
58. Ohshima K, Inoue K, Fujiwara A, Hatakeyama K, Kanto K, Watanabe Y, Muramatsu K, Fukuda Y, Ogura S, Yamaguchi K, Mochizuki T (2010) Let-7 microRNA family is selectively secreted into the extracellular environment via exosomes in a metastatic gastric cancer cell line. PLoS One 5(10):e13247
59. Pigati L, Yaddanapudi SC, Iyengar R, Kim DJ, Hearn SA, Danforth D, Hastings ML, Duelli DM (2010) Selective release of microRNA species from normal and malignant mammary epithelial cells. PLoS One 5(10):e13515

60. Villarroya-Beltri C, Gutierrez-Vazquez C, Sanchez-Cabo F, Perez-Hernandez D, Vazquez J, Martin-Cofreces N, Martinez-Herrera DJ, Pascual-Montano A, Mittelbrunn M, Sanchez-Madrid F (2013) Sumoylated hnRNPA2B1 controls the sorting of miRNAs into exosomes through binding to specific motifs. Nat Commun 4:2980
61. Santangelo L, Giurato G, Cicchini C, Montaldo C, Mancone C, Tarallo R, Battistelli C, Alonzi T, Weisz A, Tripodi M (2016) The RNA-binding protein SYNCRIP is a component of the hepatocyte exosomal machinery controlling microRNA sorting. Cell Rep 17(3):799–808
62. Squadrito ML, Baer C, Burdet F, Maderna C, Gilfillan GD, Lyle R, Ibberson M, De Palma M (2014) Endogenous RNAs modulate microRNA sorting to exosomes and transfer to acceptor cells. Cell Rep 8(5):1432–1446
63. Griffiths-Jones S (2004) The microRNA registry. Nucleic Acids Res 32(Database issue):D109–D111
64. Tay Y, Zhang J, Thomson AM, Lim B, Rigoutsos I (2008) MicroRNAs to Nanog, Oct4 and Sox2 coding regions modulate embryonic stem cell differentiation. Nature 455(7216):1124–1128
65. Schober A, Nazari-Jahantigh M, Wei Y, Bidzhekov K, Gremse F, Grommes J, Megens RT, Heyll K, Noels H, Hristov M, Wang S, Kiessling F, Olson EN, Weber C (2014) MicroRNA-126-5p promotes endothelial proliferation and limits atherosclerosis by suppressing Dlk1. Nat Med 20(4):368–376
66. Jovanovic M, Hengartner MO (2006) miRNAs and apoptosis: RNAs to die for. Oncogene 25(46):6176–6187
67. Inui M, Martello G, Piccolo S (2010) MicroRNA control of signal transduction. Nat Rev Mol Cell Biol 11(4):252–263
68. Lo Dico A, Costa V, Martelli C, Diceglie C, Rajata F, Rizzo A, Mancone C, Tripodi M, Ottobrini L, Alessandro R, Conigliaro A (2016) MiR675-5p acts on HIF-1alpha to sustain hypoxic responses: a new therapeutic strategy for glioma. Theranostics 6(8):1105–1118
69. Raimondo S, Corrado C, Raimondi L, De Leo G, Alessandro R (2015) Role of extracellular vesicles in hematological malignancies. Biomed Res Int 2015:821613
70. Zhuang G, Wu X, Jiang Z, Kasman I, Yao J, Guan Y, Oeh J, Modrusan Z, Bais C, Sampath D, Ferrara N (2012) Tumour-secreted miR-9 promotes endothelial cell migration and angiogenesis by activating the JAK-STAT pathway. EMBO J 31(17):3513–3523
71. Taverna S, Flugy A, Saieva L, Kohn EC, Santoro A, Meraviglia S, De Leo G, Alessandro R (2012) Role of exosomes released by chronic myelogenous leukemia cells in angiogenesis. Int J Cancer 130(9):2033–2043
72. Bang C, Batkai S, Dangwal S, Gupta SK, Foinquinos A, Holzmann A, Just A, Remke J, Zimmer K, Zeug A, Ponimaskin E, Schmiedl A, Yin X, Mayr M, Halder R, Fischer A, Engelhardt S, Wei Y, Schober A, Fiedler J, Thum T (2014) Cardiac fibroblast-derived microRNA passenger strand-enriched exosomes mediate cardiomyocyte hypertrophy. J Clin Investig 124(5):2136–2146
73. Halkein J, Tabruyn SP, Ricke-Hoch M, Haghikia A, Nguyen NQ, Scherr M, Castermans K, Malvaux L, Lambert V, Thiry M, Sliwa K, Noel A, Martial JA, Hilfiker-Kleiner D, Struman I (2013) MicroRNA-146a is a therapeutic target and biomarker for peripartum cardiomyopathy. J Clin Investig 123(5):2143–2154
74. Small EM, Olson EN (2011) Pervasive roles of microRNAs in cardiovascular biology. Nature 469(7330):336–342
75. Da Costa Martins PA, De Windt LJ (2012) MicroRNAs in control of cardiac hypertrophy. Cardiovasc Res 93(4):563–572
76. Lin J, Li J, Huang B, Liu J, Chen X, Chen XM, YM X, Huang LF, Wang XZ (2015) Exosomes: novel biomarkers for clinical diagnosis. ScientificWorldJournal 2015:657086
77. Shao H, Chung J, Lee K, Balaj L, Min C, Carter BS, Hochberg FH, Breakefield XO, Lee H, Weissleder R (2015) Chip-based analysis of exosomal mRNA mediating drug resistance in glioblastoma. Nat Commun 6:6999
78. Frank S, Aguirre A, Hescheler J, Kurian L (2016) A lncRNA perspective into (re)building the heart. Front Cell Dev Biol 4:128
79. Bhartiya D, Scaria V (2016) Genomic variations in non-coding RNAs: structure, function and regulation. Genomics 107(2–3):59–68

80. Wang H, Wang L, Erdjument-Bromage H, Vidal M, Tempst P, Jones RS, Zhang Y (2004) Role of histone H2A ubiquitination in Polycomb silencing. Nature 431(7010):873–878
81. Tsai MC, Manor O, Wan Y, Mosammaparast N, Wang JK, Lan F, Shi Y, Segal E, Chang HY (2010) Long noncoding RNA as modular scaffold of histone modification complexes. Science 329(5992):689–693
82. Yang L, Lin C, Liu W, Zhang J, Ohgi KA, Grinstein JD, Dorrestein PC, Rosenfeld MG (2011) ncRNA- and Pc2 methylation-dependent gene relocation between nuclear structures mediates gene activation programs. Cell 147(4):773–788
83. Yoon JH, Abdelmohsen K, Srikantan S, Yang X, Martindale JL, De S, Huarte M, Zhan M, Becker KG, Gorospe M (2012) LincRNA-p21 suppresses target mRNA translation. Mol Cell 47(4):648–655
84. Tripathi V, Ellis JD, Shen Z, Song DY, Pan Q, Watt AT, Freier SM, Bennett CF, Sharma A, Bubulya PA, Blencowe BJ, Prasanth SG, Prasanth KV (2010) The nuclear-retained noncoding RNA MALAT1 regulates alternative splicing by modulating SR splicing factor phosphorylation. Mol Cell 39(6):925–938
85. Gupta RA, Shah N, Wang KC, Kim J, Horlings HM, Wong DJ, Tsai MC, Hung T, Argani P, Rinn JL, Wang Y, Brzoska P, Kong B, Li R, West RB, van de Vijver MJ, Sukumar S, Chang HY (2010) Long non-coding RNA HOTAIR reprograms chromatin state to promote cancer metastasis. Nature 464(7291):1071–1076
86. Klattenhoff CA, Scheuermann JC, Surface LE, Bradley RK, Fields PA, Steinhauser ML, Ding H, Butty VL, Torrey L, Haas S, Abo R, Tabebordbar M, Lee RT, Burge CB, Boyer LA (2013) Braveheart, a long noncoding RNA required for cardiovascular lineage commitment. Cell 152(3):570–583
87. Grote P, Wittler L, Hendrix D, Koch F, Wahrisch S, Beisaw A, Macura K, Blass G, Kellis M, Werber M, Herrmann BG (2013) The tissue-specific lncRNA Fendrr is an essential regulator of heart and body wall development in the mouse. Dev Cell 24(2):206–214
88. Mercer TR, Dinger ME, Mattick JS (2009) Long non-coding RNAs: insights into functions. Nat Rev Genet 10(3):155–159
89. Conigliaro A, Costa V, Lo Dico A, Saieva L, Buccheri S, Dieli F, Manno M, Raccosta S, Mancone C, Tripodi M, De Leo G, Alessandro R (2015) CD90+ liver cancer cells modulate endothelial cell phenotype through the release of exosomes containing H19 lncRNA. Mol Cancer 14:155
90. Kogure T, Yan IK, Lin WL, Patel T (2013) Extracellular vesicle-mediated transfer of a novel long noncoding RNA TUC339: a mechanism of intercellular signaling in human hepatocellular cancer. Genes Cancer 4(7–8):261–272
91. Chen M, Xu R, Ji H, Greening DW, Rai A, Izumikawa K, Ishikawa H, Takahashi N, Simpson RJ (2016) Transcriptome and long noncoding RNA sequencing of three extracellular vesicle subtypes released from the human colon cancer LIM1863 cell line. Sci Rep 6:38397
92. Chistiakov DA, Chekhonin VP (2014) Extracellular vesicles shed by glioma cells: pathogenic role and clinical value. Tumour Biol 35(9):8425–8438
93. Ahadi A, Brennan S, Kennedy PJ, Hutvagner G, Tran N (2016) Long non-coding RNAs harboring miRNA seed regions are enriched in prostate cancer exosomes. Sci Rep 6:24922
94. Zhang J, Liu SC, Luo XH, Tao GX, Guan M, Yuan H, Hu DK (2016) Exosomal long noncoding RNAs are differentially expressed in the cervicovaginal lavage samples of cervical cancer patients. J Clin Lab Anal 30(6):1116–1121
95. Berrondo C, Flax J, Kucherov V, Siebert A, Osinski T, Rosenberg A, Fucile C, Richheimer S, Beckham CJ (2016) Expression of the long non-coding RNA HOTAIR correlates with disease progression in bladder cancer and is contained in bladder cancer patient urinary exosomes. PLoS One 11(1):e0147236
96. Mandel P, Metais P (1948) [Not available]. Comptes Rendus des Séances de la Société de Biologie et de Ses Filiales 142(3–4):241–243
97. Leon SA, Shapiro B, Sklaroff DM, Yaros MJ (1977) Free DNA in the serum of cancer patients and the effect of therapy. Cancer Res 37(3):646–650
98. Balaj L, Lessard R, Dai L, Cho YJ, Pomeroy SL, Breakefield XO, Skog J (2011) Tumour microvesicles contain retrotransposon elements and amplified oncogene sequences. Nat Commun 2:180

99. Guescini M, Genedani S, Stocchi V, Agnati LF (2010) Astrocytes and glioblastoma cells release exosomes carrying mtDNA. J Neural Transm (Vienna) 117(1):1–4
100. Lazaro-Ibanez E, Sanz-Garcia A, Visakorpi T, Escobedo-Lucea C, Siljander P, Ayuso-Sacido A, Yliperttula M (2014) Different gDNA content in the subpopulations of prostate cancer extracellular vesicles: apoptotic bodies, microvesicles, and exosomes. Prostate 74(14):1379–1390
101. Ronquist GK, Larsson A, Ronquist G, Isaksson A, Hreinsson J, Carlsson L, Stavreus-Evers A (2011) Prostasomal DNA characterization and transfer into human sperm. Mol Reprod Dev 78(7):467–476
102. Cai J, Wu G, Jose PA, Zeng C (2016) Functional transferred DNA within extracellular vesicles. Exp Cell Res 349(1):179–183. doi:10.1016/j.yexcr.2016.10.012. S0014-4827(16)30337-0 [pii]
103. Hill AF, Pegtel DM, Lambertz U, Leonardi T, O'Driscoll L, Pluchino S, Ter-Ovanesyan D, Nolte-'t Hoen EN (2013) ISEV position paper: extracellular vesicle RNA analysis and bioinformatics. J Extracell Vesicles 2. doi:10.3402/jev.v2i0.22859
104. Ronquist KG, Ronquist G, Carlsson L, Larsson A (2009) Human prostasomes contain chromosomal DNA. Prostate 69(7):737–743
105. Ronquist GK, Larsson A, Stavreus-Evers A, Ronquist G (2012) Prostasomes are heterogeneous regarding size and appearance but affiliated to one DNA-containing exosome family. Prostate 72(16):1736–1745
106. Waldenstrom A, Genneback N, Hellman U, Ronquist G (2012) Cardiomyocyte microvesicles contain DNA/RNA and convey biological messages to target cells. PLoS One 7(4):e34653
107. Fischer S, Cornils K, Speiseder T, Badbaran A, Reimer R, Indenbirken D, Grundhoff A, Brunswig-Spickenheier B, Alawi M, Lange C (2016) Indication of horizontal DNA Gene transfer by extracellular vesicles. PLoS One 11(9):e0163665
108. Qin Z, Ljubimov VA, Zhou C, Tong Y, Liang J (2016) Cell-free circulating tumor DNA in cancer. Chin J Cancer 35:36
109. Lee TH, Chennakrishnaiah S, Audemard E, Montermini L, Meehan B, Rak J (2014) Oncogenic ras-driven cancer cell vesiculation leads to emission of double-stranded DNA capable of interacting with target cells. Biochem Biophys Res Commun 451(2):295–301
110. Thakur BK, Zhang H, Becker A, Matei I, Huang Y, Costa-Silva B, Zheng Y, Hoshino A, Brazier H, Xiang J, Williams C, Rodriguez-Barrueco R, Silva JM, Zhang W, Hearn S, Elemento O, Paknejad N, Manova-Todorova K, Welte K, Bromberg J, Peinado H, Lyden D (2014) Double-stranded DNA in exosomes: a novel biomarker in cancer detection. Cell Res 24(6):766–769
111. Cai J, Han Y, Ren H, Chen C, He D, Zhou L, Eisner GM, Asico LD, Jose PA, Zeng C (2013) Extracellular vesicle-mediated transfer of donor genomic DNA to recipient cells is a novel mechanism for genetic influence between cells. J Mol Cell Biol 5(4):227–238
112. Cai J, Wu G, Tan X, Han Y, Chen C, Li C, Wang N, Zou X, Chen X, Zhou F, He D, Zhou L, Jose PA, Zeng C (2014) Transferred BCR/ABL DNA from K562 extracellular vesicles causes chronic myeloid leukemia in immunodeficient mice. PLoS One 9(8):e105200
113. Kahlert C, Melo SA, Protopopov A, Tang J, Seth S, Koch M, Zhang J, Weitz J, Chin L, Futreal A, Kalluri R (2014) Identification of double-stranded genomic DNA spanning all chromosomes with mutated KRAS and p53 DNA in the serum exosomes of patients with pancreatic cancer. J Biol Chem 289(7):3869–3875
114. Jin Y, Chen K, Wang Z, Wang Y, Liu J, Lin L, Shao Y, Gao L, Yin H, Cui C, Tan Z, Liu L, Zhao C, Zhang G, Jia R, Du L, Chen Y, Liu R, Xu J, Hu X (2016) DNA in serum extracellular vesicles is stable under different storage conditions. BMC Cancer 16(1):753
115. Guescini M, Guidolin D, Vallorani L, Casadei L, Gioacchini AM, Tibollo P, Battistelli M, Falcieri E, Battistin L, Agnati LF, Stocchi V (2010) C2C12 myoblasts release micro-vesicles containing mtDNA and proteins involved in signal transduction. Exp Cell Res 316(12):1977–1984
116. Zhang B, Asadi S, Weng Z, Sismanopoulos N, Theoharides TC (2012) Stimulated human mast cells secrete mitochondrial components that have autocrine and paracrine inflammatory actions. PLoS One 7(12):e49767

Chapter 3
Functional Role of Cardiovascular Exosomes in Myocardial Injury and Atherosclerosis

Maarten Vanhaverbeke, Diane Gal, and Paul Holvoet

3.1 Introduction

Exosomes are bi-lipid-membranous vesicles containing protein, lipid and nucleic acid material secreted from cells. They are smaller than shedding microvesicles and apoptotic bodies (approximately 40–100 nm for exosomes compared to 100 nm–1 μm for shedding microvesicles and 1–4 μm for apoptotic bodies) and differ by their intracellular origin [19, 42]. Exosomes are identified by their components including integrins and tetraspanins (CD63, CD89, CD81, CD9, and CD82), by maturation-related proteins (flotillin and annexin), and by heat shock proteins (HSP). Together, exosomes and microvesicles are often referred to as extracellular vesicles (EVs). They are not a result of random sampling; instead they contain selective cargo assembled through dedicated packing mechanisms [40, 44], deliver these loads to targeted cells and contain unique trafficking properties. All these mechanisms depend on the individual cell, its cellular state and different physiological, pathological and stress conditions [19, 20]. EVs mediate horizontal, paracrine transfer, delivering microRNAs (miRs), mRNA and proteins between cells of different origin, resulting in silencing or activation of signaling pathways [32]. However, the underlying mechanisms of transfer and the amount of content delivered to the cells remain controversial or unclear.

This chapter will discuss the functional roles of EVs in the prevention, repair or progression of cardiovascular disease, through the communication between several cell types in the heart and vasculature, with an emphasis on signaling pathways.

Maarten Vanhaverbeke and Diane Gal contributed equally to this work.

M. Vanhaverbeke • D. Gal • P. Holvoet, Ph.D., F.A.H.A., F.E.S.C. (✉)
Department of Cardiovascular Sciences, KU Leuven,
Herestraat 49, O&N1, POB 704, B-3000 Leuven, Belgium
e-mail: paul.holvoet@kuleuven.be

J. Xiao, S. Cretoiu (eds.), *Exosomes in Cardiovascular Diseases*, Advances in Experimental Medicine and Biology 998, DOI 10.1007/978-981-10-4397-0_3

3.2 Role of Extracellular Vesicles in Intracardial Communication

3.2.1 Cardiomyocytes, Myofibroblasts and Cardiac Injury

The heart contains cardiomyocytes (CMs), which represent only a third of all cells in the heart, endothelial cells (ECs), immune-system-related cells (macrophages), fibroblasts (FBs), smooth muscle cells (SMCs), sympathetic and parasympathetic neuronal cells, and stem cells. A tight balance between these cell types is needed to maintain the integrity of the heart [4]. Myocardial injury disrupts this integrity by inducing CM death and destroying the vasculature, triggering several effects to repair and maintain cardiac integrity, including cardiac fibrosis by activation of myofibroblasts. This activation involves a complex signaling network containing transforming growth factor (TGF)-β, endothelin-1 (EDN-1), angiotensin II (AGTII), connective tissue growth factor (CCN2), and platelet-derived growth factor (PDGF) (Fig. 3.1a). Myofibroblasts originate from several sources including quiescent tissue FBs, circulating CD34+ fibrocytes, and the phenotypic conversion of various cell types including epithelial cells and ECs. To date, little is known about the inter-organ transfer mechanisms of cardioprotection but recent reports suggest that extracellular vesicles (EVs) release may be involved [13]. In this section we will focus on EVs released from CMs, ECs, macrophages and FBs in cardiac injury and remodeling (Fig. 3.1).

3.2.2 Extracellular Vesicles from Cardiomyocytes, Endothelial Cells and Myofibroblasts

CMs and ECs have an intimate anatomical relationship that is essential for maintaining normal development and function in the heart. Regulation mechanisms of cardiac and endothelial crosstalk in situations of acute stress remain elusive. This cardiac and endothelial crosstalk may involve EVs, among them exosomes. Recently, HSP20-enriched EVs secreted by CMs acted as a novel cardiokine in regulating myocardial angiogenesis through activation of the vascular endothelial growth factor receptor (VEGFR) signaling cascade [54] (Fig. 3.1b). Indeed, overexpression of HSP20 in streptozotocin (STZ)-induced diabetic mice significantly decreased cardiac dysfunction, hypertrophy, apoptosis, fibrosis, and restored angiogenesis. This protective action against adverse cardiac remodeling involved p-Akt, survivin, and superoxide dismutase 1. In addition, HSP20 exosomes interacted with tumor susceptibility gene 101, an important regulator of cell cycle arrest and p-53 independent cell death [47]. Diabetes also significantly impaired angiogenesis by inhibiting proliferation and migration of mouse cardiac endothelial cells (CECs). Mechanistically, higher levels of microRNA (miR)-320 in exosomes of diabetic animals functionally down-regulated its target genes such as insulin growth factor

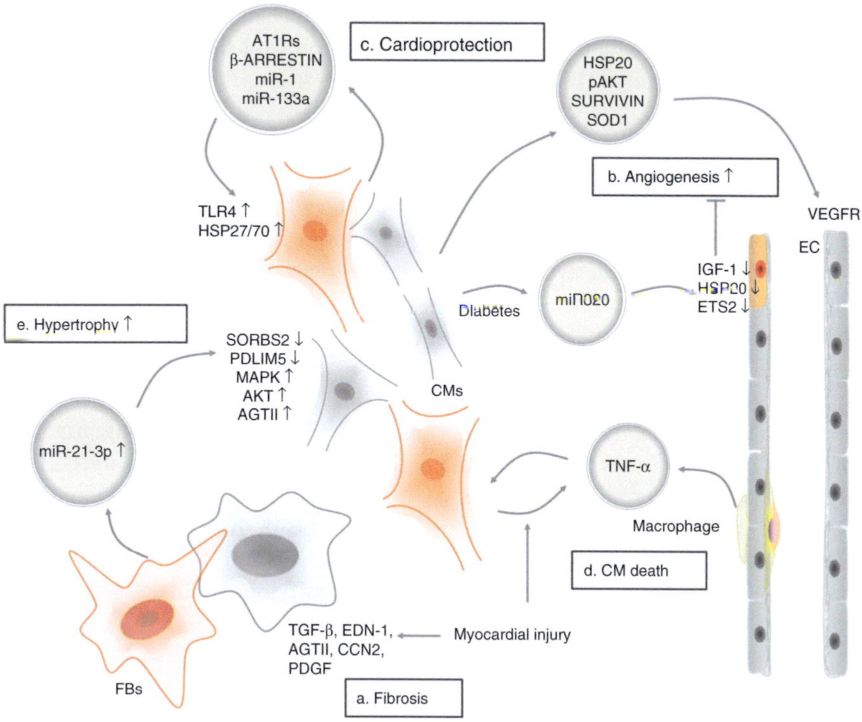

Fig. 3.1 Extracellular vesicles in cardiac remodeling: Myocardial injury induces CM death and triggers several repair mechanisms, including cardiac fibrosis. The effector cells of fibrosis are activated FBs called myofibroblasts and involves a complex signaling network containing TGF-β, EDN-1, AGTII, CCN2 and PDGF leading to fibrosis (**a**). EVs secreted from CMs were shown to be enriched in HSP20, pAKT, survivin and SOD1, and were cardioprotective and induced angiogenesis (**b**). In contrast, diabetes induces CMs to secrete EVs enriched in miR-320 impairing the expression of IGF-1, HSP20 and ETS2 in ECs resulting in impaired angiogenesis (**b**). CM-derived EVs containing A1TR, β-arrestin and miR-1 and miR133a increase TLR4/HSP27 signaling, resulting in cardioprotection (**c**). During myocardial ischemia, TNF-α is mainly released in macrophage-derived exosomes, but with persistent ischemia it can also originate from exosomes released by CMs, inducing CM death (**d**). miR-21-3p (miR-21*) in FB-derived exosomes induced CM hypertrophy by inhibiting SORBS2 and PDLIM5 (**e**). In addition, these FB-derived exosomes induced MAPK and AKT signaling resulting in intensified AGTII-induced cardiac hypertrophy. Abbreviations: *AGT* angiotensin, *AKT* AKT serine/threonine kinase 1, *A1TR* angiotensin II type I receptors, *CM* cardiomyocyte, *CCN2* connective tissue growth factor, *EC* endothelial cell, *EDN-1* endothelin-1, *FB* (myo)fibroblast, *ETS2* ETS proto-oncogene 2, transcription factor, *EV* extracellular vesicle, *HIF-1α* hypoxia inducible factor-1α, *HSP* heat shock protein, *IGF-1* insulin growth factor 1, *MAPK* mitogen-activated protein kinases, *miR* microRNA, *MQ* macrophage, *PDGF* platelet-derived growth factor, *PDLIM5* PDZ and LIM domain 5, *SOD* superoxide dismutase, *SORBS2* sorbin and SH3 domain-containing protein 2, *TGF-β* transforming growth factor, *TLR* toll like receptor, *TNF* tumor necrosis factor

(IGF-)1, HSP20 and ETS proto-oncogene 2, transcription factor (ETS2) in recipient CECs [48] (Fig. 3.1b).

In myocardial ischemia-reperfusion injury, both miR-1 and miR-133a in the exosome-rich fraction in plasma protected CMs by inhibiting cardiac hypertrophy [7]

(Fig. 3.1c). Therefore, they may be utilized to suppress maladapted hypertrophy when blood flow and energy supply is limited [26]. A pro-survival signaling pathway was activated in CMs involving toll-like receptor (TLR)-4 and various kinases, leading to activation of the cardioprotective HSP27.

In addition to secreting microvesicles and exosomes, CMs were found to secrete cytokines, chemokines, and factors like ANP, BNP, TGF-β, and tumor necrosis factor-α (TNF-α) [15, 46]. Excessive TNF-α is thought to be harmful to CMs in acute myocardial infarction (MI). During myocardial ischemia, TNF-α is mainly released in macrophage-derived exosomes, but with persistent ischemia it can also originate from exosomes released by CMs, induced directly by hypoxia and activation of hypoxia inducible factor-1α (HIF-1α) [53] (Fig. 3.1d).

Finally, FBs also have an important function in the pathophysiology of CM death, fibrosis and hypertrophy. MiR-21-3p in cardiac FB-derived exosomes induced CM hypertrophy (Fig. 3.1e). Sorbin and SH3 domain-containing protein 2 (SORBS2) and PDZ and LIM domain 5 (PDLIM5) were identified as miR-21-3p targets by proteome profiling. Silencing SORBS2 or PDLIM5 in CM and inhibition of miR-21-3p induced hypertrophy [5]. In addition, FB-derived exosomes induced mitogen-activated protein kinases (MAPKs) and Akt resulting in increased expression of the Renin Angiotensin System, thereby intensifying AGTII-induced pathological cardiac hypertrophy [29].

3.3 Vesicles from Stem Cells and Progenitor Cells

3.3.1 Cardiac Stem Cells

By definition, a stem cell is capable of self-renewal and can differentiate into at least one cell type. Embryonic stem cells (ESCs) are pluripotent stem cells which were first isolated as a small cluster of cells within mouse blastocysts [11], later from human blastocysts [38]. Hematopoietic stem cells (HSCs) differentiate into different blood cells and are $CD34^+$. Mesenchymal stem cells (MSCs) are multipotent stromal cells that can differentiate into a variety of cell types like myocytes and adipocytes [39]. In contrast to HSCs, standardization of MSCs was hampered by the lack of molecular markers to discern MSCs from FBs. However, recently, specific DNA methylation patterns were used to discriminate between MSCs and FBs, and to distinguish between MSCs from bone marrow and adipose tissue [1]. In addition to ESCs, HSCs and MSCs, a population of resident cardiac stem cells (CPCs) have been identified in the heart; but they comprise less than 1% of the cells in the heart. As yet it is not known whether the CPCs actually migrate from bone marrow to the heart, or originate from remnants of embryonic cell populations in the right atrium and right ventricle. CPCs have been sub-classified into c-kit and Scal-1. C-kit cardiogenic stem cells can differentiate into myogenic, vascular endothelial and smooth muscle lineages. Sca-1 are involved in cell signaling and cell adhesion [16]. Finally, induced pluripotent stem cells (iPSC) can generate an abundance of cells without the risk of immune rejection for cell therapy.

3.3.2 Stem Cell Derived Vesicles

MSCs have been considered to be one of the most promising candidates for regeneration of cardiac cells lost upon injury. But with age, dysregulated MSCs differentiate into dysfunctional inflammatory FBs leading to pathologic fibrosis. The phenotypic change is specific to the heart since MSCs originating from bone or FBs derived from MSCs were free of these defects [43]. In contrast with the original hypothesis that the regenerative capacity of MSCs was due to their potential to engraft, differentiate and replace damaged cardiac cells, recent studies suggested that this was primarily due to paracrine factors released from MSCs in exosomes [35]. For example, GATA binding protein 4 (GATA-4) enriched in exosomes released by MSCs at the border of an ischemic region significantly restored cardiac contractile function and reduced infarct size (Fig. 3.2). Mechanistically, these exosomes increased miR-19a in CMs, resulting in decreased expression of Phosphatase and Tensin Homolog (PTEN), a predicted target of miR-19a, and in the activation of the Akt and ERK signaling pathways [52] (Fig. 3.2). In addition to miR-19, miR profiling analysis revealed that cardiac stem cells exposed to MSC-derived exosomes secreted more miR-147, let-7i-3p, miR-503-5p, and miR-362-3p, and less miR-326-5p, miR-328a-5p, miR-207, miR-760-3pn, and miR-702-5p, associated with activation of target genes involved in angiogenesis and positive regulation of cell proliferation, cell migration, cell differentiation, and response to hypoxia [55]. They activated several signaling pathways important in wound healing (Akt, ERK, and STAT3) and induced the expression of a number of growth factors [hepatocyte growth factor (HGF), IGF1, nerve growth factor (NGF), and stromal-derived growth factor-1 (SDF1)] [37], ultimately leading to preserved cardiac performance after MI [6] (Fig. 3.2). Levels of ATP, NADH and phosphorylated-Akt and phosphorylated-inosine/guanosine kinase (GSK)-3β were increased, while phosphorylated-c-JNK was reduced, thereby decreasing oxidative stress and inflammation in ischemic/reperfused hearts [3]. In addition, the proteome of MSCs and MSC-derived exosomes, from cells cultured under expansion conditions and under ischemic tissue simulated conditions, was shown to contain key angiogenic paracrine effectors and, potentially, differentially expressed in these conditions. In total, 6342 proteins were identified in MSCs and 1927 proteins in MSC-derived exosomes. They included PDGF, epidermal growth factor (EGF), fibroblast growth factor (FGF), and most notably nuclear factor-kappaB (NFkB) signaling pathway proteins. The latter was identified as a key mediator of MSC exosome induced angiogenesis in ECs [2] (Fig. 3.2).

Exosomes derived from iPSC-derived mesenchymal stem cells (iMSC), expressing CD63, CD81, and CD9 at their surface, enhanced microvessel density and blood perfusion in mouse ischemic limbs, consistent with an attenuation of ischemic injury [18], possibly by delivery of cardioprotective miRs, including nanog-regulated miR-21 and HIF-1α-regulated miR-210 [49] (Fig. 3.2).

Furthermore, CPC-derived EVs had the same beneficial effects as their parent cells in the treatment of chronic heart failure in mice [23]. Like MSCs, CPCs

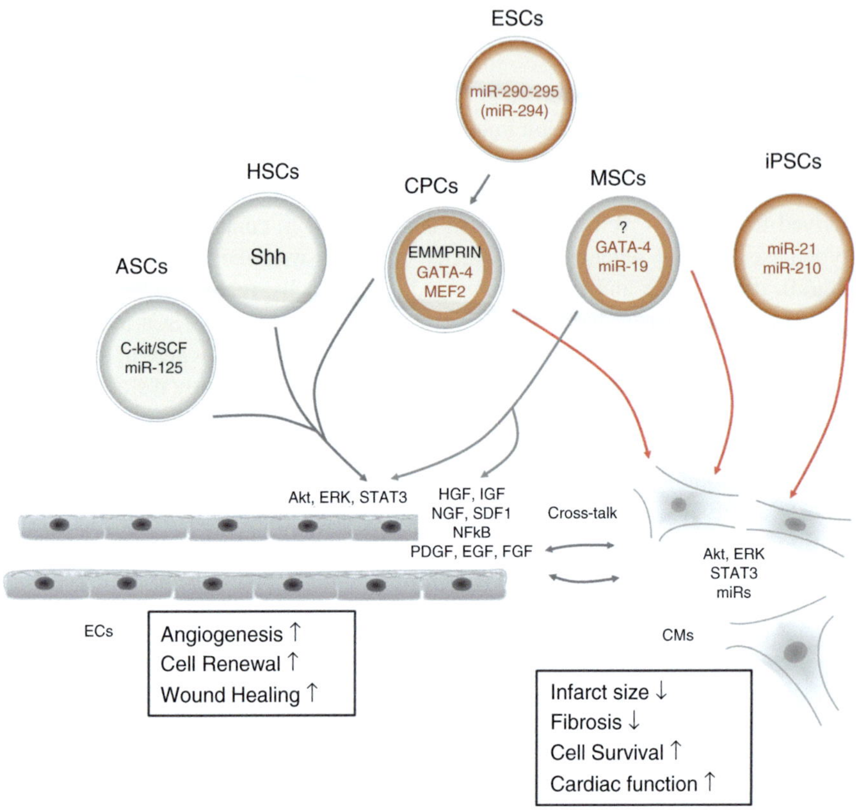

Fig. 3.2 Stem and progenitor cells secrete EVs which protect CMs. EVs released by MSCs overexpressing miR-19 and GATA-4 at the border of an ischemic region following ligation of the left anterior descending coronary artery significantly reduced infarct size and restored cardiac contractile function. Exosomes from iPSC protected against myocardial ischemia/reperfusion injury most likely by delivery of cardioprotective miRs, including nanog-regulated miR-21 and HIF-1α-regulated miR-210. CPCs release EVs containing GATA-4 and MEF2, which are important in the development of pre-cardiac cells. ESC-derived exosomes deliver ESC-specific miR-290-295 cluster and more particular miR-294 to CPCs and CMs, promoting increased survival, cell cycle progression, and proliferation in ECs and CMs. On aggregate, these EVs reduced infarct size and fibrosis and increased cell survival and cardiac function. ASC-derived exosomes carry c-kit and stem cell factor, which play a role in angiogenesis. Exosomes secreted by ASCs also promote EC angiogenesis by transferring miR-125a repressing the NOTCH ligand delta-like 4. HSCs induce angiogenesis by secreting exosomes with increased expression of Shh. CPCs also secrete exosomes with pro-angiogenic properties mediated via ERK/AKT-signaling. They contain high levels of EMMPRIN. MSCs exposed to ischemia secrete exosomes enriched in PDGF, EGF and FGF, and most notably, NFkB signaling pathway proteins inducing angiogenesis in ECs. In addition, MSC-derived exosomes activated several signaling pathways important in wound healing (AKT, ERK, and STAT3) and induced the expression of a number of growth factors (HGF, IGF1, NGF, and SDF1). On aggregate, these EVs induced angiogenesis, cell renewal and wound healing in endothelium. Abbreviations: *AKT* AKT serine/threonine kinase 1, *ASC* adipose-derived mesenchymal cell, *CPC* cardiac progenitor cell, *EGF* epidermal growth factor, *EMMPRIN* extracellular matrix metalloproteinase inducer, *ERK* extracellular signal–regulated kinase, *ESC* embryonic stem cell, *FGF* fibroblast growth factor, *GATA-4* GATA binding protein 4, *GSK* inosine/guanosine kinase, *HGF* hepatocyte growth factor, *HIF-1α* hypoxia inducible factor-1α, *HSCs* hematopoietic stem cell, *IGF-1* insulin growth factor 1, *MEF2* myocyte enhancer factor 2, *MSC* mesenchymal stem cell, *NFkB* nuclear factor-kappaB, *NGF* nerve growth factor, *PDGF* platelet-derived growth factor, *SCF* stem cell factor, *Shh* sonic hedgehog, *SDF1* stromal-derived growth factor-1, *STAT3* signal transducer and activator of transcription 3

secreted exosomes with pro-angiogenic properties mediated via ERK/Akt-signaling (Fig. 3.2). Analysis of pro-angiogenic factors revealed high levels of extracellular matrix metalloproteinase inducer (EMMPRIN) [45]. Also mouse ESC-derived exosomes augmented function in infarcted hearts by enhancing neovascularization and CM survival and by reducing fibrosis, most likely by delivery of ESC-specific miR-294 to CPCs, promoting increased survival, cell cycle progression, and proliferation [24] (Fig. 3.2).

Finally, adipose mesenchymal stem cell (ASC)-derived EVs induced in vitro vessel-like structure formation by human microvascular endothelial cells (MECs) (Fig. 3.2). Treatment of ASCs with PDGF stimulated secretion of EVs, carrying c-kit and stem cell factor, SCF, regulated by HIF-1α, inducing angiogenesis [28]. Exosomes secreted by ASCs also promoted EC angiogenesis by transferring miR-125a, which repressed the NOTCH ligand delta-like 4 [27]. Similarly, neovascularization was induced by transplantation of human HSCs to ischemic tissues in preclinical models (Fig. 3.2). These cells secreted exosomes with increased expression of the angiogenic factor sonic hedgehog (Shh), to offset age- and health-related angiogenic declines [36]. They also reduced infarct size and increased border zone capillary density compared with unmodified CD34+ cells [30].

3.4 Role of Extracellular Vesicles in Endothelial to Mesenchymal Transition

Recent studies demonstrated that the phenotypic transition of ECs into MSCs, called Endothelial to Mesenchymal Transition or EndMT, plays an important role in the pathogenesis of fibrotic disorders. During EndMT, resident ECs acquire a mesenchymal phenotype characterized by an increased ability to migrate and invade, thereby contributing to tissue remodeling and fibrosis [25, 31].

Hypoxia was found to efficiently induce human coronary artery endothelial cells (CAECs) to undergo EndMT, resulting in EndMT-derived FBs (Fig. 3.3). This process was mediated through a HIF-1α-dependent pathway, TGF/SMAD signaling pathways and DNA (cytosine-5)-methyltransferase 3A (DNMT3a)-mediated hypermethylation of Ras-Gap-like protein 1 (RASAL1) promoter and direct zinc-finger transcription factor Snail (SNAIL) induction [50]. Ultimately, this resulted in increased expression of extracellular matrix proteins such as collagen COLI and COLIII [51]. Focal myocardial fibrosis is also a structural hallmark of diabetic cardiomyopathy resulting from hyperglycemia-induced endothelial injury leading to EndMT associated with reduced expression of EC markers, such as CD31 and CD34, and increased expression of multiple mesenchymal markers, such as COLI and COLIV, and vimentin. Mir-200b reverted diabetes-associated EndMT by directly interacting with VEGF, SMAD2/3 and regulating p300-dependant histone acetylation and expression of for example extracellular matrix proteins [12].

Cardiac fibrosis does, however, not only result from a decrease in microvessels and oxygen supply but also from increased oxygen consumption by activated immune and inflammatory cells and fibroblasts leading to localized tissue hypoxia predominantly within inflammatory lesions ("inflammatory hypoxia"). Herein, HIF-1α, an oxygen-sensitive transcription factor that allows adaptation to hypoxia environments, plays an important role. Recent data suggest that the HIF-1 α-mediated metabolic shift and fibrosis is not only related to cardiovascular diseases but also to immune-related disorders [10] (Fig. 3.3). A common direct target of HIF-1α in hypoxia-induced EndMT is SNAIL [51] (Fig. 3.3).

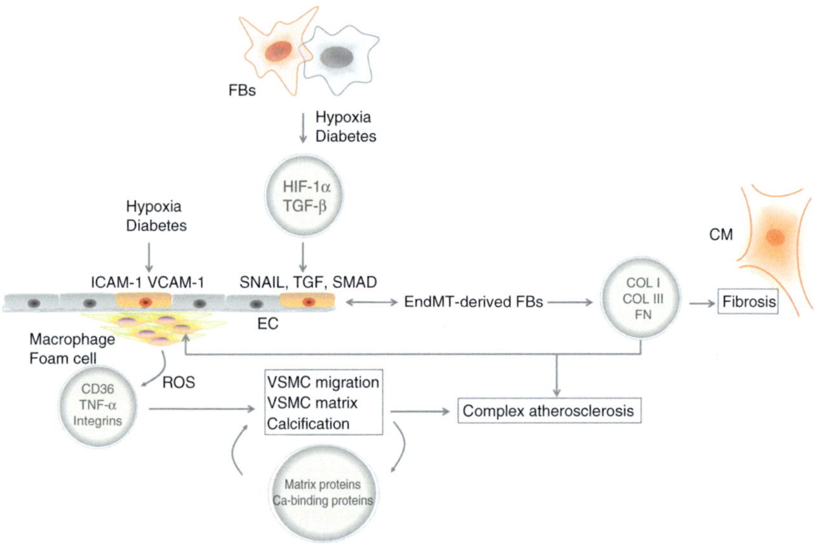

Fig. 3.3 Endothelial to Mesenchymal Transition, fibrosis and atherosclerosis. Hypoxia and diabetes induce secretion of EVs by FBs enriched in HIF-1α and TGF-β, respectively. They induce endothelial cells to undergo endothelial to mesenchymal transition (EndMT) through direct induction of SNAIL, a target of HIF-1α, and through TGF/SMAD signaling pathways and DNMT3a-mediated hypermethylation of RASAL1 promoter. EndMT results in the generation of EndMT-derived FBs with increased expression extracellular matrix proteins such as COLI, COLIII and FN. Hypoxia and diabetes cause endothelial dysfunction with increased secretion of leukocyte adhesion molecules leading to increased macrophage accumulation and oxidative stress. Macrophages secrete EVs enriched in CD36 and TNF-α which induce VSMC migration. Stimulated VSMCs secrete extracellular vesicles inducing matrix accumulation and calcification, resulting in complex atherosclerotic plaques. Collagen and fibronectin in extracellular vesicles secreted by EndMT-derived FBs exacerbate the growth of atherosclerotic plaques by inducing macrophage infiltration and making the atherosclerotic plaques more complex due to extracellular matrix deposition. Abbreviations: *COL* collagen, *DNMT3a* DNA (cytosine-5)-methyltransferase 3A, *EC* endothelial cell, *EV* extracellular vesicle, *FB* fibroblast, *FN* fibronectin, *HIF-1α* hypoxia inducible factor-1α, *ICAM-1* intercellular adhesion molecule 1, *RASAL1* Ras-Gap-like protein 1, *ROS* reactive oxygen species, *SMC* smooth muscle cells, *SNAIL* zinc-finger transcription factor Snail, *TGF-β* transforming growth factor, *TNF* tumor necrosis factor, *VCAM-1* vascular cell adhesion molecule-1

3.5 Role of Extracellular Vesicles in Atherosclerosis

Hypoxia and diabetes are known to cause endothelial dysfunction, with increased secretion of leukocyte adhesion molecules leading to increased macrophage accumulation, reactive oxygen species (ROS) and oxidative stress, ultimately leading to atherosclerosis (Fig. 3.3). The circulation of atherosclerotic patients has been shown to contain more leukocyte-derived EVs promoting vascular SMC adhesion and migration, than those of healthy participants. In addition, macrophage-derived foam cells secrete EVs which promote increased levels of vascular SMC migration and adhesion, regulating the actin cytoskeleton and focal adhesion pathways to a greater extent than macrophage-derived EVs. Western blotting revealed that foam cell-derived EVs could also promote the phosphorylation of ERK and Akt in SMCs in a time-dependent manner. Foam cell-derived EVs could enter the SMCs and transfer integrins to the surface of these cells [33].

The calcification of SMC is also mediated by regulated exosome secretion. Comparative proteomics showed that vascular SMC-derived exosomes shared components with osteoblast-derived EVs including calcium-binding and extracellular matrix proteins (Fig. 3.3). Elevated extracellular calcium was found to induce sphingomyelin phosphodiesterase 3 and the secretion of calcifying exosomes from vascular SMCs in vitro. Chemical inhibition of sphingomyelin phosphodiesterase 3 prevented SMC calcification. In vivo, EVs containing exosomes were observed in vessels from chronic kidney disease patients on dialysis, and CD63 was located where there was calcification. Importantly, factors such as TNF-α and PDGF-BB were also found to increase exosome production, leading to increased calcification of SMC [22]. Comparison between exosomes from quiescent and activated SMCs showed evidence of 29 differentially expressed proteins which are involved in cytoskeleton organization, chaperones, cell adhesion, cell signaling, metabolic pathways, vesicle trafficking and extracellular matrix production [9] (Fig. 3.3). Foam cell formation and enhanced VSMC and extracellular matrix accumulation with calcification resulted in the generation of complex atherosclerotic plaques (Fig. 3.3). The extracellular matrix proteins COL and fibronectin, induced in EndMT, were associated with increased luminal endothelial expression of intercellular adhesion molecule 1 (ICAM-1) and vascular cell adhesion molecule-1 (VCAM-1). Mechanistically, the relation between EndMT and atherosclerosis also depends on loss of endothelial fibroblast growth factor receptor 1 (FGFR1) expression and activation of endothelial TGF-β signaling [8].

3.6 Research Limitations and Future Outlook

Facing a shortage of human data we focused our review to studies containing in vivo data obtained in relevant preclinical models. Even with these restrictions, it proved to be difficult to compare the outcome of selected in vivo studies because of differences in isolation procedures of EVs (for example ultracentrifugation, size exclusion chromatography, or immunoadsorption) which may yield EVs with different sizes (not necessarily purified exosomes), use of different sets of surface markers which may yield information about cell origin of EVs and differences in extent of evaluation of cargo (miRs, proteins or none of them). Therefore standardization of separation procedures and protocols to analyze biogenesis, composition and function are needed to improve our insight in the mechanistic role of EVs [34]. Novel isolation procedures may involve microfluidics devices for on-chip isolation and quantification of circulating micro-particles [21], and microchip-based RNA extraction, amplification and RT-PCR analysis [41].

A major shortcoming of many of the previous studies is that the full content of EVs was not all analyzed or that the selection of compounds to which functional roles were attributed was biased. Indeed, many of the reviewed studies focused on one of a few numbers of compounds without displaying information about the complete cargo or specifying the reason why these compounds were selected. Furthermore, data on the signaling pathways involved in the mechanisms of action of EVs are rare and again incomplete. Therefore, an unbiased systems biology approach is needed to generate and test hypotheses about the effect of context (e.g. spatial organization of endothelial cells in relation to other cell-types like macrophages or myofibroblasts and interaction through paracrine factors) dependent on state of the disease on the functional role of specific EVs. To this end Gray and colleagues [14] proposed cue-signal-response studies using partial least square regression (PLSR) methods that study how signals (exosome content or cargo) translate cues from the secreting cell (its gene/protein expression state) to elicit a specific response in the recipient cells.

In reviewing the role of EVs in cardiovascular diseases it became obvious that, although it is generally accepted that macrophages play a crucial role in the development of cardiac fibrosis and atherosclerosis, information about the role of their precursor cells, monocytes, and exosomes derived from monocytes is limited. Recently, we found that low mitochondrial cytochrome oxidase-1, a marker of mitochondrial dysfunction, in monocyte-derived exosomes predicted the risk of future cardiovascular events in the same way as low mitochondrial cytochrome oxidase-1 in monocytes. Therefore, the role monocyte-derived exosomes should be investigated further [17].

3.7 Conclusions

The studies reviewed present evidence that extracellular vesicles secreted by CMs, ECs, immune-system-related cells (macrophages), fibroblasts (FBs) and stem cells play an important role in the regulation of endothelial cardiomyocyte and endothelial function in relation to cardiovascular diseases. However, knowledge of the underlying signaling pathways is still too sparse to identify targets for EV-mediated treatment of these diseases.

References

1. de Almeida DC, Ferreira MRP, Franzen J, Weidner CI, Frobel J, Zenke M, Costa IG, Wagner W (2016) Epigenetic classification of human mesenchymal stromal cells. Stem Cell Rep 6:168–175
2. Anderson JD, Johansson HJ, Graham CS, Vesterlund M, Pham MT, Bramlett CS, Montgomery EN, Mellema MS, Bardini RL, Contreras Z, Hoon M, Bauer G, Fink KD, Fury B, Hendrix KJ, Chedin F, EL-Andaloussi S, Hwang B, Mulligan MS, Lehtiö J, Nolta JA (2016) Comprehensive proteomic analysis of mesenchymal stem cell exosomes reveals modulation of angiogenesis via nuclear factor-KappaB signaling: MSC exosomes induce angiogenesis via NFkB pathway. Stem Cells 34:601–613
3. Arslan F, Lai RC, Smeets MB, Akeroyd L, Choo A, Aguor ENE, Timmers L, van Rijen HV, Doevendans PA, Pasterkamp G, Lim SK, de Kleijn DP (2013) Mesenchymal stem cell-derived exosomes increase ATP levels, decrease oxidative stress and activate PI3K/Akt pathway to enhance myocardial viability and prevent adverse remodeling after myocardial ischemia/reperfusion injury. Stem Cell Res 10:301–312
4. Banerjee I, Fuseler JW, Price RL, Borg TK, Baudino TA (2007) Determination of cell types and numbers during cardiac development in the neonatal and adult rat and mouse. Am J Phys Heart Circ Phys 293:H1883–H1891
5. Bang C, Batkai S, Dangwal S, Gupta SK, Foinquinos A, Holzmann A, Just A, Remke J, Zimmer K, Zeug A, Ponimaskin E, Schmiedl A, Yin X, Mayr M, Halder R, Fischer A, Engelhardt S, Wei Y, Schober A, Fiedler J, Thum T (2014) Cardiac fibroblast-derived microRNA passenger strand-enriched exosomes mediate cardiomyocyte hypertrophy. J Clin Investig 124:2136–2146
6. Bian S, Zhang L, Duan L, Wang X, Min Y, Yu H (2014) Extracellular vesicles derived from human bone marrow mesenchymal stem cells promote angiogenesis in a rat myocardial infarction model. J Mol Med 92:387–397
7. Carè A, Catalucci D, Felicetti F, Bonci D, Addario A, Gallo P, Bang M-L, Segnalini P, Gu Y, Dalton ND, Elia L, Latronico MVG, Høydal M, Autore C, Russo MA, Dorn GW, Ellingsen O, Ruiz-Lozano P, Peterson KL, Croce CM, Peschle C, Condorelli G (2007) MicroRNA-133 controls cardiac hypertrophy. Nat Med 13:613–618
8. Chen P-Y, Qin L, Baeyens N, Li G, Afolabi T, Budatha M, Tellides G, Schwartz MA, Simons M (2015) Endothelial-to-mesenchymal transition drives atherosclerosis progression. J Clin Investig 125:4514–4528
9. Comelli L, Rocchiccioli S, Smirni S, Salvetti A, Signore G, Citti L, Trivella MG, Cecchettini A (2014) Characterization of secreted vesicles from vascular smooth muscle cells. Mol Biosyst 10:1146–1152
10. Deng W, Feng X, Li X, Wang D, Sun L (2016) Hypoxia-inducible factor 1 in autoimmune diseases. Cell Immunol 303:7–15
11. Evans MJ, Kaufman MH (1981) Establishment in culture of pluripotential cells from mouse embryos. Nature 292:154–156
12. Feng B, Cao Y, Chen S, Chu X, Chu Y, Chakrabarti S (2016) miR-200b mediates endothelial-to-mesenchymal transition in diabetic cardiomyopathy. Diabetes 65:768–779
13. Giricz Z, Varga ZV, Baranyai T, Sipos P, Pálóczi K, Kittel Á, Buzás EI, Ferdinandy P (2014) Cardioprotection by remote ischemic preconditioning of the rat heart is mediated by extracellular vesicles. J Mol Cell Cardiol 68:75–78
14. Gray WD, French KM, Ghosh-Choudhary S, Maxwell JT, Brown ME, Platt MO, Searles CD, Davis ME (2015) Identification of therapeutic covariant microRNA clusters in hypoxia-treated cardiac progenitor cell exosomes using systems biology. Circ Res 116:255–263
15. Gupta S, Knowlton AA (2007) HSP60 trafficking in adult cardiac myocytes: role of the exosomal pathway. Am J Phys Heart Circ Phys 292:H3052–H3056
16. Henning RJ (2011) Stem cells in cardiac repair. Futur Cardiol 7:99–117
17. Holvoet P, Vanhaverbeke M, Bloch K, Baatsen P, Sinnaeve P, Janssens S (2016) Low MT-CO1 in monocytes and microvesicles is associated with outcome in patients with coronary artery disease. J Am Heart Assoc 5:e004207

18. Hu G, Li Q, Niu X, Hu B, Liu J, Zhou S, Guo S, Lang H, Zhang C, Wang Y, Deng Z (2015) Exosomes secreted by human-induced pluripotent stem cell-derived mesenchymal stem cells attenuate limb ischemia by promoting angiogenesis in mice. Stem Cell Res Ther 6:10
19. Huber HJ, Holvoet P (2015) Exosomes: emerging roles in communication between blood cells and vascular tissues during atherosclerosis. Curr Opin Lipidol 26:412–419
20. Hulsmans M, Holvoet P (2013) MicroRNA-containing microvesicles regulating inflammation in association with atherosclerotic disease. Cardiovasc Res 100:7–18
21. Kanwar SS, Dunlay CJ, Simeone DM, Nagrath S (2014) Microfluidic device (ExoChip) for on-chip isolation, quantification and characterization of circulating exosomes. Lab Chip 14:1891
22. Kapustin AN, Chatrou MLL, Drozdov I, Zheng Y, Davidson SM, Soong D, Furmanik M, Sanchis P, De Rosales RTM, Alvarez-Hernandez D, Shroff R, Yin X, Muller K, Skepper JN, Mayr M, Reutelingsperger CP, Chester A, Bertazzo S, Schurgers LJ, Shanahan CM (2015) Vascular smooth muscle cell calcification is mediated by regulated exosome secretion. Circ Res 116:1312–1323
23. Kervadec A, Bellamy V, El Harane N, Arakélian L, Vanneaux V, Cacciapuoti I, Nemetalla H, Périer M-C, Toeg HD, Richart A, Lemitre M, Yin M, Loyer X, Larghero J, Hagège A, Ruel M, Boulanger CM, Silvestre J-S, Menasché P, Renault NKE (2016) Cardiovascular progenitor–derived extracellular vesicles recapitulate the beneficial effects of their parent cells in the treatment of chronic heart failure. J Heart Lung Transplant 35:795–807
24. Khan M, Nickoloff E, Abramova T, Johnson J, Verma SK, Krishnamurthy P, Mackie AR, Vaughan E, Garikipati VNS, Benedict C, Ramirez V, Lambers E, Ito A, Gao E, Misener S, Luongo T, Elrod J, Qin G, Houser SR, Koch WJ, Kishore R (2015) Embryonic stem cell-derived exosomes promote endogenous repair mechanisms and enhance cardiac function following myocardial infarction. Circ Res 117:52–64
25. Kovacic JC, Mercader N, Torres M, Boehm M, Fuster V (2012) Epithelial-to-mesenchymal and endothelial-to-mesenchymal transition: from cardiovascular development to disease. Circulation 125:1795–1808
26. Kuwabara Y, Ono K, Horie T, Nishi H, Nagao K, Kinoshita M, Watanabe S, Baba O, Kojima Y, Shizuta S, Imai M, Tamura T, Kita T, Kimura T (2011) Increased microRNA-1 and microRNA-133a levels in serum of patients with cardiovascular disease indicate myocardial damage. Circ Cardiovasc Genet 4:446–454
27. Liang X, Zhang L, Wang S, Han Q, Zhao RC (2016) Exosomes secreted by mesenchymal stem cells promote endothelial cell angiogenesis by transferring miR-125a. J Cell Sci 129:2182–2189
28. Lopatina T, Bruno S, Tetta C, Kalinina N, Porta M, Camussi G (2014) Platelet-derived growth factor regulates the secretion of extracellular vesicles by adipose mesenchymal stem cells and enhances their angiogenic potential. Cell Commun Signal 12:26
29. Lyu L, Wang H, Li B, Qin Q, Qi L, Nagarkatti M, Nagarkatti P, Janicki JS, Wang XL, Cui T (2015) A critical role of cardiac fibroblast-derived exosomes in activating renin angiotensin system in cardiomyocytes. J Mol Cell Cardiol 89:268–279
30. Mackie AR, Klyachko E, Thorne T, Schultz KM, Millay M, Ito A, Kamide CE, Liu T, Gupta R, Sahoo S, Misener S, Kishore R, Losordo DW (2012) Sonic hedgehog-modified human CD34+ cells preserve cardiac function after acute myocardial infarction. Circ Res 111:312–321
31. Markwald RR, Fitzharris TP, Manasek FJ (1977) Structural development of endocardial cushions. Am J Anat 148:85–119
32. Melo SA, Sugimoto H, O'Connell JT, Kato N, Villanueva A, Vidal A, Qiu L, Vitkin E, Perelman LT, Melo CA, Lucci A, Ivan C, Calin GA, Kalluri R (2014) Cancer exosomes perform cell-independent microRNA biogenesis and promote tumorigenesis. Cancer Cell 26:707–721
33. Niu C, Wang X, Zhao M, Cai T, Liu P, Li J, Willard B, Zu L, Zhou E, Li Y, Pan B, Yang F, Zheng L (2016) Macrophage foam cell–derived extracellular vesicles promote vascular smooth muscle cell migration and adhesion. J Am Heart Assoc 5:e004099

34. Peterson MF, Otoc N, Sethi JK, Gupta A, Antes TJ (2015) Integrated systems for exosome investigation. Methods 87:31–45
35. Safari S, Malekvandfard F, Babashah S, Alizadehasl A, Sadeghizadeh M, Motavaf M (2016) Mesenchymal stem cell-derived exosomes: a novel potential therapeutic avenue for cardiac regeneration. Cell Mol Biol (Noisy-le-grand) 62:66–73
36. Sahoo S, Klychko E, Thorne T, Misener S, Schultz KM, Millay M, Ito A, Liu T, Kamide C, Agrawal H, Perlman H, Qin G, Kishore R, Losordo DW (2011) Exosomes from human CD34+ stem cells mediate their proangiogenic paracrine activity. Circ Res 109:724–728
37. Shabbir A, Cox A, Rodriguez-Menocal L, Salgado M, Badiavas EV (2015) Mesenchymal stem cell exosomes induce proliferation and migration of normal and chronic wound fibroblasts, and enhance angiogenesis in vitro. Stem Cells Dev 24:1635–1647
38. Shamblott MJ, Axelman J, Wang S, Bugg EM, Littlefield JW, Donovan PJ, Blumenthal PD, Huggins GR, Gearhart JD (1998) Derivation of pluripotent stem cells from cultured human primordial germ cells. Proc Natl Acad Sci U S A 95:13726–13731
39. Singh A, Singh A, Sen D (2016) Mesenchymal stem cells in cardiac regeneration: a detailed progress report of the last 6 years (2010–2015). Stem Cell Res Ther 7(1):82
40. Squadrito ML, Baer C, Burdet F, Maderna C, Gilfillan GD, Lyle R, Ibberson M, De Palma M (2014) Endogenous RNAs modulate microRNA sorting to exosomes and transfer to acceptor cells. Cell Rep 8:1432–1446
41. Takahashi K, Yan IK, Kim C, Kim J, Patel T (2014) Analysis of extracellular RNA by digital PCR. Front Radiat Ther Oncol 4:129
42. Théry C, Ostrowski M, Segura E (2009) Membrane vesicles as conveyors of immune responses. Nat Rev Immunol 9:581–593
43. Trial J, Entman ML, Cieslik KA (2016) Mesenchymal stem cell-derived inflammatory fibroblasts mediate interstitial fibrosis in the aging heart. J Mol Cell Cardiol 91:28–34
44. Villarroya-Beltri C, Baixauli F, Gutiérrez-Vázquez C, Sánchez-Madrid F, Mittelbrunn M (2014) Sorting it out: regulation of exosome loading. Semin Cancer Biol 28:3–13
45. Vrijsen KR, Maring JA, Chamuleau SAJ, Verhage V, Mol EA, Deddens JC, Metz CHG, Lodder K, van Eeuwijk ECM, van Dommelen SM, Doevendans PA, Smits AM, Goumans M-J, Sluijter JPG (2016) Exosomes from cardiomyocyte progenitor cells and mesenchymal stem cells stimulate angiogenesis via EMMPRIN. Adv Healthc Mater 5:2555–2565
46. Waldenström A, Gennebäck N, Hellman U, Ronquist G (2012) Cardiomyocyte microvesicles contain DNA/RNA and convey biological messages to target cells. PLoS One 7:e34653
47. Wang X, Gu H, Huang W, Peng J, Li Y, Yang L, Qin D, Essandoh K, Wang Y, Peng T, Fan G-C (2016) Hsp20-mediated activation of exosome biogenesis in cardiomyocytes improves cardiac function and angiogenesis in diabetic mice. Diabetes 65:3111–3128
48. Wang X, Huang W, Liu G, Cai W, Millard RW, Wang Y, Chang J, Peng T, Fan G-C (2014) Cardiomyocytes mediate anti-angiogenesis in type 2 diabetic rats through the exosomal transfer of miR-320 into endothelial cells. J Mol Cell Cardiol 74:139–150
49. Wang Y, Zhang L, Li Y, Chen L, Wang X, Guo W, Zhang X, Qin G, He S, Zimmerman A, Liu Y, Kim I, Weintraub NL, Tang Y (2015) Exosomes/microvesicles from induced pluripotent stem cells deliver cardioprotective miRNAs and prevent cardiomyocyte apoptosis in the ischemic myocardium. Int J Cardiol 192:61–69
50. Xu X, Tan X, Hulshoff MS, Wilhelmi T, Zeisberg M, Zeisberg EM (2016) Hypoxia-induced endothelial-mesenchymal transition is associated with RASAL1 promoter hypermethylation in human coronary endothelial cells. FEBS Lett 590:1222–1233
51. Xu X, Tan X, Tampe B, Sanchez E, Zeisberg M, Zeisberg EM (2015) Snail is a direct target of hypoxia-inducible factor 1α (HIF1α) in hypoxia-induced endothelial to mesenchymal transition of human coronary endothelial cells. J Biol Chem 290:16653–16664
52. Yu B, Kim HW, Gong M, Wang J, Millard RW, Wang Y, Ashraf M, Xu M (2015) Exosomes secreted from GATA-4 overexpressing mesenchymal stem cells serve as a reservoir of anti-apoptotic microRNAs for cardioprotection. Int J Cardiol 182:349–360

53. Yu X, Deng L, Wang D, Li N, Chen X, Cheng X, Yuan J, Gao X, Liao M, Wang M, Liao Y (2012) Mechanism of TNF-α autocrine effects in hypoxic cardiomyocytes: initiated by hypoxia inducible factor 1α, presented by exosomes. J Mol Cell Cardiol 53:848–857
54. Zhang X, Wang X, Zhu H, Kranias EG, Tang Y, Peng T, Chang J, Fan G-C (2012) Hsp20 functions as a novel cardiokine in promoting angiogenesis via activation of VEGFR2. PLoS One 7:e32765
55. Zhang Z, Yang J, Yan W, Li Y, Shen Z, Asahara T (2016) Pretreatment of cardiac stem cells with exosomes derived from mesenchymal stem cells enhances myocardial repair. J Am Heart Assoc 5:e002856

Part III
Exosomes as Biomarkers of Cardiovascular Diseases

Chapter 4
Exosomes as Diagnostic Biomarkers in Cardiovascular Diseases

Felix Jansen and Qian Li

4.1 Role and Function of Exosomes in Cardiovascular Biology

Exosomes are a subgroup of extracellular vesicles (EVs), generally ranging from 40 to 100 nm in diameter [1]. They are derived from exocytosis, through fusion of multivesicular bodies with cell membranes [1, 2]. Exosomes mediate intercellular communication by transporting biological molecules to recipient cells, playing vital roles in the regulation of vascular health [3]. Contents and quantities of exosomes are variable under different conditions and detectable in body fluids [3]. Therefore, circulating exosomes and their molecule cargos such as nucleic acids or proteins, may facilitate the diagnosis of cardiovascular diseases. Firstly, we summarize the current knowledge about the role and function of exosomes in cardiovascular biology.

4.1.1 Exosomes Participate in Cell-to-Cell Communication

Many types of cardiac cells are able to release exosomes, such as cardiomyocytes, fibroblasts, endothelial cells, cardiac progenitor cells (CPCs) and even stem cells. Numerous studies suggest that exosomes play important roles in cardiac cell-cell communication under physical and pathological conditions [1, 2]. Exosome-mediated cellular cross-talk relies on their capability to transport biomolecules from cell to cell [3].

Felix Jansen and Qian Li contributed equally to this work.

F. Jansen (✉) • Q. Li
Medizinische Klinik und Poliklinik II, Universitätsklinikum Bonn,
Sigmund-Freud-Str. 25, 53105 Bonn, Germany
e-mail: felix.jansen@ukb.uni-bonn.de

J. Xiao, S. Cretoiu (eds.), *Exosomes in Cardiovascular Diseases*, Advances in Experimental Medicine and Biology 998, DOI 10.1007/978-981-10-4397-0_4

Exosomes carry biological cargos derived from their parent cells including mRNA, microRNA (miRNA), other non-coding RNA, DNA, cytoplasmic and membrane proteins, growth factors, cytokines, lipids and others [4, 5]. After released from the cells of origin, exosomes can be taken up by adjacent or even distant cells [1, 4].

Previous studies have shown that the intercellular transfer of functional biomolecules via exosomes affects function and phenotype of the recipient cell [6]. For example, miRNA from exosomes control gene expression by binding to mRNA in the target cells inducing mRNA degradation [7]. In this context, a recent study reported that exosomes derived from CPCs reduced cardiac hypertrophy and cardiomyocyte apoptosis via exosome based miRNA-133-transfer from CPCs to cardiomyocytes [8]. Endothelial exosomes were shown to transfer miRNA-214 to target cells and mediate angiogenesis [9]. MiRNA-133 in exosomes from CPCs reduced cardiac hypertrophy and cardiomyocyte apoptosis [8]. Another exosomal miRNA derived from CPCs, miRNA-451, was demonstrated to protect cardiomyocytes from oxidative stress [10]. Also, miRNA-143/145 in EVs were reported to deliver atheroprotective messages to smooth muscle cells [11].

Moreover, it was shown that the interaction between exosomes and target cells functions in a ligand-receptor mediated way [12]. Given the distinct identification between exosome surface ligands and recipient cell receptors, exosomes specifically interact with distinct recipient cells. One study exploring exosome-target cell interaction found that exosomes derived from antigen presenting cells express major histocompatibility complex (MHC) class I and II molecules, which enable exosomes to interact with CD8+ and CD4+ T cells [12].

4.1.2 Exosomes Content and Function Reflect the Condition of the Cell of Origin

Recent studies show that exosome compositions depend on the status of the original cells at the time of exosomes biogenesis [7, 13]. During this process, signaling proteins, RNAs or other molecules expressed in parent cells can be selectively loaded into exosomes, though the specific mechanisms are still elusive [14]. In vitro experiments indicated that biological contents in exosomes depend on the pretreatment with diverse stressors. De Jong et al. showed that in endothelial cell-derived exosomes, RNA and protein content reflects the effects of cellular stress induced by hypoxia, inflammation or hyperglycemia [13]. Hypoxia stimulation induced expression of lysyl oxidase like-2 and TNF-α treatment promoted ICAM-1 expression in exosomes [13]. Short-term exposure to hyperglycemia, however, only altered exosome protein composition without affecting its RNA expression [13].

In experimental conditions of myocardial injury conditions, increasing evidences show that both contents and quantities of exosomes display significant alterations [15, 16]. Chen et al. demonstrated that miRNA-451 enriched in exosomes from CPCs show anti-apoptotic effects in ischemia/reperfusion-induced cardiomyocyte cell damage [10]. Deddens and colleagues assessed EVs and circulating miRNAs in a porcine model of ischemia/reperfusion injury, suggesting that the amount of EVs increase 2.5 h after ischemia. However, alterations of circulating miRNA levels were detectable only after 1 h [16]. In addition, they found that miRNA-133b, -208b, and -499 were selectively increased in plasma-derived EVs [16].

As exosome composition reflects the status of the releasing cell, they provide a biological illustration of the individual health status [17, 18]. Therefore, exosomes may be regarded as a potential fingerprint of diseases.

4.1.3 Exosomes act as a Protective Carrier

Exosomes contain biomolecules and can be secreted by different cell types into body fluids, e.g. plasma, urine, semen, breast milk [5, 19]. As exosomes consists of a lipid bilayer, they represent an efficient protection barrier for the intravesicular molecules [20]. For example, due to the enzymes in plasma or other body fluids, naked RNAs or proteins without protection may be easily degraded, resulting in unstable detection results [20, 21]. Exosomes, with the help of the vesicle structure, protect their contents from being affected by external environments. Consequently, previous studies have shown associations between exosomes biological contents and diverse cardiovascular diseases, which is a crucial prerequisite for biomarkers [22]. In summary, given their function as carrier and protection device for intravesicular biological contents, increasing evidences suggest exosomes and their contents as reliable biomarkers for cardiovascular diseases.

4.2 Exosome Based Biomarkers as Diagnostic Tool in Cardiovascular Diseases

The contents and quantities of exosomes are variable under different pathological cardiovascular conditions. Therefore, exosomes may function as novel diagnostic biomarkers. Exosomes incorporated miRNAs and proteins are the most widely investigated components. In addition, other cargos such as lipids can also serve as potential biomarkers.

4.2.1 Exosomes Based miRNA

Extracellular circulating miRNAs can be detected in body fluids, including blood. Furthermore, exosomes and other EVs are a main source of circulating miRNAs. Compared to freely circulating miRNAs, the majority of plasma miRNAs are concentrated in exosomes and bound to RNA-binding proteins [23, 24]. Therefore, exosomes-incorporated circulating miRNAs hold great potential as a novel diagnostic biomarker for cardiac diseases. However, the investigation of exosomal miRNAs as biomarker for cardiac diseases is still in its infancy. In the following paragraphs, we summarize the current knowledge about exosomes-incorporated miRNAs as diagnostic biomarkers in diverse cardiovascular pathologies.

4.2.1.1 Exosomes as Biomarkers for Acute Coronary Syndromes (ACS) and Myocardial Infarction (MI)

Injured cardiomyocytes release exosomes which are enriched with cardiac-specific miRNAs, such as miRNA-1, miRNA-133a [25, 26]. Accoringly, alterations of exosomal cardiac-specific miRNAs can be detected in circulation during AMI or ACS. Kuwabara et al. included 29 patients with ACS and 42 healthy controls [26]. They compared the levels of miRNA-1 and miRNA-133a in serum samples between the two groups [26]. Results indicated that levels of miRNA-1 and miRNA-133a were up-regulated in patients with ACS compared with control group [26]. Notably, they suggested that miR-133a could be released into circulation via active secretion in exosomes [26].

Compared with established biomarkers for cardiac ischemia such as cardiac troponins (troponin I and troponin T), expression levels of exosomal miRNAs change much faster and earlier in the circulation [27]. Additionally, for diagnosing AMI, one study showed that detection of circulating cardiac-specific miRNAs had higher sensitivity and specificity than troponin T [28]. The time course of miRNA release into the circulation has been explored in several studies. miRNA-1 and miRNA-133a showed the highest plasma levels 2.5 h after the onset of symptoms in MI patients [26, 27].

4.2.1.2 Exosomes as Biomarkers for Heart Failure (HF)

Brain natriuretic peptide (BNP) and N-terminal probrain natriuretic peptide (NT-proBNP) are classical biomarkers for diagnosing HF. Additionally, evidences suggest that specific exosomes-associated miRNAs are differentially regulated in the failing heart [29–31], suggesting their potential complementary role in the diagnosis of HF. Matsumoto S et al. suggested that exosomes-bound miRNAs were predictive indicators of ischemic HF in post-AMI patients [31]. In their study, a panel of 377 miRNAs were examined from serum of registry patients who developed HF after acute AMI [31]. Results show that circulating levels of p53-responsive miRNAs, including miR-192, miR-194 and miRNA-34a, increase markedly in the patients with HF [31]. Moreover, they showed that levels of miRNA-192, miRNA-194, and miRNA-34a were highly enriched in the exosome fraction [31]. They confirmed their findings in a validation cohort of 21 patients, suggesting that exosomes based circulating p53-responsive miRNAs (miRNA-192, miRNA-194 and miRNA-34a) may be regarded as predictors of ischemic HF that develops after AMI [31].

Peripartum cardiomyopathy (PPCM) associated HF has a high mortality, which lacks of specific diagnosis biomarkers. Recently, exosomes based circulating miRNA-146a is reported to be a potential hallmark for diagnosing PPCM associated HF [32]. A clinical study conducted by Halkein et al. included 30 patients with dilated CM and 38 patients with acute PPCM [32]. Patients with acute PPCM showed significantly increased levels of circulating miRNA-146a [32]. Interestingly, the level of circulating exosomal miRNA-146a decreased in the patients with acute PPCM after undergoing the standard therapy of HF, suggesting miRNA-146a might reflect a promising biomarker for PPCM associated acute HF [32]. Moreover, many studies demonstrated that circulating miRNA-92 has a close connection with cardiovascular disease [33, 34]. Goren et al. examined its concentration in the serum of HF and control groups, showing higher level of miRNA-92 in the exosomal fraction of the HF patients [35]. In summary, specific exosomes based circulating miRNAs are specifically regulated under different pathogenetic conditions of HF.

4.2.1.3 Exosomes as Biomarkers for Metabolic Diseases (Diabetes, Metabolic Syndrome)

Metabolic diseases, in particular type 2 diabetes mellitus (T2DM) and metabolic syndrome (MetS) increase the risk to develop cardiovascular diseases. T2DM is characterized by abnormal hyperglycaemia and insulin resistance. In patients with T2DM, the quantities of circulating exosomes are increased [36, 37], Additionally, the changes of exosomal contents are also indicators for T2DM. Wang et al. suggest that cardiomyocytes can secret exosomes containing higher levels of miRNA-320 under diabetic conditions, a prominent regulator of endothelial dysfunction [38]. Kong et al. and Karolina et al. respectively demonstrated that circulating miRNA-29a, miRNA-30d, and miRNA-146a were strongly related to T2DM [39, 40]. Moreover, in a large prospective cohort of 822 individuals, the diagnostic ability of circulating miRNA-126 was successfully established in diabetic patients compared to a non-diabetic control group [41–43].

It is generally accepted that MetS includes insulin resistance, central obesity, high triglycerides, dyslipidaemia and hypertension [44]. The substantial issue with diagnosing MetS is to identify unique biomarkers, especially blood-based ones. Circulating miRNAs, carried by exosomes or other EVs, are currently explored and potentially represent novel biomarkers for MetS [45, 46]. Karolina et al. compared EVs based circulating miRNAs between patients with MetS, hypertension and healthy controls. Results showed that circulating levels of miRNA-17, miRNA-197, miRNA-509-5p, miRNA-92a and miRNA-320a significantly increased in MetS patients [47]. In patients with T2DM, only miRNA-320a was found to have higher levels compared to non-diabetic patients [47]. In patients with hypertension, higher levels of miRNA-197, miRNA-92a and lower levels of miRNA-17, miRNA-509-5p and miRNA-320a were reported [47]. In summary, the changes of specific miRNAs in the circulation are reflections of different pathological conditions in MetS.

4.2.2 Exosomes Based Proteins

In addition to non-coding RNAs, exosomes are also enriched with bioactive proteins such as chemokines, heat shock proteins and growth factors, all mediating signal transduction. Many studies discussed the change of protein cargos in EVs under hypoxic conditions, which is a substantial pathological process in CAD. DeJong et al. proved that exosomes secreted by cells under hypoxic conditions in vitro contain fibronectin, collagen, and lysyl-oxidase-like 2 (LOXL2) [13]. Yu X et al. showed that exosomes derived from hypoxic cardiomyocytes mediate TNF-α production [48]. Besides, Pironti G et al. investigated exosomal proteins under pressure overload. They demonstrated that exosomes derived from cardiomyocytes contain up-regulated levels of angiotensin II (AngII) type 1 receptor (AT1R) [49]. AT1R plays a vital part in maintaining blood pressure and heart function [49]. The results were from both in vitro experiments under hypotonic conditions(143 mOsm/kg; osmotic stretch) and in vivo studies using a model of pressure overload [49]. All these studies suggest that specific proteins enriched in exosomes can be detected in the circulation in cardiac diseases, which needs further validation in clinical studies.

De Hoog et al. performed multivariate logistic regression analysis between ACS and non-ACS patients [50]. They compared three selected proteins from plasma EVs, including polygenic immunoglobulin receptor (pIgR), cystatin C and complement factor C5a (C5a) [50]. Results suggested that the three selected proteins were independently associated with ACS and show gender differences [50]. Cheowet al. compared plasma protein cargos in EVs between MI and stable angina patients and conducted quantitative proteomics profiling analysis [51]. In this study, 252 EVs based circulating proteins were significantly modulated in patients with MI compared to patients with stable angina [51]. Furthermore, they selected six up-regulated proteins after MI which reflected key factors in MI progression [51]. Specifically, Complement C1q subcomponent subunit A (C1QA) and Complement C5 (C5) are involved in post-infarct pathways of complement activation. Apoliporotein D (APOD) and Apolipoprotein C-III (APOCC3) participate in lipoprotein metabolism, Platelet glycoprotein Ib alpha chain (GP1BA) and platelet basic protein (PPBP) represent surrogate parameters for platelet activation [51]. This is the first study exploring EV based proteins in plasma as diagnostic biomarkers for MI. However, large, prospective studies are urgently needed to assess the potential role of exosomes or EVs protein contents in diagnosing coronary artery diseases.

4.2.3 Other Contents in Exosomes

In addition to non-coding RNAs and proteins, exosomal components such as lipids, mRNAs as well as DNAs may also serve as biomarkers for cardiac diseases. Aswad et al. conducted lipiomic analysis of exosomes in a lipid-induced insulin resistant mouse model. They demonstrated that exosomes derived from palmitate-treated cells are enriched in palmitate [52]. Furthermmore, exosomal palmitate participates in modifying muscle homeostasis, suggesting exosomes may mediate cell-cell communication by transferring lipids [52]. Waldenström et al. analyzed total RNAs in cardiomyocytes derived exosomes and compared their profiles with original cells [14]. Results show that 1520 mRNAs are similar and shared by exosomes and their parent cells [14]. Remarkably, most of the mRNAs are related with biological mechanisms, including gene expression changes [14].

4.3 Clinical Application and Future Perspectives

Exosomal cargos, including non-coding RNAs, proteins or lipids, are potential diagnostic biomarkers for cardiac diseases. However, many issues remain unsolved, which have to be addressed, before exosome contents might be helpful in the clinical routine diagnosing cardiovascular diseases. Although current studies have given valuable insights into the formation of exosomes, the exact mechanisms of cargo selection and packaging remain unknown. Additionally, the interactions between exosomal contents, such as circulating miRNAs, and target cells have not been completely understood. Furthermore, exosomes are only one kind of carrier for extracellular intravesicular bioactive molecules. Other EVs including microparticles and apoptotic bodies are also protective carriers of biological contents in the circulation. Besides, ideal biomarkers should be unique and sensitive for certain disease. However, the same variation of exosomal quantities or cargos can be observed under different pathological conditions. Therefore, more efforts should be put into identifying specific exosomes based biomarkers. Ultimately, previous studies about exosomal biomarkers have limitations. Some of them fail to take time-dependent exosomes release kinetics into consideration; It is known that exosomal cargos depend on the pathological stage of diseases [16]. Most studies just analyzed exosomes number and contents at one time point without continuous measurements; The majority of studies only a selected number of miRNAs or proteins were analyzed, which are not able to fully cover all the differentially expressed exosomal contents.

To address the mentioned limitations and further explore the potential use of exosome-associated biomarkers for cardiac disease, we may take more factors into account. Firstly, for the sake of enhancing specificity and sensitivity of biomarkers, it is a potential way to combine multiple candidate biomarkers together to diagnose certain disease. Given different kinds of miRNAs have distinct release kinetics under MI, a combination of miRNA-208a, miRNA-133, miRNA-1, and miRNA-499-5p in one test could potentially identify patients with MI in a broader time range [27]. The combinations of exosomal and traditional biomarkers may also enhance their diagnostic potential. On the other hand, beyond some small-scale studies by individual research groups, large randomized clinical trials are needed to address the mentioned limitations, which are necessary with carefully selected control groups. The mass of data gained from experimental and clinical science also calls for methods to deal with it, such as high quality meta-analysis. Additionally, it is a tough task to balance good purity and high quantity of exosomal cargos derived from circulation. Therefore, clinical applications require efficient detection protocols and methods, which may depend on the progression of novel technology, e.g. innovation of detection kits.

4.4 General Conclusion

Exosomes play important roles in the development and progression of cardiovascular diseases through modulating intercellular communication between different cardiac cell types. Exosomes numbers and its cargos are reflections of the (patho) physiological status of their parent cells, which provides great potential to use them for diagnosing cardiac diseases. To establish exosomes as biomarkers in cardiovascular diseases, convinced clinical trials as well as efficient and standardized detection methods are prerequisites.

References

1. Simons M, Raposo G (2009) Exosomes—vesicular carriers for intercellular communication. Curr Opin Cell Biol 21(4):575–581
2. Das S, Halushka MK (2015) Extracellular vesicle microRNA transfer in cardiovascular disease. Cardiovasc Pathol 24(4):199–206
3. Amabile N, Rautou PE, Tedgui A, Boulanger CM (2010) Microparticles: key protagonists in cardiovascular disorders. Semin Thromb Hemost 36(8):907–916
4. Owens AP III, Mackman N (2011) Microparticles in hemostasis and thrombosis. Circ Res 108(10):1284–1297
5. Khalyfa A, Gozal D (2014) Exosomal miRNAs as potential biomarkers of cardiovascular risk in children. J Transl Med 12:162
6. Pfeifer P, Werner N, Jansen F (2015) Role and function of microRNAs in extracellular vesicles in cardiovascular biology. Biomed Res Int 2015:161393
7. Loyer X, Vion AC, Tedgui A, Boulanger CM (2014) Microvesicles as cell-cell messengers in cardiovascular diseases. Circ Res 114(2):345–353
8. Izarra A, Moscoso I, Levent E, Canon S, Cerrada I, Diez-Juan A, Blanca V, Nunez-Gil IJ, Valiente I, Ruiz-Sauri A, Sepulveda P, Tiburcy M, Zimmermann WH, Bernad A (2014) miR-133a enhances the protective capacity of cardiac progenitors cells after myocardial infarction. Stem Cell Rep 3(6):1029–1042
9. Lawson C, Vicencio JM, Yellon DM, Davidson SM (2016) Microvesicles and exosomes: new players in metabolic and cardiovascular disease. J Endocrinol 228(2):R57–R71
10. Chen L, Wang Y, Pan Y, Zhang L, Shen C, Qin G, Ashraf M, Weintraub N, Ma G, Tang Y (2013) Cardiac progenitor-derived exosomes protect ischemic myocardium from acute ischemia/reperfusion injury. Biochem Biophys Res Commun 431(3):566–571
11. Boon RA, Hergenreider E, Dimmeler S (2012) Atheroprotective mechanisms of shear stress-regulated microRNAs. Thromb Haemost 108(4):616–620
12. Thery C, Duban L, Segura E, Veron P, Lantz O, Amigorena S (2002) Indirect activation of naive CD4+ T cells by dendritic cell-derived exosomes. Nat Immunol 3(12):1156–1162
13. de Jong OG, Verhaar MC, Chen Y, Vader P, Gremmels H, Posthuma G, Schiffelers RM, Gucek M, van Balkom BW (2012) Cellular stress conditions are reflected in the protein and RNA content of endothelial cell-derived exosomes. J Extracell Vesicles 1. doi:10.3402/jev.v1i0.18396

14. Waldenstrom A, Genneback N, Hellman U, Ronquist G (2012) Cardiomyocyte microvesicles contain DNA/RNA and convey biological messages to target cells. PLoS One 7(4):e34653
15. Iaconetti C, Sorrentino S, De Rosa S, Indolfi C (2016) Exosomal miRNAs in heart disease. Physiology (Bethesda) 31(1):16–24
16. Deddens JC, Vrijsen KR, Colijn JM, Oerlemans MI, Metz CH, van der Vlist EJ, Nolte-'t Hoen EN, den Ouden K, Jansen Of Lorkeers SJ, van der Spoel TI, Koudstaal S, Arkesteijn GJ, Wauben MH, van Laake LW, Doevendans PA, Chamuleau SA, Sluijter JP (2016) Circulating extracellular vesicles contain miRNAs and are released as early biomarkers for cardiac injury. J Cardiovasc Transl Res 9(4):291–301
17. Stremersch S, De Smedt SC, Raemdonck K (2016) Therapeutic and diagnostic applications of extracellular vesicles. J Control Release 244(Pt B):167–183
18. Yellon DM, Davidson SM (2014) Exosomes: nanoparticles involved in cardioprotection? Circ Res 114(2):325–332
19. Tual-Chalot S, Leonetti D, Andriantsitohaina R, Martinez MC (2011) Microvesicles: intercellular vectors of biological messages. Mol Interv 11(2):88–94
20. Valadi H, Ekstrom K, Bossios A, Sjostrand M, Lee JJ, Lotvall JO (2007) Exosome-mediated transfer of mRNAs and microRNAs is a novel mechanism of genetic exchange between cells. Nat Cell Biol 9(6):654–659
21. Dear JW, Street JM, Bailey MA (2013) Urinary exosomes: a reservoir for biomarker discovery and potential mediators of intrarenal signalling. Proteomics 13(10–11):1572–1580
22. Keller S, Sanderson MP, Stoeck A, Altevogt P (2006) Exosomes: from biogenesis and secretion to biological function. Immunol Lett 107(2):102–108
23. Gallo A, Tandon M, Alevizos I, Illei GG (2012) The majority of microRNAs detectable in serum and saliva is concentrated in exosomes. PLoS One 7(3):e30679
24. Arroyo JD, Chevillet JR, Kroh EM, Ruf IK, Pritchard CC, Gibson DF, Mitchell PS, Bennett CF, Pogosova-Agadjanyan EL, Stirewalt DL, Tait JF, Tewari M (2011) Argonaute2 complexes carry a population of circulating microRNAs independent of vesicles in human plasma. Proc Natl Acad Sci U S A 108(12):5003–5008
25. Li C, Pei F, Zhu X, Duan DD, Zeng C (2012) Circulating microRNAs as novel and sensitive biomarkers of acute myocardial infarction. Clin Biochem 45(10–11):727–732
26. Kuwabara Y, Ono K, Horie T, Nishi H, Nagao K, Kinoshita M, Watanabe S, Baba O, Kojima Y, Shizuta S, Imai M, Tamura T, Kita T, Kimura T (2011) Increased microRNA-1 and microRNA-133a levels in serum of patients with cardiovascular disease indicate myocardial damage. Circ Cardiovasc Genet 4(4):446–454
27. D'Alessandra Y, Devanna P, Limana F, Straino S, Di Carlo A, Brambilla PG, Rubino M, Carena MC, Spazzafumo L, De Simone M, Micheli B, Biglioli P, Achilli F, Martelli F, Maggiolini S, Marenzi G, Pompilio G, Capogrossi MC (2010) Circulating microRNAs are new and sensitive biomarkers of myocardial infarction. Eur Heart J 31(22):2765–2773
28. Sayed AS, Xia K, Yang TL, Peng J (2013) Circulating microRNAs: a potential role in diagnosis and prognosis of acute myocardial infarction. Dis Markers 35(5):561–566
29. Divakaran V, Mann DL (2008) The emerging role of microRNAs in cardiac remodeling and heart failure. Circ Res 103(10):1072–1083
30. Small EM, Olson EN (2011) Pervasive roles of microRNAs in cardiovascular biology. Nature 469(7330):336–342
31. Matsumoto S, Sakata Y, Suna S, Nakatani D, Usami M, Hara M, Kitamura T, Hamasaki T, Nanto S, Kawahara Y, Komuro I (2013) Circulating p53-responsive microRNAs are predictive indicators of heart failure after acute myocardial infarction. Circ Res 113(3):322–326
32. Halkein J, Tabruyn SP, Ricke-Hoch M, Haghikia A, Nguyen NQ, Scherr M, Castermans K, Malvaux L, Lambert V, Thiry M, Sliwa K, Noel A, Martial JA, Hilfiker-Kleiner D, Struman I (2013) MicroRNA-146a is a therapeutic target and biomarker for peripartum cardiomyopathy. J Clin Investig 123(5):2143–2154
33. Jansen F, Yang X, Proebsting S, Hoelscher M, Przybilla D, Baumann K, Schmitz T, Dolf A, Endl E, Franklin BS, Sinning JM, Vasa-Nicotera M, Nickenig G, Werner N (2014) MicroRNA expression in circulating microvesicles predicts cardiovascular events in patients with coronary artery disease. J Am Heart Assoc 3(6):e001249

34. Natarelli L, Schober A (2015) MicroRNAs and the response to injury in atherosclerosis. Hamostaseologie 35(2):142–150
35. Goren Y, Kushnir M, Zafrir B, Tabak S, Lewis BS, Amir O (2012) Serum levels of microRNAs in patients with heart failure. Eur J Heart Fail 14(2):147–154
36. Jayachandran M, Litwiller RD, Lahr BD, Bailey KR, Owen WG, Mulvagh SL, Heit JA, Hodis HN, Harman SM, Miller VM (2011) Alterations in platelet function and cell-derived microvesicles in recently menopausal women: relationship to metabolic syndrome and atherogenic risk. J Cardiovasc Transl Res 4(6):811–822
37. Diamant M, Nieuwland R, Pablo RF, Sturk A, Smit JW, Radder JK (2002) Elevated numbers of tissue-factor exposing microparticles correlate with components of the metabolic syndrome in uncomplicated type 2 diabetes mellitus. Circulation 106(19):2442–2447
38. Wang X, Huang W, Liu G, Cai W, Millard RW, Wang Y, Chang J, Peng T, Fan GC (2014) Cardiomyocytes mediate anti-angiogenesis in type 2 diabetic rats through the exosomal transfer of miR-320 into endothelial cells. J Mol Cell Cardiol 74:139–150
39. Karolina DS, Armugam A, Tavintharan S, Wong MT, Lim SC, Sum CF, Jeyaseelan K (2011) MicroRNA 144 impairs insulin signaling by inhibiting the expression of insulin receptor substrate 1 in type 2 diabetes mellitus. PLoS One 6(8):e22839
40. Kong L, Zhu J, Han W, Jiang X, Xu M, Zhao Y, Dong Q, Pang Z, Guan Q, Gao L, Zhao J, Zhao L (2011) Significance of serum microRNAs in pre-diabetes and newly diagnosed type 2 diabetes: a clinical study. Acta Diabetol 48(1):61–69
41. Tijsen AJ, Pinto YM, Creemers EE (2012) Circulating microRNAs as diagnostic biomarkers for cardiovascular diseases. Am J Physiol Heart Circ Physiol 303(9):H1085–H1095
42. Jansen F, Yang X, Hoelscher M, Cattelan A, Schmitz T, Proebsting S, Wenzel D, Vosen S, Franklin BS, Fleischmann BK, Nickenig G, Werner N (2013) Endothelial microparticle-mediated transfer of MicroRNA-126 promotes vascular endothelial cell repair via SPRED1 and is abrogated in glucose-damaged endothelial microparticles. Circulation 128(18):2026–2038
43. Jansen F, Wang H, Przybilla D, Franklin BS, Dolf A, Pfeifer P, Schmitz T, Flender A, Endl E, Nickenig G, Werner N (2016) Vascular endothelial microparticles-incorporated microRNAs are altered in patients with diabetes mellitus. Cardiovasc Diabetol 15:49
44. O'Neill S, O'Driscoll L (2015) Metabolic syndrome: a closer look at the growing epidemic and its associated pathologies. Obes Rev 16(1):1–12
45. O'Neill S, Bohl M, Gregersen S, Hermansen K, O'Driscoll L (2016) Blood-based biomarkers for metabolic syndrome. Trends Endocrinol Metab 27(6):363–374
46. Rottiers V, Naar AM (2012) MicroRNAs in metabolism and metabolic disorders. Nat Rev Mol Cell Biol 13(4):239–250
47. Karolina DS, Tavintharan S, Armugam A, Sepramaniam S, Pek SL, Wong MT, Lim SC, Sum CF, Jeyaseelan K (2012) Circulating miRNA profiles in patients with metabolic syndrome. J Clin Endocrinol Metab 97(12):E2271–E2276
48. Yu X, Deng L, Wang D, Li N, Chen X, Cheng X, Yuan J, Gao X, Liao M, Wang M, Liao Y (2012) Mechanism of TNF-alpha autocrine effects in hypoxic cardiomyocytes: initiated by hypoxia inducible factor 1alpha, presented by exosomes. J Mol Cell Cardiol 53(6):848–857
49. Pironti G, Strachan RT, Abraham D, Mon-Wei Yu S, Chen M, Chen W, Hanada K, Mao L, Watson LJ, Rockman HA (2015) Circulating exosomes induced by cardiac pressure overload contain functional angiotensin II type 1 receptors. Circulation 131(24):2120–2130
50. de Hoog VC, Timmers L, Schoneveld AH, Wang JW, van de Weg SM, Sze SK, van Keulen JK, Hoes AW, den Ruijter HM, de Kleijn DP, Mosterd A (2013) Serum extracellular vesicle protein levels are associated with acute coronary syndrome. Eur Heart J Acute Cardiovasc Care 2(1):53–60
51. Cheow ES, Cheng WC, Lee CN, de Kleijn D, Sorokin V, Sze SK (2016) Plasma-derived extracellular vesicles contain predictive biomarkers and potential therapeutic targets for myocardial ischemic (MI) injury. Mol Cell Proteomics 15(8):2628–2640
52. Aswad H, Forterre A, Wiklander OP, Vial G, Danty-Berger E, Jalabert A, Lamaziere A, Meugnier E, Pesenti S, Ott C, Chikh K, El-Andaloussi S, Vidal H, Lefai E, Rieusset J, Rome S (2014) Exosomes participate in the alteration of muscle homeostasis during lipid-induced insulin resistance in mice. Diabetologia 57(10):2155–2164

Chapter 5
Exosomes-Based Biomarkers for the Prognosis of Cardiovascular Diseases

Yihua Bei, Pujiao Yu, Dragos Cretoiu, Sanda Maria Cretoiu, and Junjie Xiao

5.1 Background

Cardiovascular diseases (CVDs), such as coronary artery diseases, heart failure, and stroke, have a high prevalence and annually increasing incidence with high mortality and morbidity. In 2012, approximately 17.5 million people died from CVDs and this number is predicted to increase up to 23 million by 2030. Although treatment for CVDs has made great progress in the past decades, the 5-year survival rate for CVDs patients fails to be considerably improved [1–3]. With the increasing heterogeneity and complexity observed in the progression of CVDs, the need for specific and accurate diagnosis of disease state and molecular monitoring of disease progression have become more urgent. Besides that, to identify a biomarker with high sensitivity and specificity for assessing the prognosis of CVDs is also necessary for optimizing personalized treatment and reducing mortality.

Exosomes, in some studies also called extracellular vesicles (EVs), are capable of carrying signaling molecules including mRNA, miRNA, and proteins, and can serve as a platform of complex intercellular communications [4, 5]. Exosomes have been proved to be accessible in nearly all body fluids including blood [6], urine [7, 8],

---

Y. Bei • J. Xiao (✉)
Cardiac Regeneration and Ageing Lab, School of Life Science, Shanghai University, 333 Nan Chen Road, Shanghai 200444, China
e-mail: junjiexiao@shu.edu.cn

P. Yu
Department of Cardiology, Tongji Hospital, Tongji University School of Medicine, Shanghai 200065, China

D. Cretoiu • S.M. Cretoiu
Victor Babes National Institute of Pathology, Bucharest 050096, Romania

Division of Cellular and Molecular Biology and Histology, Carol Davila University of Medicine and Pharmacy, Bucharest 050474, Romania

J. Xiao, S. Cretoiu (eds.), *Exosomes in Cardiovascular Diseases*, Advances in Experimental Medicine and Biology 998, DOI 10.1007/978-981-10-4397-0_5

saliva, ascites [9], and bronchoalveolar lavage [10, 11]. The vast repertoire of proteins and nucleic acids that can be packaged within exosomes appear to reflect the extensive, diverse, and complex signaling potential of these vesicles. The roles of exosomes have been investigated in many different areas such as immunology, pregnancy, cancer, neurodegenerative diseases, and cardiovascular diseases, and increasing studies have focused on its potential in diagnostic and prognostic monitoring in recent years [12–16]. Along with numerous studies, it has been proved that exosomes could contain diverse biological contents and might be reflective of disease state, thus making them potential candidates for non-invasive biomarkers [17].

Exosome count is a simple and intuitional index in prognostic monitoring of diseases as exosomes derived from different cells could show up first in quantity under pathological conditions. For example, higher levels of CD24-positive exosomes were reported to indicate poor prognosis and reduce patients' survival rate with ovarian cancer [18]. Stage III and IV melanoma patients showed increased levels of plasmatic caveolin-1 and CD63-positive exosomes and exosomes associated with caveolin-1 displayed a specificity of 96.3% and a sensitivity of 69% [19]. Meanwhile, analysis of contents within exosomes, such as proteins or miRNAs, is also a critical observed index and this index has already been widely applied in the field of cancer research. For example, different tumors are characterized by distinct and specific miRNA profile [20], and exosomal miRNAs have been suggested as diagnostic and prognostic indicators for lung cancer, esophageal squamous cell carcinoma, prostate cancer, breast cancer, glioblastoma, ovarian cancer, and other cancer types. These miRNAs are also correlated with the stage and degree of cancer progression [21–25]. Exosomal miR-1290 and miR-375 were reported as prognostic markers in castration-resistant prostate cancer [26]. Exosomal miR-34 was suggested to be a predictive biomarker for response to docetaxel with clinical relevance to progression of prostate cancer [27]. miR-195 levels isolated from plasma exosomes are proved to be higher in breast tumor patients. miR-195 and let-7a levels decreased upon tumor removal, showing a possible application as a prognostic biomarker as well [28]. Besides that, identification of cancer aggressiveness is also an important marker in prognosis. Levels of EpCAM and CD24 present in exosomes were correlated with the aggressiveness of ovarian cancer and cytoplasmic localization of CD24 occurred in tumors with high invasive potential [29]. Exosomes isolated from metastatic cells were proved to be capable of making primary tumor aggressive by permanently converting bone marrow progenitors in melanoma [30]. Besides the studies of exosomes as biomarkers in cancer research, the potential of exosomes in the prognosis of CVDs has also been increasingly reported. Here we will summarize the advance of exosome-based biomarkers in CVDs, particularly by focusing on their potential in the prognosis of CVDs.

5.2 Exosomes-Based Biomarkers for the Prognosis of Coronary Artery Diseases

Coronary artery diseases (CADs), including stable angina, unstable angina, myocardial infarction, and sudden cardiac death, are among the most common causes of death in the world [31]. Treatments in acute phase of CADs, such as coronary bypass surgery, balloon dilatation of coronary vessels and percutaneous transluminal coronary angioplasty can alleviate the initial cardiac damage and decrease the mortality dramatically [32]. Treatments during the chronic phase are also eager. Atherosclerosis is a major pathological change of most CAD patients during the chronic phase, which could even be present at a young age. Identification of the people at high risk for an adverse cardiovascular event is challenging in a background of atherosclerosis already existing for decades. Meanwhile, tissue repair after the acute phase of CADs is also important for cardiac functional recovery.

Various methods of assessments have been used in individuals with CADs in order to predict or stratify their risk of mortality and also to provide a personalized treatment for patients. As an example, the Killip classification is a classic system used in the clinic to assess individuals with myocardial infarction and it is also well acknowledged that individuals with a low Killip class are less likely to die within the first 30 days after their myocardial infarction than those with a high one [33]. The Canadian Cardiovascular Society grading of angina pectoris (also referred to as the CCS Angina Grading Scale or the CCS functional Classification of Angina) is another classic assessment for patients with CADs. Increasing CCS class was proved to be associated with increased long-term mortality, and it has been reported that 8-year mortality rates were 20.5%, 24.1%, 40.4%, and 35.3% among CCS class I, II, III, and IV patients, respectively [34].

In addition to these cardiovascular risk stratification tools using established risk factors, several circulating biomarkers have also been shown to be associated with adverse cardiovascular outcomes. These prognostic biomarkers include high-sensitivity cardiac troponin (hs-cTn), high-sensitivity C-reactive protein (hs-CRP), and creatine kinase MB (CK-MB) [35]. Several studies have shown that the baseline troponin elevation was associated with a poor outcome after percutaneous coronary intervention (PCI) and an isolated elevation of cTnT was a predictor of long-term risk of death [36, 37]. However, none of these biomarkers have been included in the clinical guidelines yet. The progress of CADs always accompany with apoptosis of cardiomyocytes, activation of pro-inflammatory cytokines, and release of various signaling molecules while counts of exosomes and regulation of exosomal contents are also followed with these changes.

5.2.1 *Proteins from Exosomes in Coronary Artery Diseases*

After 1 year of follow-up under controlled diet and drug treatment, patients with CADs from the study arms of the PREDIMED trial that did not have a future cardiovascular event showed reduced exosomes shedding from lymphocytes ($CD3^+$/$CD45^+$) and smooth muscle cells (α-SMA^+) [38]. These two kinds of circulating exosomes have been shown to add prognostic value to Framingham Risk Score (FRS) in the cluster model for cardiovascular event prediction. The ROC curve analyses demonstrated that AUC increased from 0.548 ± 0.087 [95% CI: 0.377–0.719] ($P = 0.585$) to 0.748 ± 0.078 [95% CI: 0.596–0.900] ($P = 0.006$), which indicated that these vesicles display a higher predictive value for cardiovascular event than the commonly used FRS [38]. In another heterogeneous population of 488 patients with CADs, CD144-positive EVs were found to be a reliable predictor for cardiovascular events. The incidences of cardiovascular death and acute coronary syndrome were significantly higher in the high-EVs group than in the low-EVs group and Kaplan-Meier analysis based on high and low levels of biomarkers also showed a significantly higher probability of cardiovascular events in presence of this kind of EVs during the follow [39]. Likewise, after a 6-year follow-up, the level of $CD31^+$/Annexin V^+ (endothelial) EVs were proved to be an independent risk factor for major adverse cardiac events (MACE) and cardiovascular death in patients with CADs. Inclusion of these EV levels into a classical risk factor model increased predictive value (c-statistic from 0.637 to 0.702, $P = 0.03$) [40].

Cellular stress conditions can be reflected in the protein content of exosomes [41]. A study based on proteomics evaluated the risk of EV protein levels on the occurrence of new vascular events in patients with clinically vascular manifest in a large cohort (n = 1060). With a long-term follow-up (median 6.4 years), the EV Cystatin C, Serpin F2, Serpin G1, and CD14 levels were identified to be related to an increased risk for cardiovascular events and death. The Cystatin C and CD14 levels were related to an elevated risk for vascular events (hazard ratio (HR): 1.27; 95%CI: 1.07–1.52 and HR: 1.32; 95%CI: 1.12–1.55, respectively) and all-cause mortality (HR: 1.41; 95%CI:1.18–1.69 and HR: 1.36; 95%CI: 1.15–1.62, respectively). These HRs were corrected for age, gender, and estimated glomerular filtration rate (eGFR) [42].

5.2.2 miRNAs from Exosomes in Coronary Artery Diseases

Injured cardiomyocytes were shown to release miRNAs via exosomes and many evidence suggested that EV-packaged miRNAs might represent functional mediators in CADs. Hergenreider et al. described an atheroprotective communication between endothelial cells and vascular smooth muscle cells via endothelial cell-derived exosomes in a miR-143/145 dependent way [43]. It was also reported that cardiac progenitor cells (CPCs)-derived miR-146a enriched exosomes were cardioprotective in myocardial infarction [44]. These findings suggest that exosomal miRNAs might have a potential as prognostic biomarkers in CADs. A study analyzing the records of patients with myocardial infarction registered in the OACIS reported that serum levels of p53-responsive miRNAs (including miR-192, miR-194, and miR-34a) particularly inside EVs were highly related to the development of heart failure [45]. Furthermore, a trail enrolling 181 patients with stable CAD quantified 10 miRNAs involved in the regulation of vascular performance both in plasma and circulating EVs by reverse transcription polymerase chain reaction (RT-PCR). Interestingly, after a median follow-up duration of 6.1 years, increased expression of miR-126 and miR-199a in circulating EVs were found to be significantly associated with a lower MACE rate, while none of the plasma miRNA expression were predictive of cardiovascular events [46].

5.3 Exosomes-Based Biomarkers for the Prognosis of Heart Failure

Chronic heart failure(HF) is the end stage of many CVDs. The frequencies of emerging cases of HF arise annually worldwide. A reliable and simple tool that enables physicians to have a realistic expectation of the prognosis and to guide treatment options is a major challenge in the management of HF. European Society of Cardiology and the American Heart Association/American Colleges of Cardiology have published a series of guidelines for the diagnosis and treatments of patients with HF [47, 48] and have also recommended several biomarkers with high predictive value, such as natriuretic peptides, pentraxin-3, galectin-3, and cardiac specific troponin [49]. Although these methods and biomarkers have some advantages, there are still limitations in risk stratification among HF patients. The New York Heart Association (NYHA) Functional Classification is a classical way to evaluate the extent of HF and remains arguably the most important prognostic method in routine clinical use. But this assessment still has its limitations such as challenge of consistently classifying patients in class II or III. Moreover, relying on patients' subjective statements instead of objective condition might greatly influence the accuracy of results [50]. B-type natriuretic peptide (BNP) is among the most studied and validated biomarkers used in chronic HF. They are used as aids in the diagnosis and assessment of severity of HF. But in some clinical trails, it was found that measurement of BNP did not conclusively affect hospital mortality rates or had no apparent effects on clinical outcomes, making its prognostic value debatable [51]. The progression of HF accompanies cardiac remodeling and vascular dysfunction. Exosomes are reported to play important roles in regulating all these pathological processes, indicating its potential in assessing patients with HF [52, 53].

5.3.1 Proteins from Exosomes in Heart Failure

It is well known that a degeneration of vascular integrity and endothelial function, imbalanced angiogenesis, and inflammation critically contribute to the development of HF. The levels of circulating EVs derived from endothelial cells were previously reported to be closely related to vascular endothelial dysfunction. A cohort of 169 HF patients with NYHA class I or more were studied for measurement of the plasma endothelium-derived EVs. The endothelium-derived EV levels were found to be significantly increased with NYHA functional class [EVs median (range): healthy, 0.325 (0.164–0.354) × 10^6/mL; NYHA I, 0.484 (0.426–0.575); NYHA II, 0.646 (0.439–0.795); and NYHA III/IV, 0.786 (0.569–1.026), $P < 0.001$]. The Kaplan-Meier analysis demonstrated that the high-EVs group was associated with a significantly higher probability of cardiovascular events [54]. Likewise, a cohort of 388 patients with chronic HF during 3 years showed that $CD144^+/CD31^+$/Annexin V^+EVs, and $CD31^+$/AnnexinV^+ EVs remained statistically significant for cumulative endpoint. Notably, combining endothelial apoptotic EVs ($CD144^+$/$CD31^+$/Annexin V^+EVs and $CD31^+$/AnnexinV^+ EVs) to the standard ABC model(NYHA class, decreased LVEF < 45%, NT-pro-BNP, and hs-CRP) may further improve the relative integrated discrimination indices(IDI) for cumulative endpoint by 11.4% and 10.5% respectively [53]. In another study, Berezin et al. reported that a higher ratio of endothelial-derived apoptotic microparticles ($CD31^+$/Annexin V^+EVs) to mononuclear progenitor cells (MPCs) is related to all-cause mortality in patients with chronic HF [55]. These vesicles may help determine HF patients at high risk and make biomarker-guided therapies possible for HF patients.

5.3.2 *miRNAs from Exosomes in Heart Failure*

Matusmoto et al. reported that several p53-responsive miRNAs, including miR-192, miR-194, and miR-34a, were elevated in the serum of HF patients enrolled in the Osaka Acute Coronary Insufficiency study [45]. Remarkably, these three miRNAs were found to be highly enriched in the exosome fraction in HF patients as compared to controls [56]. Knockdown of all these three miRNAs enhanced cell survival after treatment with doxorubicin, while overexpression of these miRNAs decreased cell viability *in vitro*, suggesting that higher level of these miRNAs may lead to the deterioration of heart function.

5.4 Exosomes-Based Biomarkers for the Prognosis of Stroke

Stroke, which causes part of the brain not functioning properly, is a major cause of death and disability worldwide and also represents one important branch of CVDs [32, 57, 58]. About half of the patients who have a stroke history live less than 1 year [59]. Intracranial stenosis is one of the important causes of stroke and also critically contributes to the pathogenesis of cerebral vascular diseases. It is highly necessary to give a risk stratification to patients with intracranial vascular diseases which might help improve functional recovery after stroke and avoid the second hit from stroke recurrence.

Several modifiable risk factors have been combined into validated clinical prediction tools such as the Framingham Stroke Risk Score (FSRS), CHA2DS2-VASc risk scores, and the National Institutes of Health Stroke Scale (NIHSS) [60–62]. Actually, these scores are not routinely used in clinic because of their complexity. Thus, combining the biomarkers into these risk scores might help discriminate the individuals at future risk for cardiovascular events. Recently, some unique RNA expression changes have been observed in peripheral blood after stroke and a large case-control study was used to identify novel RNA biomarkers of stroke. Peripheral blood MCEMP1 gene expression was proved to be able to predict 1-month disability and morality after stroke [63]. Importantly, exosomes can also be synthesized and released from brain cells [64], serving as important intercellular players and participating in the pathological processes during stroke and neural injury [65]. Furthermore, exosomes are able to pass blood brain barrier (BBB) and can also be detected in the cerebrospinal fluid (CSF) [64, 66], making them possible to be ideal biomarkers to reflect the pathological progresses of cerebral vascular diseases.

5.4.1 Proteins from Exosomes in Stroke

The levels of platelet-derived EVs are reported to predict shorter event-free survival in patients with prior stroke [67]. EV protein levels of Cystatin C and CD14 are related to an elevated risk for vascular events in patients with coronary arterial diseases. Moreover, these two proteins are also proved to be associated with the progression of cerebral atrophy in patients with manifest vascular disease [68]. Similarly, a proteomics analysis was performed based on EVs enriched plasma of patients with lacunar cerebral infarction (LACI) and different outcome events, such as cognitive decline and recurrent vascular events. A total of 45 patients following a non-disabling LACI along with 17 matched control subjects were enrolled and myelin basic protein, integrin alpha-IIb, talin-1, and filamin-A and proteins of coagulation cascade (fibrinogen alpha chain and fibrinogen beta chain) were up-regulated while albumin was down-regulated in patients with recurrent vascular event or cognitive decline without any recurrence of vascular events [69].

5.4.2 miRNAs from Exosomes in Stroke

Distinct miRNA expression patterns have been reported in various stroke pathogenic processes, including hyperlipidaemia, hypertension and plaque rupture [70], and atherosclerosis [71]. Circulating miRNAs have also been demonstrated to be candidates as biomarkers for stroke. In a recent study, 65 patients with acute ischemic stroke (AIS) at the acute stage and 66 non-stroke volunteers were enrolled and serum exosome miRNAs were analyzed by qPCRs. Serum exosomal miR-9 and miR-124 levels were found to be significantly higher in stroke patients. These two exosomal miRNAs were also positively correlated with National Institutes of Health Stroke Scale(NIHSS) scores, infarct volumes, and serum IL-6 level. Thus, circulating exosomal miR-9 and miR-124 might be promising biomarkers for AIS diagnosis, though their potential as biomarkers in the prediction of post stroke complications remains to be further investigated [72].

5.5 Exosomes-Based Biomarkers for the Prognosis of Other CVDs

5.5.1 *Hypertension*

Hypertension affects 16–37% of the population globally [73]. The angiotensin II type I receptor (AT1R) is a key receptor in cardiovascular system and plays a key role in hypertension [74]. Exosomes were demonstrated to be important mediators for transferring functional AT1Rs as evidenced in cellular stretch *in vitro* model [75]. In mice with cardiac pressure overload, it was further demonstrated that cardiomyocytes are the major source of exosomes containing AT1R [75]. These exosomes are secreted into circulation upon pressure overload and are taken up by skeletal muscles and resistance vessels. The transfer of AT1Rs may increase circulating AngII levels and further aggravate cardiac function during blood pressure overload [75]. Besides that, the adenosine 2A receptors and dopamine receptors can be released with EVs and transferred to other cells retaining their abilities to increase blood pressure and promote cardiac remodeling [76]. All these findings leave potential for using exosomes as prognostic biomarkers for assessing hypertension patients.

5.5.2 *Cardiac Arrhythmia*

Cardiac arrhythmia is a group of disorders with the electrical conduction system of the heart which affects millions of people. Atrial fibrillation (AF) is one of the major types of arrhythmia and will also increase the risk of HF, kidney failure, coronary heart diseases, and death [77]. Improved techniques have identified various biomarkers including vWF, D-dimer, and natriuretic peptides in the prediction of AF and related outcomes [78]. With the understanding of the pathogenesis of AF, more and more biomarkers have been identified to improve the predictive power of risk stratification of AF. EVs are associated with AF due to their direct or indirect contribution to the noxious amplification loops [79] and may also have a predictive function in AF [80, 81]. Based on a specific monoclonal antibody AD-1, serum levels of EVs were found to be increased in nonvalvular atrial fibrillation, showing EVs-bound IL-1β as an independent predictor of platelet activation [82]. Besides that, high levels of platelet-derived EVs were proved to have a significantly direct relationship with the severity of rheumatic mitral stenosis in patients with valvular AF [83].

5.5.3 *Cardiomyopathy*

Cardiomyopathy is a group of diseases that affect the heart muscle, including hypertrophic cardiomyopathy (HCM), dilated cardiomyopathy (DCM), restrictive cardiomyopathy (RCM), arrhythmogenic right ventricular dysplasia (ARVD), and broken heart syndrome. Notably, DCM is a genetically heterogeneous disease with multifactorial pathogenesis, which is responsible for about 30% cases of congestive HF [84]. Cardiac stem cell-derived exosomes have previously been proved to be capable of improving cardiac function and reducing apoptosis and fibrosis in mice with doxorubicin-induced DCM [85]. Interestingly, cardiac-specific overexpression of HSP20 could attenuate cardiac dysfunction and adverse remodeling in mice with diabetic cardiomyopathy, through an increase in the number as well as a change of composite cargo in cardiomyocyte-derived exosomes [86]. Moreover, a study recruiting 13 patients with advanced congestive HF (NYHA III/IV) due to chronic inflammatory DCM found that immunoadsorption was able to improve endothelial function by reducing the release of EVs and endothelial-derived EVs into the circulation [87]. These findings indicate that EVs are probably involved in the pathological processes of cardiomyopathy and may also have a potential in reflecting the outcome or prognostics of cardiomyopathy patients, though further studies are highly needed.

5.5.4 *Valvular Heart Diseases*

Valvular heart disease is one important branch of CVDs which may progress to HF eventually without prompt treatment [88]. Treatment of valvular heart disease relies on surgery to improve outcomes, such as surgical valve repair or prosthetic valves implantation [88]. Although these treatments can improve symptoms immediately, patients are still necessary to be evaluated for long-term recovery. A study with 6 months follow-up including 60 patients with severe mitral regurgitation showed that miR-133a, miR-199a-3p, and miR-221 were upregulated in patients with improved heart function and upregulation of miR-590-5p and miR-25 was associated with improved right ventricle function [89].

5.5.5 *Pulmonary Arterial Hypertension*

Pulmonary arterial hypertension (PAH) is a condition characterized by increased pulmonary arterial blood pressure and remodeling of distal pulmonary arterial circulation. With a persistent increase in pulmonary vascular resistance, patients will ultimately develop right ventricular failure and death [90]. PAH is a heterogeneous disease which may be linked to other diseases, such as connective tissue disease, portal hypertension, and congenital heart disease [91]. The prognosis of PAH has an untreated median survival of 2–3 years after diagnosis [92]. Exosomes have been demonstrated to participate in the development of PAH. The circulating exosomes isolated from the monocrotaline (MCT)-treated mice were able to induce pulmonary hypertension (PH) in healthy mice, while mesenchymal stem cell-derived exosomes can blunt the development of hypoxic PH. The exosomes isolated from MCT-treated mice and idiopathic PAH patients contained increased levels of miR-19b, miR-20a, miR-20b, and miR-145 [93] . These dysregulated exosomal miRNAs may have a great potential for the diagnosis and prognosis of PAH, though large cohort studies are warranted.

5.6 Perspective

With the development of surgical and medical treatments, the acute therapeutic effect for CVD patients have been largely improved, however the long-term outcome and survival remains poor. Because of the deteriorated cardiac function or second hit from the relapse, patients with previous heart diseases have very high mortality and are vulnerable to be disabled. It also leads to a new problem that patients with chronic CVDs are increased sharply, thus bringing serious burden to the society and families. Individualized treatment according to proper prognostic assessment is a key point in the management of CVD patients. Effective and accurate assessment for prognosis is important for guiding treatment for CVD patients.

Exosomes play crucial roles in cell-to-cell communications and widely participate in various physiological and pathological progresses of human body. Exosomes have relatively stable structure which can protect their cargos from destruction and are present in almost all biological fluids, making them reliable biomarkers, which may reflect disease progression earlier and more precisely. Although great efforts have been made in basic and clinical studies to learn about the roles of exosomes in CVDs, the use of exosomes as biomarkers for CVDs is still years away. The review here shows the beginning of a new era into the exploration of exosomes as novel prognositic biomarkers to predict the outcome and survival of CVD patients (Table 5.1 and Fig. 5.1). However, multicenter studies and large cohorts of patients are still warranted. Moreover, compared to the present prognostic biomarkers that have been well-established for CVDs, if exosome-based biomarkers achieve additionally benefits remains to be determined.

Table 5.1 Prognostic values of extracellular vesicles in cardiovascular diseases

Diseases	Effector molecules	Biosample	Method	Function	Reference
Coronary artery diseases	$CD3^+/CD45^+$ and α-SMA^+ EVs	Plasma	Flow cytometry	Add prognostic value to FRS and display a predictive value for cardiovascular event	[38]
	$CD144^+$ EVs	Plasma	Flow cytometry	Predictor for cardiovascular events	[39]
	$CD31^+$/Annexin V^+ EVs	Plasma	Flow cytometry	Predictor for MACE and cardiovascular death	[40]
	EVs containing Cystatin C, Serpin F2, Serpin G1, and CD14	Plasma	Multiplex immunoassay	Predictor for cardiovascular event and cardiovascular death	[42]
	Exosomes containing miR-126/miR-199a	Plasma	qPCR	Predictor for a reduced risk of MACE	[46]
Heart failure	Endothelium-derived EVs	Plasma	Flow cytometry	Potential biomarker of endothelial dysfunction in HF risk stratification	[54]
	$CD144^+/CD31^+$/Annexin V^+ EVs, and $CD31^+$/AnnexinV^+EVs	Plasma	Flow cytometry	Predictor for poor prognosis in cardiac failure	[53]
	$CD31^+$/Annexin V^+ EVs	Plasma	Flow cytometry	Predictor for all-cause mortality in HF patients	[55]
	Exosomes containing miR-192/miR-194/miR-34a	Serum	qPCR	Highly related to heart failure development after myocardial infarction	[56]

(continued)

Table 5.1 (continued)

Diseases	Effector molecules	Biosample	Method	Function	Reference
Stroke	Platelet-derived EVs	Plasma	Flow cytometry	Predictor for shorter event-free survival after stroke	[67]
	Exosomes containing Cystatin C and CD14 EVs containing upregulated myelin basic protein and proteins of coagulation cascade and focal adhesion, and downregulated albumin	Plasma Plasma	Multiplex immunoassay iTRAQ	Predictor for vascular events and progression of cerebral atrophy Prognostic biomarker candidates of lacunar infarction	[68] [69]
	Exosomes containing miR-9/miR-124	Serum	Nanoparticle-tracking analysis, qPCR	Upregulated in stroke patients and correlated with NIHSS scores, infarct volumes, and serum IL-6 level	[72]
Hypertension	Exosomes containing AT1R	Serum	Electron microscopy and radio ligand receptor binding assay	Increase circulating AngII level and aggravate cardiac function during blood pressure overload	[75]
Cardiac arrhythmia	AD-1$^+$ EVs containing IL-1β	Serum	ELISA	Predictor for platelet activation in nonvalvular atrial fibrillation	[82]
Valvular heart diseases	Exosomes containing miR-133a/miR-199a-3p/miR-221/miR-590-5p/miR-25	Serum	qPCR	Predictor for functional recovery after MitraClip repair of severe mitral regurgitation	[89]
Pulmonary arterial hypertension	Exosomes containing miR-19b, miR-20a, miR-20b, and miR-145	Plasma	qPCR	Potential predictor for pulmonary hypertension	[93]

FRS Framingham Risk Score, *MACE* major adverse cardiac events, *HF* heart failure, *NIHSS* the National Institutes of Health Stroke Scale, *AT1R* angiotensin II type I receptor

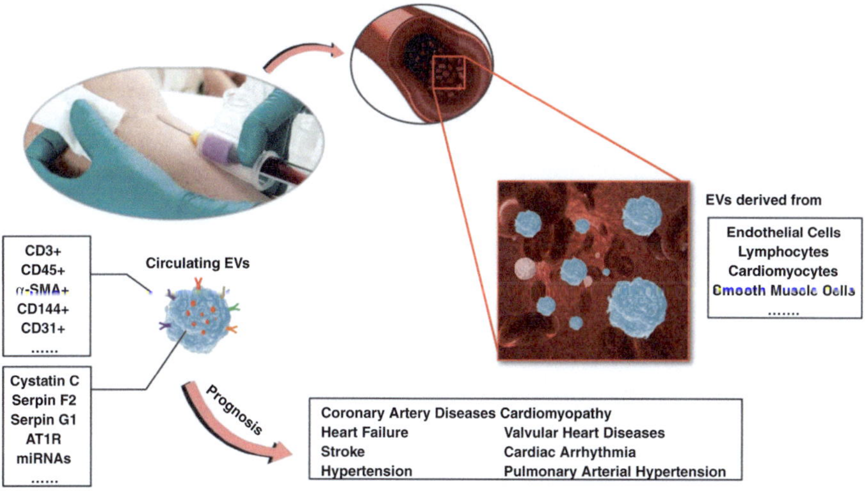

Fig. 5.1 Exosomes-based biomarkers for the prognosis of cardiovascular diseases

Acknowledgements This work was supported by the grants from National Natural Science Foundation of China (81570362, 91639101 and 81200169 to J.J. Xiao and 81400647 to Y. Bei), and the development fund for Shanghai talents (to J.J. Xiao).

Competing Financial Interests The authors declare no competing financial interests.

References

1. Mathers CD, Loncar D (2006) Projections of global mortality and burden of disease from 2002 to 2030. PLoS Med 3(11):e442–e442
2. Mendis S, Lindholm LH, Mancia G, Whitworth J, Alderman M, Lim S, Heagerty T (2007) World Health Organization (WHO) and International Society of Hypertension (ISH) risk prediction charts: assessment of cardiovascular risk for prevention and control of cardiovascular disease in low and middle-income countries. J Hypertens 25(8):1578–1582
3. WHO, Federation WH, Organization WS (2012) Global atlas on cardiovascular disease prevention and control
4. Simpson RJ, Jensen SS, Lim JW (2008) Proteomic profiling of exosomes: current perspectives. Proteomics 8(19):4083–4099
5. Valadi H, Ekstrom K, Bossios A, Sjostrand M, Lee JJ, Lotvall JO (2007) Exosome-mediated transfer of mRNAs and microRNAs is a novel mechanism of genetic exchange between cells. Nat Cell Biol 9(6):654–659
6. Caby MP, Lankar D, Vincendeau-Scherrer C, Raposo G, Bonnerot C (2005) Exosomal-like vesicles are present in human blood plasma. Int Immunol 17(7):879–887
7. Pisitkun T, Shen RF, Knepper MA (2004) Identification and proteomic profiling of exosomes in human urine. Proc Natl Acad Sci U S A 101(36):13368–13373
8. Hoorn EJ, Pisitkun T, Zietse R, Gross P, Frokiaer J, Wang NS, Gonzales PA, Star RA, Knepper MA (2005) Prospects for urinary proteomics: exosomes as a source of urinary biomarkers. Nephrology (Carlton) 10(3):283–290

9. Koga K, Matsumoto K, Akiyoshi T, Kubo M, Yamanaka N, Tasaki A, Nakashima H, Nakamura M, Kuroki S, Tanaka M, Katano M (2005) Purification, characterization and biological significance of tumor-derived exosomes. Anticancer Res 25(6a):3703–3707
10. Andre F, Schartz NE, Movassagh M, Flament C, Pautier P, Morice P, Pomel C, Lhomme C, Escudier B, Le Chevalier T, Tursz T, Amigorena S, Raposo G, Angevin E, Zitvogel L (2002) Malignant effusions and immunogenic tumour-derived exosomes. Lancet 360(9329):295–305
11. Bard MP, Hegmans JP, Hemmes A, Luider TM, Willemsen R, Severijnen LA, van Meerbeeck JP, Burgers SA, Hoogsteden HC, Lambrecht BN (2004) Proteomic analysis of exosomes isolated from human malignant pleural effusions. Am J Respir Cell Mol Biol 31(1):114–121
12. Mincheva-Nilsson L, Baranov V (2010) The role of placental exosomes in reproduction. Am J Reprod Immunol 63(6):520–533
13. Chaput N, Thery C (2011) Exosomes: immune properties and potential clinical implementations. Semin Immunopathol 33(5):419–440
14. Silverman JM, Reiner NE (2011) Exosomes and other microvesicles in infection biology: organelles with unanticipated phenotypes. Cell Microbiol 13(1):1–9
15. Schorey JS, Bhatnagar S (2008) Exosome function: from tumor immunology to pathogen biology. Traffic 9(6):871–881
16. Vella LJ, Sharples RA, Nisbet RM, Cappai R, Hill AF (2008) The role of exosomes in the processing of proteins associated with neurodegenerative diseases. Eur Biophys J 37(3):323–332
17. Tickner JA, Urquhart AJ, Stephenson SA, Richard DJ, O'Byrne KJ (2014) Functions and therapeutic roles of exosomes in cancer. Front Oncol 4:127
18. Runz S, Keller S, Rupp C, Stoeck A, Issa Y, Koensgen D, Mustea A, Sehouli J, Kristiansen G, Altevogt P (2007) Malignant ascites-derived exosomes of ovarian carcinoma patients contain CD24 and EpCAM. Gynecol Oncol 107(3):563–571
19. Logozzi M, De Milito A, Lugini L, Borghi M, Calabro L, Spada M, Perdicchio M, Marino ML, Federici C, Iessi E, Brambilla D, Venturi G, Lozupone F, Santinami M, Huber V, Maio M, Rivoltini L, Fais S (2009) High levels of exosomes expressing CD63 and caveolin-1 in plasma of melanoma patients. PLoS One 4(4):e5219
20. Choi D, Lee TH, Spinelli C, Chennakrishnaiah S, D'Asti E, Rak J (2017) Extracellular vesicle communication pathways as regulatory targets of oncogenic transformation. Semin Cell Dev Biol 67:11–22
21. Wang WT, Chen YQ (2014) Circulating miRNAs in cancer: from detection to therapy. J Hematol Oncol 7:86
22. Rabinowits G, Gercel-Taylor C, Day JM, Taylor DD, Kloecker GH (2009) Exosomal microRNA: a diagnostic marker for lung cancer. Clin Lung Cancer 10(1):42–46
23. Lau C, Kim Y, Chia D, Spielmann N, Eibl G, Elashoff D, Wei F, Lin YL, Moro A, Grogan T, Chiang S, Feinstein E, Schafer C, Farrell J, Wong DT (2013) Role of pancreatic cancer-derived exosomes in salivary biomarker development. J Biol Chem 288(37):26888–26897
24. Nilsson J, Skog J, Nordstrand A, Baranov V, Mincheva-Nilsson L, Breakefield XO, Widmark A (2009) Prostate cancer-derived urine exosomes: a novel approach to biomarkers for prostate cancer. Br J Cancer 100(10):1603–1607
25. Corcoran C, Friel AM, Duffy MJ, Crown J, O'Driscoll L (2011) Intracellular and extracellular microRNAs in breast cancer. Clin Chem 57(1):18–32
26. Huang X, Yuan T, Liang M, Du M, Xia S, Dittmar R, Wang D, See W, Costello BA, Quevedo F, Tan W, Nandy D, Bevan GH, Longenbach S, Sun Z, Lu Y, Wang T, Thibodeau SN, Boardman L, Kohli M, Wang L (2015) Exosomal miR-1290 and miR-375 as prognostic markers in castration-resistant prostate cancer. Eur Urol 67(1):33–41
27. Corcoran C, Rani S, O'Driscoll L (2014) miR-34a is an intracellular and exosomal predictive biomarker for response to docetaxel with clinical relevance to prostate cancer progression. Prostate 74(13):1320–1334
28. Heneghan HM, Miller N, Lowery AJ, Sweeney KJ, Newell J, Kerin MJ (2010) Circulating microRNAs as novel minimally invasive biomarkers for breast cancer. Ann Surg 251(3):499–505
29. Aktas IY, Bugdayci M, Usubutun A (2012) Expression of p16, p53, CD24, EpCAM and calretinin in serous borderline tumors of the ovary. Turk Patoloji Derg 28(3):220–230

30. Peinado H, Aleckovic M, Lavotshkin S, Matei I, Costa-Silva B, Moreno-Bueno G, Hergueta-Redondo M, Williams C, Garcia-Santos G, Ghajar C, Nitadori-Hoshino A, Hoffman C, Badal K, Garcia BA, Callahan MK, Yuan J, Martins VR, Skog J, Kaplan RN, Brady MS, Wolchok JD, Chapman PB, Kang Y, Bromberg J, Lyden D (2012) Melanoma exosomes educate bone marrow progenitor cells toward a pro-metastatic phenotype through MET. Nat Med 18(6):883–891
31. Wong ND (2014) Epidemiological studies of CHD and the evolution of preventive cardiology. Nat Rev Cardiol 11(5):276–289
32. Mortality GBD, Causes of Death C (2015) Global, regional, and national age–sex specific all-cause and cause-specific mortality for 240 causes of death, 1990–2013: a systematic analysis for the Global Burden of Disease Study 2013. Lancet 385(9963):117–171
33. Killip T, Kimball JT (1967) Treatment of myocardial infarction in a coronary care unit. Am J Cardiol 20(4):457–464
34. Kaul P, Naylor CD, Armstrong PW, Mark DB, Theroux P, Dagenais GR (2009) Assessment of activity status and survival according to the Canadian Cardiovascular Society angina classification. Can J Cardiol 25(7):e225–e231
35. Ndrepepa G, Colleran R, Braun S, Cassese S, Hieber J, Fusaro M, Kufner S, Ott I, Byrne RA, Husser O, Hengstenberg C, Laugwitz KL, Schunkert H, Kastrati A (2016) High-sensitivity Troponin T and mortality after elective percutaneous coronary intervention. J Am Coll Cardiol 68(21):2259–2268
36. Prasad A, Rihal CS, Lennon RJ, Singh M, Jaffe AS, Holmes DR (2008) Significance of periprocedural myonecrosis on outcomes after percutaneous coronary intervention. An analysis of preintervention and postintervention troponin T levels in 5487 patients. Circ Cardiovasc Interv 1(1):10–19
37. Miller WL, Garratt KN, Burritt MF, Lennon RJ, Reeder GS, Jaffe AS (2006) Baseline troponin level: key to understanding the importance of post-PCI troponin elevations. Eur Heart J 27(9):1061–1069
38. Chiva-Blanch G, Suades R, Crespo J, Vilahur G, Arderiu G, Padro T, Corella D, Salas-Salvado J, Aros F, Martinez-Gonzalez MA, Ros E, Fito M, Estruch R, Badimon L (2016) CD3(+)/CD45(+) and SMA-alpha(+) circulating microparticles are increased in individuals at high cardiovascular risk who will develop a major cardiovascular event. Int J Cardiol 208:147–149
39. Nozaki T, Sugiyama S, Koga H, Sugamura K, Ohba K, Matsuzawa Y, Sumida H, Matsui K, Jinnouchi H, Ogawa H (2009) Significance of a multiple biomarkers strategy including endothelial dysfunction to improve risk stratification for cardiovascular events in patients at high risk for coronary heart disease. J Am Coll Cardiol 54(7):601–608
40. Sinning JM, Losch J, Walenta K, Bohm M, Nickenig G, Werner N (2011) Circulating CD31+/Annexin V+ microparticles correlate with cardiovascular outcomes. Eur Heart J 32(16):2034–2041
41. de Jong OG, Verhaar MC, Chen Y, Vader P, Gremmels H, Posthuma G, Schiffelers RM, Gucek M, van Balkom BW (2012) Cellular stress conditions are reflected in the protein and RNA content of endothelial cell-derived exosomes. J Extracell Vesicles 1. doi:10.3402/jev.v1i0.18396
42. Kanhai DA, Visseren FL, van der Graaf Y, Schoneveld AH, Catanzariti LM, Timmers L, Kappelle LJ, Uiterwaal CS, Lim SK, Sze SK, Pasterkamp G, de Kleijn DP (2013) Microvesicle protein levels are associated with increased risk for future vascular events and mortality in patients with clinically manifest vascular disease. Int J Cardiol 168(3):2358–2363
43. Hergenreider E, Heydt S, Treguer K, Boettger T, Horrevoets AJ, Zeiher AM, Scheffer MP, Frangakis AS, Yin X, Mayr M, Braun T, Urbich C, Boon RA, Dimmeler S (2012) Atheroprotective communication between endothelial cells and smooth muscle cells through miRNAs. Nat Cell Biol 14(3):249–256
44. Barile L, Lionetti V, Cervio E, Matteucci M, Gherghiceanu M, Popescu LM, Torre T, Siclari F, Moccetti T, Vassalli G (2014) Extracellular vesicles from human cardiac progenitor cells inhibit cardiomyocyte apoptosis and improve cardiac function after myocardial infarction. Cardiovasc Res 103(4):530–541
45. Matsumoto S, Sakata Y, Suna S, Nakatani D, Usami M, Hara M, Kitamura T, Hamasaki T, Nanto S, Kawahara Y, Komuro I (2013) Circulating p53-responsive microRNAs are predictive indicators of heart failure after acute myocardial infarction. Circ Res 113(3):322–326

46. Jansen F, Yang X, Proebsting S, Hoelscher M, Przybilla D, Baumann K, Schmitz T, Dolf A, Endl E, Franklin BS, Sinning JM, Vasa-Nicotera M, Nickenig G, Werner N (2014) MicroRNA expression in circulating microvesicles predicts cardiovascular events in patients with coronary artery disease. J Am Heart Assoc 3(6):e001249
47. Hunt SA, Abraham WT, Chin MH, Feldman AM, Francis GS, Ganiats TG, Jessup M, Konstam MA, Mancini DM, Michl K, Oates JA, Rahko PS, Silver MA, Stevenson LW, Yancy CW, Antman EM, Smith Jr SC, Adams CD, Anderson JL, Faxon DP, Fuster V, Halperin JL, Hiratzka LF, Hunt SA, Jacobs AK, Nishimura R, Ornato JP, Page RL, Riegel B (2005) ACC/AHA 2005 guideline update for the diagnosis and management of chronic heart failure in the adult—summary article: a report of the American College of Cardiology/American Heart Association Task Force on practice guidelines (Writing Committee to update the 2001 guidelines for the evaluation and management of heart failure). J Am Coll Cardiol 46(6):1116–1143
48. McMurray JJV, Adamopoulos S, Anker SD, Auricchio A, Bohm M, Dickstein K, Falk V, Filippatos G, Fonseca C, Gómez Sánchez MA, Jaarsma T, Kober L, Lip GYH, Maggioni AP, Parkhomenko A, Pieske BM, Popescu BA, Ronnevik PK, Rutten FH, Schwitter J, Seferovic P, Stepinska J, Trindade PT, Voors AA, Zannad F, Zeiher A (2013) ESC guidelines for the diagnosis and treatment of acute and chronic heart failure 2012. Revista Portuguesa de Cardiologia (English Edition) 32(7–8):641–642
49. Dubin R, Li Y, Ix JH, Shlipak MG, Whooley M, Peralta CA (2012) Associations of pentraxin-3 with cardiovascular events, incident heart failure, and mortality among persons with coronary heart disease: data from the Heart and Soul Study. Am Heart J 163(2):274–279
50. Raphael C, Briscoe C, Davies J, Whinnett ZI, Manisty C, Sutton R, Mayet J, Francis DP (2007) Limitations of the New York Heart Association functional classification system and self-reported walking distances in chronic heart failure. Heart 93(4):476–482
51. Schneider HG, Lam L, Lokuge A, Krum H, Naughton MT, De Villiers Smit P, Bystrzycki A, Eccleston D, Federman J, Flannery G, Cameron P (2009) B-type natriuretic peptide testing, clinical outcomes, and health services use in emergency department patients with dyspnea: a randomized trial. Ann Intern Med 150(6):365–371
52. Waldenstrom A, Ronquist G (2014) Role of exosomes in myocardial remodeling. Circ Res 114(2):315–324
53. Berezin AE, Kremzer AA, Martovitskaya YV, Samura TA, Berezina TA (2015) The predictive role of circulating microparticles in patients with chronic heart failure. BBA Clin 3:18–24
54. Nozaki T, Sugiyama S, Sugamura K, Ohba K, Matsuzawa Y, Konishi M, Matsubara J, Akiyama E, Sumida H, Matsui K, Jinnouchi H, Ogawa H (2010) Prognostic value of endothelial microparticles in patients with heart failure. Eur J Heart Fail 12(11):1223–1228
55. Berezin AE, Kremzer AA, Samura TA, Martovitskaya YV, Malinovskiy YV, Oleshko SV, Berezina TA (2015) Predictive value of apoptotic microparticles to mononuclear progenitor cells ratio in advanced chronic heart failure patients. J Cardiol 65(5):403–411
56. Evans S, Mann DL (2013) Circulating p53-responsive microRNAs as predictive biomarkers in heart failure after acute myocardial infarction: the long and arduous road from scientific discovery to clinical utility. Circ Res 113(3):242–244
57. Lackland DT, Roccella EJ, Deutsch AF, Fornage M, George MG, Howard G, Kissela BM, Kittner SJ, Lichtman JH, Lisabeth LD, Schwamm LH, Smith EE, Towfighi A (2014) Factors influencing the decline in stroke mortality. A statement from the American Heart Association/American Stroke Association. Stroke 45(1):315–353
58. Lo EH, Dalkara T, Moskowitz MA (2003) Mechanisms, challenges and opportunities in stroke. Nat Rev Neurosci 4(5):399–414
59. Donnan GA, Fisher M, Macleod M, Davis SM (2008) Stroke. Lancet 371(9624):1612–1623
60. Harrison SL, de Craen AJ, Kerse N, Teh R, Granic A, Davies K, Wesnes KA, den Elzen WP, Gussekloo J, Kirkwood TB, Robinson L, Jagger C, Siervo M, Stephan BC (2017) Predicting risk of cognitive decline in very old adults using three models: the Framingham stroke risk profile; the cardiovascular risk factors, aging, and dementia model; and oxi-inflammatory biomarkers. J Am Geriatr Soc 65(2):381–389. doi:10.1111/jgs.14532

61. Christophersen IE, Yin X, Larson MG, Lubitz SA, Magnani JW, McManus DD, Ellinor PT, Benjamin EJ (2016) A comparison of the CHARGE-AF and the CHA2DS2-VASc risk scores for prediction of atrial fibrillation in the Framingham Heart Study. Am Heart J 178:45–54
62. Fischer U, Arnold M, Nedeltchev K, Brekenfeld C, Ballinari P, Remonda L, Schroth G, Mattle HP (2005) NIHSS score and arteriographic findings in acute ischemic stroke. Stroke 36(10):2121–2125
63. Raman K, O'Donnell MJ, Czlonkowska A, Duarte YC, Lopez-Jaramillo P, Penaherrera E, Sharma M, Shoamanesh A, Skowronska M, Yusuf S, Pare G (2016) Peripheral blood MCEMP1 gene expression as a biomarker for stroke prognosis. Stroke 47(3):652–658
64. Fruhbeis C, Frohlich D, Kuo WP, Kramer-Albers EM (2013) Extracellular vesicles as mediators of neuron-glia communication. Front Cell Neurosci 7:182
65. Zhang ZG, Chopp M (2016) Exosomes in stroke pathogenesis and therapy. J Clin Investig 126(4):1190–1197
66. Zhuang X, Xiang X, Grizzle W, Sun D, Zhang S, Axtell RC, Ju S, Mu J, Zhang L, Steinman L, Miller D, Zhang H-G (2011) Treatment of brain inflammatory diseases by delivering exosome encapsulated anti-inflammatory drugs from the nasal region to the brain. Mol Ther 19(10):1769–1779
67. Lee YJ, Jy W, Horstman LL, Janania J, Reyes Y, Kelley RE, Ahn YS (1993) Elevated platelet microparticles in transient ischemic attacks, lacunar infarcts, and multiinfarct dementias. Thromb Res 72(4):295–304
68. Kanhai DA, de Kleijn DPV, Kappelle LJ, Uiterwaal C, van der Graaf Y, Pasterkamp G, Geerlings MI, Visseren FLJ (2014) Extracellular vesicle protein levels are related to brain atrophy and cerebral white matter lesions in patients with manifest vascular disease: the SMART-MR study. BMJ Open 4(1)
69. Koomen JM, Datta A, Chen CP, Sze SK (2014) Discovery of prognostic biomarker candidates of lacunar infarction by quantitative proteomics of microvesicles enriched plasma. PLoS One 9(4):e94663
70. Rink C, Khanna S (2011) MicroRNA in ischemic stroke etiology and pathology. Physiol Genomics 43(10):521–528
71. Cipollone F, Felicioni L, Sarzani R, Ucchino S, Spigonardo F, Mandolini C, Malatesta S, Bucci M, Mammarella C, Santovito D, de Lutiis F, Marchetti A, Mezzetti A, Buttitta F (2011) A unique microRNA signature associated with plaque instability in humans. Stroke 42(9):2556–2563
72. Ji Q, Ji Y, Peng J, Zhou X, Chen X, Zhao H, Xu T, Chen L, Xu Y (2016) Increased brain-specific MiR-9 and MiR-124 in the serum exosomes of acute ischemic stroke patients. PLoS One 11(9):e0163645
73. Campbell NR, Lackland DT, Lisheng L, Niebylski ML, Nilsson PM, Zhang XH (2015) Using the Global Burden of Disease Study to assist development of nation-specific fact sheets to promote prevention and control of hypertension and reduction in dietary salt: a resource from the World Hypertension League. J Clin Hypertens (Greenwich) 17(3):165–167
74. Zou Y, Akazawa H, Qin Y, Sano M, Takano H, Minamino T, Makita N, Iwanaga K, Zhu W, Kudoh S, Toko H, Tamura K, Kihara M, Nagai T, Fukamizu A, Umemura S, Iiri T, Fujita T, Komuro I (2004) Mechanical stress activates angiotensin II type 1 receptor without the involvement of angiotensin II. Nat Cell Biol 6(6):499–506
75. Pironti G, Strachan RT, Abraham D, Mon-Wei Yu S, Chen M, Chen W, Hanada K, Mao L, Watson LJ, Rockman HA (2015) Circulating exosomes induced by cardiac pressure overload contain functional angiotensin II type 1 receptors. Circulation 131(24):2120–2130
76. Guescini M, Leo G, Genedani S, Carone C, Pederzoli F, Ciruela F, Guidolin D, Stocchi V, Mantuano M, Borroto-Escuela DO, Fuxe K, Agnati LF (2012) Microvesicle and tunneling nanotube mediated intercellular transfer of g-protein coupled receptors in cell cultures. Exp Cell Res 318(5):603–613
77. Zoni-Berisso M, Lercari F, Carazza T, Domenicucci S (2014) Epidemiology of atrial fibrillation: European perspective. Clin Epidemiol 6:213–220
78. Munger TM, LQ W, Shen WK (2014) Atrial fibrillation. J Biomed Res 28(1):1–17

79. Jesel L, Abbas M, Toti F, Cohen A, Arentz T, Morel O (2013) Microparticles in atrial fibrillation: a link between cell activation or apoptosis, tissue remodelling and thrombogenicity. Int J Cardiol 168(2):660–669
80. Choudhury A, Chung I, Blann AD, Lip GY (2007) Elevated platelet microparticle levels in nonvalvular atrial fibrillation: relationship to p-selectin and antithrombotic therapy. Chest 131(3):809–815
81. Chirinos JA, Castrellon A, Zambrano JP, Jimenez JJ, Jy W, Horstman LL, Willens HJ, Castellanos A, Myerburg RJ, Ahn YS (2005) Digoxin use is associated with increased platelet and endothelial cell activation in patients with nonvalvular atrial fibrillation. Heart Rhythm 2(5):525–529
82. Wang H, Yan HM, Tang MX, Wang ZH, Zhong M, Zhang Y, Deng JT, Zhang W (2010) Increased serum levels of microvesicles in nonvalvular atrial fibrillation determinated by ELISA using a specific monoclonal antibody AD-1. Clin Chim Acta 411(21–22):1700–1704
83. Azzam H, Zagloul M (2009) Elevated platelet microparticle levels in valvular atrial fibrillation. Hematology 14(6):357–360
84. Maron BJ, Towbin JA, Thiene G, Antzelevitch C, Corrado D, Arnett D, Moss AJ, Seidman CE, Young JB (2006) Contemporary definitions and classification of the cardiomyopathies. An American Heart Association Scientific Statement from the Council on Clinical Cardiology, Heart Failure and Transplantation Committee; Quality of Care and Outcomes Research and Functional Genomics and Translational Biology Interdisciplinary Working Groups; and Council on Epidemiology and Prevention. Circulation 113(14):1807–1816
85. Vandergriff AC, de Andrade JB, Tang J, Hensley MT, Piedrahita JA, Caranasos TG, Cheng K (2015) Intravenous cardiac stem cell-derived exosomes ameliorate cardiac dysfunction in doxorubicin induced dilated cardiomyopathy. Stem Cells Int 2015:960926
86. Wang X, Gu H, Huang W, Peng J, Li Y, Yang L, Qin D, Essandoh K, Wang Y, Peng T, Fan GC (2016) Hsp20-mediated activation of exosome biogenesis in cardiomyocytes improves cardiac function and angiogenesis in diabetic mice. Diabetes 65(10):3111–3128
87. Bulut D, Scheeler M, Niedballa LM, Miebach T, Mugge A (2011) Effects of immunoadsorption on endothelial function, circulating endothelial progenitor cells and circulating microparticles in patients with inflammatory dilated cardiomyopathy. Clin Res Cardiol 100(7):603–610
88. Bonow RO, Carabello BA, Chatterjee K, de Leon AC, Faxon DP, Freed MD, Gaasch WH, Lytle BW, Nishimura RA, O'Gara PT, O'Rourke RA, Otto CM, Shah PM, Shanewise JS (2006) ACC/AHA 2006 guidelines for the Management of Patients with valvular heart disease. A report of the American College of Cardiology/American Heart Association Task Force on Practice Guidelines (Writing Committee to revise the 1998 Guidelines for the Management of Patients with Valvular Heart Disease): developed in collaboration with the Society of Cardiovascular Anesthesiologists: endorsed by the Society for Cardiovascular Angiography and Interventions and the Society of Thoracic Surgeons. Circulation 114(5):e84–e231
89. Kasner M, Gast M, Galuszka O, Stroux A, Rutschow S, Wang X, Dohmen P, Skurk C, Landmesser U, Poller W, Gross M (2016) Circulating exosomal microRNAs predict functional recovery after MitraClip repair of severe mitral regurgitation. Int J Cardiol 215:402–405
90. Lohani O, Colvin KL, Yeager ME (2015) Biomarkers for pediatric pulmonary arterial hypertension: challenges and recommendations. Paediatr Respir Rev 16(4):225–231
91. McLaughlin VV, Archer SL, Badesch DB, Barst RJ, Farber HW, Lindner JR, Mathier MA, McGoon MD, Park MH, Rosenson RS, Rubin LJ, Tapson VF, Varga J (2009) ACCF/AHA 2009 expert consensus document on pulmonary hypertension a report of the American College of Cardiology Foundation Task Force on Expert Consensus Documents and the American Heart Association developed in collaboration with the American College of Chest Physicians; American Thoracic Society, Inc.; and the Pulmonary Hypertension Association. J Am Coll Cardiol 53(17):1573–1619
92. Hoeper MM, Bogaard HJ, Condliffe R, Frantz R, Khanna D, Kurzyna M, Langleben D, Manes A, Satoh T, Torres F, Wilkins MR, Badesch DB (2013) Definitions and diagnosis of pulmonary hypertension. J Am Coll Cardiol 62(25, Suppl):D42–D50
93. Aliotta JM, Pereira M, Wen S, Dooner MS, Del Tatto M, Papa E, Goldberg LR, Baird GL, Ventetuolo CE, Quesenberry PJ, Klinger JR (2016) Exosomes induce and reverse monocrotaline-induced pulmonary hypertension in mice. Cardiovasc Res 110(3):319–330

Part IV
Pathological Effects of Exosomes

Chapter 6
Exosomes as New Intercellular Mediators in Development and Therapeutics of Cardiomyocyte Hypertrophy

Qi Huang and Benzhi Cai

6.1 Introduction

With the dramatic changes of human lifestyle and diet, the mortality and morbidity of cardiovascular diseases is increasing year by year. Cardiovascular diseases such as coronary heart disease and severe heart failure remain the main cause of people death all over the world. Now there is a growing body of evidence that indicates that cardiac hypertrophy is the precursor lesions and independent risk factors of coronary heart disease, heart failure, sudden cardiac death and other heart diseases. In particularly, pathological myocardial hypertrophy leads to the impairment of cardiac function and is a major determinant of these heart diseases.

6.2 Cardiac Hypertrophy

6.2.1 Classification of Cardiac Hypertrophy

Cardiomyocyte is a highly differentiated terminal cells, and functions to pumping the blood into the whole body. Hypertrophic growth of cardiomyocyte is closely associated with a variety of neurohumoral factors [1], activation of intracellular signaling pathways and alteration of functional proteins. Myocardial hypertrophy is usually categorized into two types: physiology myocardial hypertrophy and pathology myocardial hypertrophy. Physiological myocardial hypertrophy is

Q. Huang • B. Cai (✉)
Department of Pharmacy, The Affiliated Second Hospital of Harbin Medical University, 157# Baojian Road, Nangang District, Harbin 150081, Heilongjiang Province, China

Department of Pharmacology, College of Pharmacy, Harbin Medical University, Harbin 150081, Heilongjiang Province, China
e-mail: caibenzhi@163.com

J. Xiao, S. Cretoiu (eds.), *Exosomes in Cardiovascular Diseases*, Advances in Experimental Medicine and Biology 998, DOI 10.1007/978-981-10-4397-0_6

reversible and mostly caused by increased cardiac load from exercise, pregnancy and other physiological factors. The functional and structural changes occurring in physiological cardiac hypertrophy include the enlarged horizontal axis of cardiomyocyte, the increased size of cardiomyocyte, the improved ventricular diastolic compliance, the stronger reserve capacity and contractility of heart, as well as the higher cardiac efficiency [2–4]. These structural and functional changes lead to effective cardiac output and compliance with physiological limits. So, to some extent physiological cardiac hypertrophy shows beneficial and effective cardiac output in healthy people similar as in sports persons.

Pathological myocardial hypertrophy is caused by a long-term and chronic stress stimulation, or certain diseases such as primary or secondary hypertension, valvular heart disease and coronary artery diseases, which is considered as compensatory adoption of hearts in response to oxidant stress and its overload. Long-term pathophysiological stimuli lead to the changes of contraction ability of cardiac muscles, and further form the substrate of pathological heart hypertrophy. At the early stage, myocardial hypertrophy does a favor to maintain normal cardiac function. However, at the late stage, it will result in the increase of myocardial oxygen consumption and the decrease of cardiac compliance and its contraction ability, finally leading to decompensated myocardial hypertrophy that is predisposed to heart failure and sudden cardiac death [5]. Pathological cardiac hypertrophy is usually irreversible at the late stage [6]. There are typical structural, functional and molecular alterations in pathological cardiac hypertrophy including the enlarged longitudinal axis of cardiomyocyte and length of muscle, the more myocardial sarcomeres, the excessive deposition of collagen and myocardial fibrosis, the lower cardiac efficiency, the increased cardiomyocyte apoptosis, the upregulation of brain natriuretic peptide (BNP) and β-myosin heavy chain (β-MHC), and downregulation of α-myosin heavy chain (α-MHC) and SERCA2a (a sarcoplasmic endoplasmic reticulum calcium (Ca^{2+}) ATPase) [2]. So in pathological cardiac hypertrophy, the muscles of heart are under the condition of oxidant stress which leads to lower cardiac output and serious pathological changes in structure.

6.2.2 *Mechanism of Cardiac Hypertrophy*

The detailed mechanisms underlying cardiac hypertrophy has not yet been fully elucidated, but a large number of studies have shown that cardiac hypertrophy is closely related to the activation of multiple signaling pathways induced by various stimuli, such as mitogen-activated protein kinase (MAPK) signaling pathway, Ca^{2+} and its dependent signaling pathways, Wnt signaling pathway and microRNAs (miRNAs), JAK-STAT signaling pathway, AMPK signaling pathway and so on [7–13]. Mechanical stretch is thought to be one of the most important initial factors in cardiac hypertrophy [9–11, 13]. Mechanical tension directly stimulates cell growth, and also promotes the synthesis and secretion of various endocrine factors in myocardial tissues such as Ang II, catecholamines, insulin-like growth factor 1 (IGF-1), nitric oxide (NO), endothelin, etc. In addition, RASS also is an important mediator of blood pressure in human body. Ang II binds to ATR1 receptor in vascular smooth muscle cells, and promotes the contraction of vascular smooth muscle through Ca^{2+}-phospholipid-dependent protein kinase pathway, resulting in the increase of blood pressure and cardiac load [9]. The increased secretion of catecholamine after pressure overload is able to aggravate the damage and induces the apoptosis of cardiomyocytes. Catecholamine also promotes myocardial protein synthesis, collagen accumulation, myocardial fibrosis, and eventually lead to cardiac hypertrophy [10]. It has been proved that Norepinephrine (NE) increases protein synthesis and gives rise to cardiac hypertrophy [13]. Endothelin-1 is a potent vasoconstrictor peptide and plays an important role in regulating cardiovascular function. Endothelin-1 has been shown to promote the proliferation of cardiac fibroblasts and induce myocardial hypertrophy [11]. WNT signaling pathway is silenced under the normal condition, but is activated during pressure load. It was demonstrated that blocking of WNT/frizzled signaling could reduce stress-mediated cardiac hypertrophy [12]. Recent studies have shown that miRNAs are involved in the regulation of cardiac hypertrophy. Van Rooij et al. analyzed 186 miRNAs and found that 7 of them were downregulated and 21 of them were upregulated under pathological stress loading [7]. Sayed D et al. found that the expression of miR-1 in myocardium was significantly decreased after transverse aortic constriction (TAC) [14]. It suggests that miR-1 plays a essential role in the development of myocardial hypertrophy. Increasing evidence suggests that activation of AMPK can protect against myocardial ischemia and limit cardiac hypertrophy caused by various factors [15]. The level of IGF-1 was positively correlated with myocardial hypertrophy, indicating that insulin-like growth factor 1 (IGF-1) is also involved in hypertensive hypertrophy [16, 17].

6.3 Exosomes

6.3.1 The Concept and Characteristics of Exosomes

In 1983, the exosomes were firstly found in reticulocytes of sheep, and it is a small vesicle secreted by reticulocytes [18]. In 1987 Johnstone named it "exosome" [19]. Nowadays, the exosomes are defined as nanomembrane vesicles with a diameter of 50–100 nm and originated from multivesicular bodies (MVBs). A variety of cells such as epithelial cells, T cells, B cells, reticulocytes, mast cells, platelets, monocytes and tumor cells have the ability to release the exosomes [20, 21]. With extracellular environment stimuli, cell membrane inwardly fold to form a closed vesicle which is called early endosomes, and then gradually mature into late endosomes. The late endosomes containing ILVs are also called as multivesicular bodies (MVBs). The extracellular release of MVBs after fused with plasma membrane is known as exosomes [22, 23]. Exosomes were shown to be cup-shaped morphology by electron microscopy which has been shown for a majority of isolated exosomes

[24]. Moreover, exosomes are distinct from microvesicles (100–1000 nm in diameter) and apoptotic bodies (1–4 μm in diameter) [25]. The exosomes contain a lot of biological active substances including cytokines, proteins, lipids, mRNA, miRNA and ribosomal RNA, and these substances contributes to the biological functions of exosomes [25, 26]. For instance, miR-21 and miR-29a secreted from nonsmall cell lung cancer (NSCLC) cells were found recruited in tumor microenvironment, which favors cancer growth and dissemination [27].

6.3.2 Biological Function of Exosomes

Recent studies have shown that exosomes participate in cellular communication, migration, proliferation, differentiation and angiogenesis. These findings support exosomes as important messengers between cells. A growing body of evidence suggests that exosomes help to transfer mRNA and miRNA to adjacent cells and thus play an important role in cellular communications. Let-7, miR-1, miR-15, miR-16, miR-181 and miR-375 were found in the exosomes derived from mouse and human mast cell lines. These mRNAs and miRNAs can be further transported to target cells and produce biological actions [28]. Macrophages induced the secretion of miR-150 in the exosomes from THP-1 cells, which stimulates the migration of endothelial cells (HMVEC) by inhibiting the translation of target gene c-Myb in endothelial cells [29]. It was also reported that the exosomes are capable of transferring proteins among cells. The cardiomyocytes with the higher levels of Hsp20, p-Akt, survivin, and SOD1, promote endothelial cell proliferation in a paracrine and autocrine manner under high glucose conditions [30].

6.4 Roles of Exosomes in Cardiac Hypertrophy

6.4.1 The Exosomes from Cardiosphere-Derived Cells on Cardiac Hypertrophy After MI

Cardiosphere-derived cells (CDCs) are one kind of stem cells isolated from biopsy specimens of patients. Autologous CDCs transplantation has been shown to regenerate damaged hearts after MI in patients [31]. But its precise mechanism remained unclear. Recent studies showed that porcine CDCs-derived exosomes plays protective role in myocardial remodeling after myocardial infarction (Table 6.1). They found that CDCs exosomes not only reduced myocardial fibrosis in injection sites, but also had a global anti-fibrotic effect. In addition to decreasing fibrosis, CDCs exosomes also prevented cardiomyocyte hypertrophy associated with adverse remodeling. In the peri-infarct area, hypertrophic growth of cardiomyocyte was significantly inhibited, but myocardial hypertrophy in the remote zone was not

Table 6.1 Roles of exosomes in cardiac hypertrophy

Exosomes	Cells	microRNAs	Functions
Exosomes [32]	CDCs	Unknown	Improved ventricular remodeling, reduced scarring, promoted angiogenesis and enhanced cardiomyocyte proliferation
Exosomes [33]	CFs	Unknown	Activate MAPK, ERK, p38, Akt and JNK to promote the synthesis and release of Ang II and cause cardiomyocyte hypertrophy
Exosomes [34]	CFs	miR-21*	Reduce the protein expression of SORBS2 and PDLIM5 and cause cardiomyocyte hypertrophy
Exosomes [30]	CMCs	Unknown	Exosomes containing high levels of Hsp20 are resistant to myocardial hypertrophy and other myocardial injury in diabetes mellitus
Exosomes [38]	CPCs	miR-21	Counter cardiomyocytes apoptosis by targeting PDCD4
Exosomes [42]	MSCs	miR-22	Reduce the apoptosis of CMCs by targeting Mecp2
Exosomes [43]	MSCs	miR-221	Inhibit the apoptosis of CMCs
Exosomes [45]	ADs	miR-200a	Reduce TSC1 and subsequent mTOR activation of cardiomyocytes and cause cardiomyocyte hypertrophy

inhibited after CDCs exosomes treatment. At the same time, it has been also found that treatment with CDCs exosomes contributes to angiogenesis characterized by more arterioles in the marginal and infarcted areas of MI. In general, exosomes treatment improved ventricular remodeling, reduced scarring, promoted angiogenesis, enhanced cardiomyocyte proliferation in porcine models of acute (AMI) and convalescent myocardial infarction (CMI) [32].

6.4.2 Cardiac Fibroblasts-Derived Exosomes on Cardiac Hypertrophy

It was recently reported that Ang II-induced pathological cardiac hypertrophy is mediated by exosomes secreted by cardiac fibroblasts which act on cardiomyocytes through paracrine action (Table 6.1). The exosomes from cardiac fibroblasts upregulated the expression of renin, Agt, AT1R and AT2R, and downregulated ACE2 expression in cultured neonatal cardiomyocytes. Besides, CF exosomes is able to activate MAPK, ERK, p38, Akt and JNK to promote the synthesis and release of Ang II, which is associated with EGFR and spp1. Thus, CF exosomes activate RAS in cardiomyocytes and promotes the production as well as secretion of Ang II, which leads to cardiomyocyte hypertrophy [33]. Additionally, it was found that CF exosomes are rich in miR-21*, and miR-21* in the exosomes leads to cardiomyocyte hypertrophy. Further studies showed that overexpression of miR-21* can reduce the protein expression of SORBS2 (sorbin and SH3 domain containing 2) and PDLIM5 (PDZ and LIM domain 5) in cardiomyocytes, while silencing of SORBS2 and PDLIM5 in cardiomyocytes induced cardiac hypertrophy. So, miR-21* in the exosomes can enter cardiomyocytes and cause cardiomyocyte hypertrophy [34].

6.4.3 Cardiomyocytes-Derived Exosomes on Cardiac Hypertrophy After MI

Diabetic cardiomyopathy is a leading cause of death of patients with diabetes. The reduction of heat shock protein (Hsp) expression in diabetes mellitus is a major contributor to organ damage. It was recently reported that diabetic cardiomyocytes release harmful exosomes with low level of Hsp20. Cardiac-specific overexpression of Hsp20 in transgenic mice confirmed that the exosomes from cardiomyocytes contribute to the Hsp20-mediated myocardial remodeling of diabetic mice. The exosomes derived from cardiomyocytes in Type-2 diabetic rats were able to inhibit the proliferation, migration and tube formation of myocardial endothelial cells. In addition, the exosomes containing higher levels of Hsp20 may prevent cardiac hypertrophy, cardiomyocyte apoptosis, fibrosis and microvascular rarefaction, thereby improving myocardial systolic function. These results suggest that exosomes containing high levels of Hsp20 are resistant to myocardial hypertrophy and other myocardial injury in diabetes mellitus [30]. Therefore, Hsp20 engineered exosomes are likely to become a new therapeutic way for diabetic cardiomyopathy.

6.4.4 Cardiac Progenitor Cell-Derived Exosomes on Cardiac Hypertrophy

Many studies have pointed out that exosomes secreted by CPCs improve the function of injured heart, and participate in cardiac protection and repair [35, 36]. Ischemia-reperfusion may lead to the irreversible structural damage leading to organ dysfunction, whereas this process is strongly associated with the increased production of oxygen free radicals [37]. *In vitro* experiments showed that oxidative stress may increase the release of exosomes from CPCs, and the exosomes produced by CPCs with or without H_2O_2 treatment both inhibited H_2O_2-induced apoptosis of H9C2, but the exosomes derived from H_2O_2-treated CPCs were more effective in preventing the apoptosis of H9C2. Further studies showed that the level of miR-21 in the exosomes is increased by oxidative stress, and miR-21 inhibits H9C2 apoptosis by targeting Programmed Cell Death 4 (PDCD4) [38]. So, under oxidative stress condition, transplanted CPCs are able to promote secretion of exosomes enriched in miR-21 and counter cardiomyocytes apoptosis by targeting

PDCD4 pathway. However, the direct relationship between CPC-derived exosomes and cardiac hypertrophy has not been yet elucidated. It is well known that the inhibition of apoptosis attenuates the decline of contraction function of cardiomyocyte and the compensatory hypertrophy of surviving cardiomyocytes [39, 40]. Thus, it can be proposed that CPCs-derived exosomes exerts protective effects on cardiac hypertrophy after MI.

6.4.5 Mesenchymal Stem Cells Derived Exosomes on Cardiac Hypertrophy

The exosomes derived from mesenchymal stem cells (MSCs) were able to suppress myocardial remodeling and improve heart functions by anti-apoptotic, anti-cardiac remodeling, anti-inflammatory and anti-vascular remodeling activities [20]. Its detailed mechanisms include increasing ATP levels, reducing oxidative stress, activating the PI3K/Akt pathway, etc. [41]. It was recently reported that miR-22 from the exosomes derived from MSCs can reduce the apoptosis of CMCs by targeting methyl CpG binding protein 2 (Mecp2) after ischemia/reperfusion injury [42] (Table 6.1). It was also proved that exosomes derived from MSCs can protect against CMCs apoptosis by releasing miR-221, which is associated with the inhibition of p53 upregulation of PUMA (apoptosis modulator, belonging to the Bcl-2 protein family of subclasses) [43]. Thus, the exosomes derived from MSCs have potential to become a novel therapeutic approach for myocardial infarction.

6.4.6 Adipose Cells-Derived Exosomes on Cardiac Hypertrophy

Adipose tissue is an important endocrine organ, which mediates insulin sensitivity, blood pressure, endothelial function, and inflammatory response. It has been showed that rosiglitazone (RSG) is able to treat diabetes, accompanied by adverse effects on cardiovascular system [44, 45]. But its molecular mechanisms remained incompletely clear. Recently it was uncovered that RSG was able to activate the PPARγ (a member of the nuclear hormone receptor superfamily) signaling pathway in adipocytes, which increases the expression and secretion of miR-200a [45] (Table 6.1). Notably, miR-200a is expressed abundantly in adipose tissue but hardly expressed in blood vessels, heart or skeletal muscle. Co-cultured adipocytes secret the exosomes containing a large number of miR-200a to reduce TSC1 (a repressor of mTOR signaling) and subsequent mTOR activation of cardiomyocytes, which leads to cardiomyocyte hypertrophy [45].

6.5 Outlook

A large body of evidence suggests that the exosomes secreted by a variety of cells are inextricably linked to cellular physiological and pathological conditions. As a novel important carrier for intercellular communications, exosomes play important roles in regulating cellular proliferation, apoptosis, differentiation and growth of many types of cells. At present, the studies about exosomes mainly focused on the angiogenesis, metastasis and early diagnosis of cancers as well as biological functions of stem cells. The role of exosomes in cardiovascular disease has not been elucidated completely yet, and requires more further investigations to clarify it. It is no doubt that the discovery of exosomes will not only provide a safe and convenient method for early diagnosis of cardiovascular diseases, but also be developed as new therapeutic strategy for heart conditions. Moreover, it also opens a brand-new idea for the future development of drug carriers using the exosomes due to its unique biological structure and function.

References

1. Widiapradja A, Chunduri P, Levick SP (2017) The role of neuropeptides in adverse myocardial remodeling and heart failure. Cell Mol Life Sci 74(11):2019–2038. doi:10.1007/s00018-017-2452-x
2. McMullen JR, Jennings GL (2007) Differences between pathological and physiological cardiac hypertrophy: novel therapeutic strategies to treat heart failure. Clin Exp Pharmacol Physiol 34(4):255–262
3. Baker JS, Grace F, Kilgore L, Smith DJ, Norris SR, Gardner AW, Ringseis R, Eder K, Shephard RJ, Kokkinos P (2012) Physiological cardiac hypertrophy: encyclopedia of exercise medicine in health and disease. Springer Berlin Heidelberg p 711. doi: 10.1007/978-3-540-29807-6_2877
4. Bernardo BC, Weeks KL, Pretorius L, McMullen JR (2010) Molecular distinction between physiological and pathological cardiac hypertrophy: experimental findings and therapeutic strategies. Pharmacol Ther 128(1):191–227
5. Berenji K, Drazner MH, Rothermel BA, Hill JA (2005) Does load-induced ventricular hypertrophy progress to systolic heart failure? Physiol Heart Circ Physiol 289:H8–H16
6. Opie LH, Commerford PJ, Gersh BJ, Pfeffer MA (2006) Controversies in ventricular remodelling. Lancet 367(9507):356–367
7. van Rooij E, Sutherland LB, Liu N, Williams AH, McAnally J, Gerard RD, Richardson JA, Olson EN (2006) A signature pattern of stress-responsive microRNAs that can evoke cardiac hypertrophy and heart failure. Proc Natl Acad Sci U S A 103(48):18255–18260
8. Steinberg SF (2000) Many pathways to cardiac hypertrophy. J Mol Cell Cardiol 32(8):1381–1384
9. Zablocki D, Sadoshima J (2013) Solving the cardiac hypertrophy riddle: the angiotensin II-mechanical stress connection. Circ Res 113(11):1192–1195
10. Gava AL, Balarini CM, Peotta VA, Abreu GR, Cabral AM, Vasquez EC, Meyrelles SS (2012) Baroreflex control of renal sympathetic nerve activity in mice with cardiac hypertrophy. Auton Neurosci 170(1–2):62–65
11. Bupha-Intr THK (2012) Role of endothelin in the induction of cardiac. PLoS One 7:e43179

12. van de Schans VA, van den Borne SW, Strzelecka AE, Janssen BJ, van der Velden JL, Langen RC, Wynshaw-Boris A, Smits JF, Blankesteijn WM (2007) Interruption of Wnt signaling attenuates the onset of pressure overload-induced cardiac hypertrophy. Hypertension 49(3):473–480
13. Yamazaki T, Komuro I, Yazaki Y (1998) Signalling pathways for cardiac hypertrophy. Cell Signal 10:693–698
14. Sayed D, Hong C, Chen IY, Lypowy J, Abdellatif M (2007) MicroRNAs play an essential role in the development of cardiac hypertrophy. Circ Res 100(3):416–424
15. Kim AS, Miller EJ, Young LH (2009) AMP-activated protein kinase: a core signalling pathway in the heart. Acta Physiol (Oxf) 196(1):37–53
16. Modesti PA, Vanni S, Bertolozzi I, Cecioni I, Lumachi C, Perna AM, Boddi M, Gensini GF (2004) Different growth factor activation in the right and left ventricles in experimental volume overload. Hypertension 43(1):101–108
17. Horio T, Kamide K, Takiuchi S, Yoshii M, Miwa Y, Matayoshi T, Yoshihara F, Nakamura S, Tokudome T, Miyata T, Kawano Y (2010) Association of insulin-like growth factor-1 receptor gene polymorphisms with left ventricular mass and geometry in essential hypertension. J Hum Hypertens 24(5):320–326
18. Pan BT, Johnstone RM (1983) Fate of the transferrin receptor during maturation of sheep reticulocytes in vitro: selective externalization of the receptor. Cell Death Dis 33:967–978
19. Johnstone RM, Adam M, Hammond JR, Orr L, Turbide C (1987) Vesicle formation during reticulocyte maturation. Association of plasma membrane activities with released vesicles (exosomes). Biol Chem 262:9412–9420
20. Huang L, Ma W, Ma Y, Feng D, Chen H, Cai B (2015) Exosomes in mesenchymal stem cells, a new therapeutic strategy for cardiovascular diseases? Int J Biol Sci 11(2):238–245
21. Mathivanan S, Ji H, Simpson RJ (2010) Exosomes: extracellular organelles important in intercellular communication. J Proteome 73(10):1907–1920
22. Colombo M, Raposo G, Thery C (2014) Biogenesis, secretion, and intercellular interactions of exosomes and other extracellular vesicles. Annu Rev Cell Dev Biol 30:255–289
23. Fevrier B, Raposo G (2004) Exosomes: endosomal-derived vesicles shipping extracellular messages. Curr Opin Cell Biol 16(4):415–421
24. Thery C, Boussac M, Veron P, Ricciardi-Castagnoli P, Raposo G, Garin J, Amigorena S (2001) Proteomic analysis of dendritic cell-derived exosomes: a secreted subcellular compartment distinct from apoptotic vesicles. J Immunol 166(12):7309–7318
25. Jenjaroenpun P, Kremenska Y, Nair VM, Kremenskoy M, Joseph B, Kurochkin IV (2013) Characterization of RNA in exosomes secreted by human breast cancer cell lines using next-generation sequencing. PeerJ 1:e201
26. Lai RC, Chen TS, Lim SK (2011) Mesenchymal stem cell exosome: a novel stem cell-based therapy for cardiovascular disease. Regen Med 6:481–492
27. Fabbria M, Paonea A, Calorea F, Gallia R, Gaudioa E, Santhanama R, Lovata F, Faddaa P (2012) MicroRNAs bind to Toll-like receptors to induce prometastatic inflammatory response. Proc Natl Acad Sci U S A 109:E2110–E2116
28. Valadi H, Ekstrom K, Bossios A, Sjostrand M, Lee JJ, Lotvall JO (2007) Exosome-mediated transfer of mRNAs and microRNAs is a novel mechanism of genetic exchange between cells. Nat Cell Biol 9(6):654–659
29. Zhang Y, Liu D, Chen X, Li J, Li L, Bian Z, Sun F, Lu J, Yin Y, Cai X, Sun Q, Wang K, Ba Y, Wang Q, Wang D, Yang J, Liu P, Xu T, Yan Q, Zhang J, Zen K, Zhang CY (2010) Secreted monocytic miR-150 enhances targeted endothelial cell migration. Mol Cell 39(1):133–144
30. Wang X, Haitao G, Huang W, Peng J, Li Y, Yang L, Qin D, Essandoh K, Wang Y, Peng T (2016) Hsp20-mediated activation of exosome biogenesis in cardiomyocytes improves cardiac function and angiogenesis in diabetic mice. Diabetes 65:3111–3128
31. Johnston PV, Sasano T, Mills K, Evers R, Lee ST, Smith RR, Lardo AC, Lai S, Steenbergen C, Gerstenblith G, Lange R, Marban E (2009) Engraftment, differentiation, and functional benefits of autologous cardiosphere-derived cells in porcine ischemic cardiomyopathy. Circulation 120(12):1075–1083, 1077 p following 1083

32. Gallet R, Dawkins J, Valle J, Simsolo E, de Couto G, Middleton R, Tseliou E, Luthringer D, Kreke M, Smith RR, Marban L, Ghaleh B, Marban E (2016) Exosomes secreted by cardiosphere-derived cells reduce scarring, attenuate adverse remodelling, and improve function in acute and chronic porcine myocardial infarction. Eur Heart J. doi:10.1093/eurheartj/ehw240
33. Lyu L, Wang H, Li B, Qin Q, Qi L, Nagarkatti M, Nagarkatti P, Janicki JS, Wang XL, Cui T (2015) A critical role of cardiac fibroblast-derived exosomes in activating renin angiotensin system in cardiomyocytes. J Mol Cell Cardiol 89(Pt B):268–279
34. Bang C, Batkai S, Dangwal S, Gupta SK, Foinquinos A, Holzmann A, Just A (2014) Cardiac fibroblast–derived microRNA passenger strand-enriched exosomes mediate cardiomyocyte hypertrophy. J Clin Investig 124:2136–2146
35. Sluijter JP, van Rooij E (2015) Exosomal microRNA clusters are important for the therapeutic effect of cardiac progenitor cells. Circ Res 116(2):219–221
36. Chen L, Wang Y, Pan Y, Zhang L, Shen C, Qin G, Ashraf M, Weintraub N, Ma G, Tang Y (2013) Cardiac progenitor-derived exosomes protect ischemic myocardium from acute ischemia/reperfusion injury. Biochem Biophys Res Commun 431(3):566–571
37. Cannon RO III (2005) Mechanisms, management and future directions for reperfusion injury after acute myocardial infarction. Nat Clin Pract Cardiovasc Med 2(2):88–94
38. Xiao J, Pan Y, Li XH, Yang XY, Feng YL, Tan HH, Jiang L, Feng J, Yu XY (2016) Cardiac progenitor cell-derived exosomes prevent cardiomyocytes apoptosis through exosomal miR-21 by targeting PDCD4. Cell Death Dis 7(6):e2277
39. Persoon-Rothert M, van der Wees KGC, van der Laarse A (2002) Mechanical overload-induced apoptosis: a study in cultured neonatal ventricular myocytes and fibroblasts. Mol Cell Biochem 241:115–124
40. Zhang D, Gaussin V, Taffet GE, Belaguli NS, Yamada M, Schwartz RJ, Michael LH (2000) TAK1 is activated in the myocardium after pressure overload and is sufficient to provoke heart failure in transgenic mice. Nat Med 6:556–563
41. Arslan F, Lai RC, Smeets MB, Akeroyd L, Choo A, Aguor EN, Timmers L, van Rijen HV, Doevendans PA, Pasterkamp G, Lim SK, de Kleijn DP (2013) Mesenchymal stem cell-derived exosomes increase ATP levels, decrease oxidative stress and activate PI3K/Akt pathway to enhance myocardial viability and prevent adverse remodeling after myocardial ischemia/reperfusion injury. Stem Cell Res 10(3):301–312
42. Feng Y, Huang W, Wani M, Yu X, Ashraf M (2014) Ischemic preconditioning potentiates the protective effect of stem cells through secretion of exosomes by targeting Mecp2 via miR-22. PLoS One 9:e88685
43. Yu B, Gong M, Wang Y, Millard RW, Pasha Z, Yang Y, Ashraf M, Xu M (2013) Cardiomyocyte protection by GATA-4 gene engineered mesenchymal stem cells is partially mediated by translocation of miR-221 in microvesicles. PLoS One 8(8):e73304
44. Son NH, Park TS, Yamashita H, Yokoyama M, Huggins LA, Okajima K, Homma S, Szabolcs MJ, Huang LS, Goldberg IJ (2007) Cardiomyocyte expression of PPARgamma leads to cardiac dysfunction in mice. J Clin Investig 117(10):2791–2801
45. Fang X, Stroud MJ, Ouyang K, Fang L, Zhang J, Dalton ND, Gu Y, Wu T, Peterson KL, Huang HD, Chen J, Wang N (2016) Adipocyte-specific loss of PPARgamma attenuates cardiac hypertrophy. JCI Insight 1(16):e89908

Chapter 7
Dual Behavior of Exosomes in Septic Cardiomyopathy

Valter Vinícius Silva Monteiro, Jordano Ferreira Reis, Rafaelli de Souza Gomes, Kely Campos Navegantes, and Marta Chagas Monteiro

7.1 Introduction

Sepsis is one of the main causes of admission of patients to Intensive Care Units, which is related to a high morbidity and mortality rate of the patients affected by this syndrome [1]. It is characterized as a potentially fatal organ dysfunction caused by a dysregulated and exacerbated response of the host to an infection [2]. When nonfatal, sepsis may induce a series of comorbidities, which may lead to death in the following years after the disease. The main comorbidity due to sepsis is the septic cardiomyopathy, present in about 25% of the cases [3].

Sepsis can be caused by a series of microorganisms, such as, bacteria, fungi or viruses. The most common pathogen associated with sepsis are the gram-negative bacteria, followed by gram-positive [3]. The pathophysiological process of sepsis occurs after a generalized infection in which the immune system exacerbates after exposure to molecules that have high affinity with the immune system, the superantigens [4]. The lipopolysaccharide (LPS) present in the gram-negative bacteria and the Peptidoglycan, presents in the gram-positives are the main endotoxins related to the septic process [3, 4].

V.V.S. Monteiro • J.F. Reis
School of Pharmacy, Health Science Institute, Federal University of Pará/UFPA, Belém, PA 66075900, Brazil

R. de Souza Gomes • K.C. Navegantes • M.C. Monteiro (✉)
Pharmaceutical Science Post-Graduation Program, Health Science Institute, Federal University of Pará/UFPA, Belém, PA 66075900, Brazil
e-mail: martachagas2@yahoo.com.br

J. Xiao, S. Cretoiu (eds.), *Exosomes in Cardiovascular Diseases*, Advances in Experimental Medicine and Biology 998, DOI 10.1007/978-981-10-4397-0_7

7.2 Sepsis Induced Cardiomyopathy

The initial phase of sepsis is characterized by exacerbated inflammation, probably related to the binding of superantigens to major histocompatibility complex (MHC) and T-cell receptors (TCR), as well as the activation of pattern recognition receptors (PRRs) such as Toll-like receptors (TLR) [4, 5]. These processes will trigger the activation of various cells of the immune system, such as macrophages, dendritic cells and T-cells, leading to a cytokine storm that will induce an uncontrolled inflammatory process [6, 7]. These cytokines and activation of the receptors will trigger suppression or activation of various pathways in cells systems [8].

The main inflammatory pathway activated by this binding is the nuclear transcription factor kappa B (NF-κB), that will activate the gene transcription of inflammatory cytokines, mainly Tumor necrosis factor (TNF)-α and Interleukin (IL)-1 [4, 9]. These cytokines will induce the production of other proinflammatory cytokines, such as IL-6 and IL-8, in addition to the induction of Reactive Oxygen Species (ROS), especially the Nitric Oxide (NO) [8, 10, 11].

The high concentrations of inflammatory cytokines and ROS will provoke an increased tissue damage. Inflammatory cytokines, especially IL-1β, IL-6 and TNF-α, will increase the migration of neutrophils [12]. However, these neutrophils will remain in the bloodstream for a longer time because of their delayed apoptosis, and it is possible to find several stages of maturation of these cells in patients with sepsis [4, 13].

In consequence of the great amount and the different stages of maturation of these neutrophils, they will act unspecifically [12]. This process can cause high endothelial damage, especially in the vascular endothelium, which will lead to series of dysfunctions, including structural and molecular dysfunction and also will lead to a myocardial dysfunction [14].

7.2.1 Decrease of Myofibril Response to Ca^{2+}

Ion Calcium (Ca^{2+}) is a molecule of extreme importance in the function of the cardiac system [15]. Highly related to the cardiac contraction, the Ca^{2+} interacts with the troponin C complex, in which this interaction leads to a conformational change in the filaments of Troponin I, leading to the release of actin and promoting its binding to the myosin, leading to the fiber contraction [16].

Some studies have shown that during the course of sepsis, the myofilaments response to calcium is reduced, probably due to the increase in the phosphorylation of Troponin I [17]. In addition to the direct action on muscular contraction, some studies have shown that endotoxins and cytokines, such as IL-1β, promote the decrease of Ca^{2+} influx into cells, probably due to changes in the conformational structure of calcium channels [18–21]. Furthermore, in sepsis models the concentration of Ryanodine receptors is decreased, which will significantly reduce the release of Ca^{2+} by the sarcoplasmic reticulum (SR) [22, 23].

7.2.2 Mitochondrial Dysfunction

The circulatory system, especially the heart, is a system that needs a great amount of energy to operate, because of its constant functioning and the strength exerted to pump the blood [24]. Due to this function, the cardiomyocytes present a high concentration of mitochondria within their cytoplasm, to provide the necessary Adenosine Triphosphate (ATP) for its normal functionality [25, 26].

The main mechanism in which mitochondrial dysfunction occurs is due to lipid peroxidation. This mechanism occurs because the heart cells have a low level of antioxidants, been more prone to oxidative stress [27, 28]. In sepsis, the inflammatory cascade will produce a high concentration of ROS, especially NO, that will interact with the mitochondrial macromolecules, causing a change or complete loss of its function [29, 30]. The change of the function will cause a decrease in the production of ATP, consequently causing loss of cardiomyocyte function, that will lead to the activation of cell death pathways, such as apoptosis and necrosis especially in the cardiac cells [30, 31].

7.2.3 Downregulation of β-Adrenergic Receptor

Beta(β)-adrenergic receptors are important receptors present in most tissues [32]. In cardiac tissue, these receptors release the Ca^{2+} from the SR leading to contraction of the cardiac muscle, through the production of cyclic adenosine monophosphate (cAMP) by the adenylate cyclase [33, 34]. Being one of the most important receptors in the performance of cardiac function [32].

In patients with septic shock, the number of these receptors is decreased, as is the concentration of adenylate cyclase [35, 36]. In addition, other studies show the decrease of stimulant G proteins and the increased expression of inhibitory G protein [37, 38]. These processes can lead to the malfunction of the cardiac muscles and increase the probability of the development of septic cardiomyopathy [39].

7.2.4 Other Mechanisms

The main mechanism of septic induced cardiomyopathy is summarized in Fig. 7.1. In addition to these mechanisms other mechanisms have been suggested as inducers of septic cardiomyopathy, such as the presence of TLR in cardiomyocytes inducing local activation of NF-κB [40], the expression of microRNA involved in the regulation of cytokine expression [41] and in the influence of exosomes related to the aggravation of cardiomyopathy that will be further discussed in this chapter.

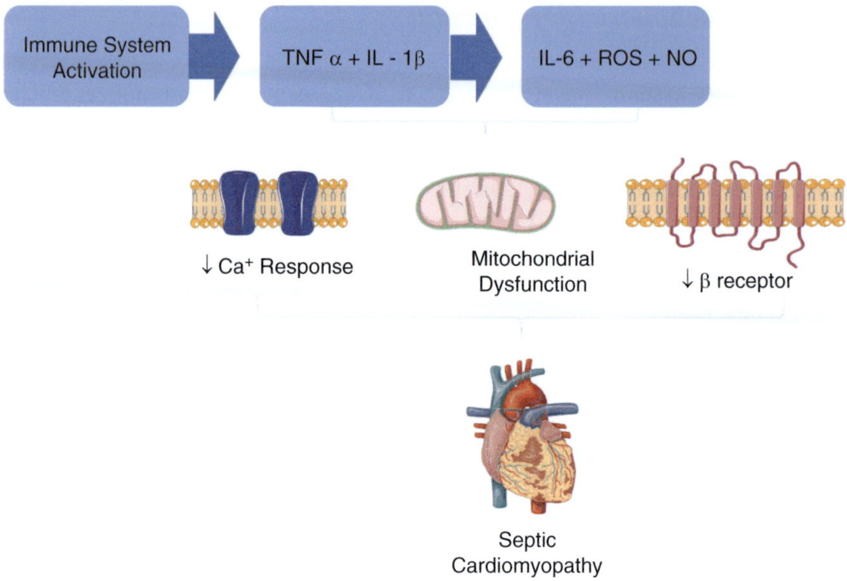

Fig. 7.1 The interaction super antigens with receptors of immune system, cause its activation in an uncontrolled manner, mainly through the NF-κB pathway, which will stimulate the transcription of TNF-α and IL-1β, these cytokines will stimulate immune system cells to produce other proinflammatory cytokines and also, the production of ROS, especially NO. The large amount of cytokines and ROS will trigger a series of dysfunctions in the cardiovascular system, mainly the Decrease in Ca^{2+} response, Mitochondrial dysfunction and Decrease in β-adrenergic receptor response. These combined processes will lead to a series of dysfunctions, causing Septic Cardiomyopathy

7.3 Role of Exosomes in Septic Cardiomyopathy

As stated in previous chapters, exosomes are small cell-derived vesicles ranging from 30 to 100 nm originate from multivesicular bodies. Exosomes released from leukocytes, platelets and dendritic cells containing major histocompatibility complex components and small amounts of phosphatidylserine are common at low levels in peripheral blood [42]. However, in thrombotic conditions and severe traumas, such as septic cardiomyopathy derived from septic shock, those exosomes appear in increased concentrations, which can be correlated with a role in physiologic/pathologic balance [43]. Much of those exosomes contain procoagulant and proinflammatory proteins [44, 45].

The role of exosomes in septic cardiomyopathy can be divided in two main mechanisms, the NADPH and the microRNA-223, as will be explained in the next sections.

7.3.1 Mechanisms NADPH-Dependent

It is known until this moment that endothelial cells apoptosis is a mechanism mainly controlled by the caspase-3 cascade activation by tumor necrosis factor, which exposure is then regulated by superoxide generation, showing that reagent oxidative species (ROS) may be involved deeply in the regulation of vascular cell survival and death as it was reviewed elsewhere [46]. Some studies have showed that platelets have the innate ability to produce ROS, mainly superoxide, to regulate their functions [47–49].

A research developed by Janiszewski and colleagues using flow cytometry analysis of microparticles obtained from septic patients and healthy individuals showed a surface containing CD42b and CD61 similar to exosomes and suggestive of platelet origin. Those exosomes also displayed the $p22^{phox}$ and $gp91^{phox}$ subunits of phagocyte-simile NADPH oxidase and exhibited intrinsic ROS production [50]. It was performed to investigate if, in septic shock, exosome could produce ROS through NADPH oxidase, showing that ROS in platelets may be produced in another via than the phospholipase A2 [51] discovered before. That same work showed a 60% increase of exosomes in plasma of septic patients relating to healthy plasma indicating the importance of platelet-derived exosomes in the pathogenesis of cardiomyopathy dysfunction.

In order to verify the theory of the pathogenesis proposed by Janiszewski, that the excess of exosomes produced by platelets in septic shock caused the cardiac inotropic dysfunction, Azevedo and collaborators collected samples from 55 patients in septic shock and 12 healthy patients to prepare exosome samples and compare [52]. They came to the conclusion that exosomes from septic patients significantly decreased positive and negative derivatives of left ventricular pressure in isolated rabbit hearts or developed tension and its first positive derivative in papillary muscles [52].

On the same research, the authors could notice that platelet-derived exosomes also contained both constitutive and inducible nitric oxide synthase (NOS), 3 and 2, respectively, being both already related to myocardial dysfunction in sepsis [53–55]. The same conclusions were drawn by Ferdinandy and collaborators, where they demonstrated that infusion of cytokines such as TNF-α or IL-1-β is associated with impairment of cardiac mechanical function in isolated hearts, accompanied by increased myocardial activities of NOS2 and NADH oxidase [56].

Azevedo and peers also pointed that peroxynitrite and by-products of the reactions of NO and ROS, which was investigated after by Gambim and colleagues. In this research, they showed that exosomes from septic patients could induce a decrease in myocardial dysfunction in isolated rabbit hearts and in papillary muscle preparations. They showed those exosomes contained a high percentage of NOS, which would contribute to increase of myocardial NO production. The authors were

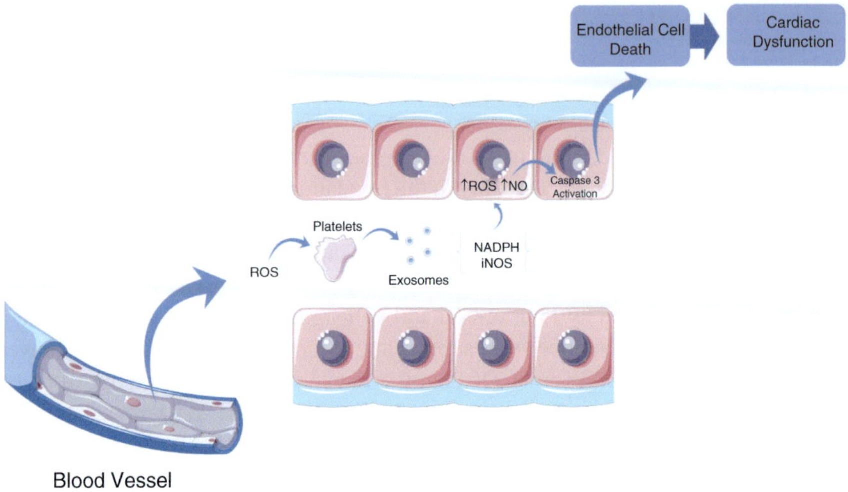

Fig. 7.2 Sepsis-generated ROS activates platelets to release exosomes containing high levels of Nicotinamide adenine dinucleotide phosphate (NADPH) and Inducible nitric oxide synthase(iNOS), which inside the endothelium increases the production and concentrations of Nitric Oxide (NO) and Reactive oxygen species (ROS), that will lead to caspase-3 activation and apoptosis of these cells

able to conclude that the NO must be a pathway for septic exosome-derived myocardial dysfunction [57]. They also noticed the high amounts of Protein Disulfide Isomerase (PDI), a chaperone associated with protein transport in the cytosol to the membrane, and its association with NADPH activation. Other groups have also showed that caspase-3 apoptosis pathway could be initiated via by-products of NO metabolism, mainly peroxynitrite [58]. In fact, it has been shown that NO can modulate apoptosis in a variety of other tissues [59, 60]. The mechanisms of platelet-derived exosomes are summarized in Fig. 7.2.

7.3.2 Mechanisms miRNA-223 Dependent

microRNAs (miRNAs) are small non-coding fragments of RNA with only the regulatory function. They regulate a specific protein transcription as they generally interact with the UTR of many mRNA, which is the three-prime untranslated region, impeding their translation and controlling the protein output in many different cells, being known that a single miRNA can globally regulates the expression of hundreds of proteins [61]. miRNAs are generally formed through transcription by RNA polymerase II as a long precursor RNA (pre-miRNA). This pre-miRNA is a double-stranded RNA that will be further transformed by RNase III endonuclease in a smaller miRNA (miRNA duplex), being one strand 3′ and the other 5′. Also, it has become clearer that in some cases both strands are able to bind to the UTR via the RNA-induced silencing complex (RISC) [62, 63].

Dysregulation of miRNAs is associated with a good assortment of human diseases, including cancer, diabetes, obesity, viral infection and cardiovascular disorders [64–66]. In the context of sepsis, some studies have correlated the organ failure and mortality to miRNA reduced levels in human individuals in septic shock. One of them, performed by Wang and collaborators enrolled 214 septic patients and quantified some miRNAs, being six of them downregulated in non-survival patients: miR-223, miR-15a, miR-16, miR-122, miR-193* and miR-483-5p [67]. The importance of miRNAs in sepsis was a target of a review from Benz and colleagues, where they put in perspective the importance of extracellular microRNAs as biomarkers for sepsis [68].

In the sphere of septic cardiomyopathy, the microRNA, the one that is the most targeted from studies is miR-223, a oncomiR related to the onset of colon, breast, ovarian and prostate cancers [64], and it was first mentioned to have any relationship with cardiac diseases by Taïbi and peers, where they showed its important role played in the regulation of several pathways [69]. However, the first time it was mentioned to regulate sepsis was in a study from Wang and collaborators (2014), where they showed its importance in aggravation of myocardial depression [70].

Wang and collaborators, in order to see the relevance of miR-223 in cardiac dysfunction under sepsis, performed an experiment comparing both wild type and miR-223 and its duplex miR-223*-knockout mice in a sepsis model [70]. Their data showed that the knocked-out mice had an exacerbation in the cardiac dysfunction and increased lethality. Also, they assessed the production of inflammatory cytokines in the mice hearts under sepsis, mainly TNF-α, IL-6 and IL-1β and their collected data showed an increase of production of TNF-α and IL-1β in cardiomyocytes, which suggests that the loss of miR-223 aggravates the inflammatory response in endothelial cells and cardiomyocytes [70].

In the same research previously described, the authors noticed that miR-223/-223* negatively regulate the STAT3, a transcription factor that is implicated with a variety of autoimmune diseases [71, 72], and the Sema3A expression in mouse hearts. A study from Ieda and peers has shown that Sema3A plays a critical role in heart rate control [73] and Wang group concluded that excess of Sema3A is related to the sepsis-induced inflammation and cardiac dysfunction [70].

Another important research regarding the effect of miR-223 and the physiopathology of septic cardiomyopathy was the one from Tabet and colleagues, where they noticed the action of extracellular miR-223 transported via HDL [74]. However, the authors also mentioned that this transportation could have been dealt by exosomes. In this research, the authors proved that miR-223 is directly responsible to the downregulation of ICAM-1, an adhesion protein related to migration of immune cells that enhance the inflammatory process in the tissue, in human coronary artery endothelial cells [74]. They also suggested that the miR-223 may be transferred from macrophages in the context of inflammatory diseases, mainly sepsis, in order to antagonize the inflammation [74].

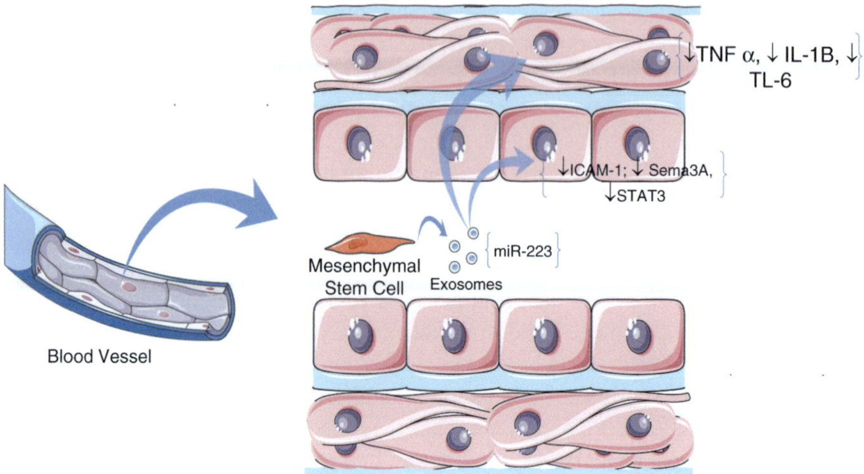

Fig. 7.3 Mesenchymal stem cells-derived exosomes contain miR-223, which downregulates Intercellular Adhesion Molecule (ICAM)-1, Signal transducer and activator of transcription (STAT)3 and Semaphorin (Sema)3A in the endothelium. In the cardiomyocytes, miR-223 acts downregulating TNF-α, IL-6 and IL-1β. Both of them have a cardio protection effect

Even though the effects of circulating extracellular miR-223 were known in the septic cardiomyopathy, the first time that it was showed that the miR-223 playing this major in the cardiac dysfunction was exosome-derived was the study from Wang and colleagues (2015), where they showed that mesenchymal stem cells (MSC)-derived exosomes containing miR-223 elicited cardio protection in polymicrobial sepsis [75]. Even though MSC are not found in the heart tissues, they showed that administration of miR-223-KO MSCs does not improve animal survival and cardiac function, which means that MSC-derived exosomes containing miR-223 may play a systemic role in septic cardiomyopathy [75]. The mechanisms of miR-223 derived from mesenchymal stem cells are compiled in Fig. 7.3.

7.4 Concluding Remarks

The cardiomyopathy and cardiac dysfunction that occurs in septic patients might be explained by the fact that platelet-derived exosomes increase in concentration in the bloodstream of sepsis patients, leading to a higher rate of endothelial damage that will further propagate to the cardiomyocytes in the vicinity. Also, MSC-derived exosomes contain miR-223, which is a miRNA associated with cardio protection, are found in less quantities in the blood of septic shock patients, accounting even more to a higher degree of cardiac dysfunction and increased mortality in patients with sepsis.

Conflicts of Interest The authors declare no conflicts of interest in relation to this article.

References

1. Dellinger R, Levy M, Rhodes A (2013) Surviving Sepsis Campaign: international guidelines for management of severe sepsis and septic shock, 2012. Intensive Care 41(2):580–637
2. Singer M, Deutschman CS, Seymour CW, Shankar-Hari M, Annane D, Bauer M, Bellomo R, Bernard GR, Chiche J-D, Coopersmith CM, Hotchkiss RS, Levy MM, Marshall JC, Martin GS, Opal SM, Rubenfeld GD, van der Poll T, Vincent J-L, Angus DC (2016) The third international consensus definitions for sepsis and septic shock (sepsis-3). JAMA 315(8):801–810
3. Dombrovskiy VY, Martin AA, Sunderram J, Paz HL (2007) Rapid increase in hospitalization and mortality rates for severe sepsis in the United States: a trend analysis from 1993 to 2003. Crit Care Med 35(5):1244–1250
4. Remick DG (2007) Pathophysiology of sepsis. Am J Pathol 170(5):1435–1444
5. Fallach R, Shainberg A, Avlas O, Fainblut M, Chepurko Y, Porat E, Hochhauser E (2010) Cardiomyocyte Toll-like receptor 4 is involved in heart dysfunction following septic shock or myocardial ischemia. J Mol Cell Cardiol 48(6):1236–1244
6. Boomer JS, Green JM, Hotchkiss RS (2014) The changing immune system in sepsis: is individualized immuno-modulatory therapy the answer? Virulence 5(1):45–56
7. Jacobi J (2002) Pathophysiology of sepsis. Am J Health Syst Pharm 59(Suppl 1):S3–S8
8. Turner MD, Nedjai B, Hurst T, Pennington DJ (2014) Cytokines and chemokines: at the crossroads of cell signalling and inflammatory disease. Biochim Biophys Acta, Mol Cell Res 1843(11):2563–2582
9. Abraham E (2003) Nuclear factor-kappaB and its role in sepsis-associated organ failure. J Infect Dis 187(Suppl):S364–S369
10. Nakae H, Motoyama S, Kurosawa S, Inaba H (1999) The effective removal of proinflammatory cytokines by continuous hemofiltration with a polymethylmethacrylate membrane following severe burn injury: report of three cases. Surg Today 29(8):762–765
11. Shimaoka M, Park EJ (2008) Advances in understanding sepsis. Eur J Anaesthesiol Suppl 42:146–153
12. de Oliveira S, Rosowski EE, Huttenlocher A (2016) Neutrophil migration in infection and wound repair: going forward in reverse. Nat Rev Immunol 16(6):378–391
13. Drifte G, Dunn-Siegrist I, Tissieres P, Pugin J (2013) Innate immune functions of immature neutrophils in patients with sepsis and severe systemic inflammatory response syndrome. Crit Care Med 41(3):820–832
14. Bhagat K (1998) Endothelial function and myocardial infarction. Cardiovasc Res 39(2):312–317
15. Szent-Gyorgyi AG (1975) Calcium regulation of muscle contraction. Biophys J 15(7):707–723
16. Wakabayashi T (2015) Mechanism of the calcium-regulation of muscle contraction—in pursuit of its structural basis. Proc Jpn Acad Ser B Phys Biol Sci 91(7):321–350
17. Tavernier B, Li JM, El-Omar MM, Lanone S, Yang ZK, Trayer IP, Mebazaa A, Shah AM (2001) Cardiac contractile impairment associated with increased phosphorylation of troponin I in endotoxemic rats. FASEB J 15(2):294–296
18. Abi-Gerges N, Tavernier B, Mebazaa A, Faivre V, Paqueron X, Payen D, Fischmeister R, Mery PF (1999) Sequential changes in autonomic regulation of cardiac myocytes after in vivo endotoxin injection in rat. Am J Respir Crit Care Med 160(4):1196–1204

19. Liu S, Schreur KD (1995) G protein-mediated suppression of L-type Ca2+ current by interleukin-1 beta in cultured rat ventricular myocytes. Am J Phys 268(2 Pt 1):C339–C349
20. Tsai T-Y, Lou S-L, Wong K-L, Wang M-L, Su T-H, Liu Z-M, Yeh L-J, Chan P, Gong C-L, Leung Y-M (2015) Suppression of Ca(2+) influx in endotoxin-treated mouse cerebral cortex endothelial bEND.3 cells. Eur J Pharmacol 755:80–87
21. Zhong J, Hwang TC, Adams HR, Rubin LJ (1997) Reduced L-type calcium current in ventricular myocytes from endotoxemic guinea pigs. Am J Phys 273(5 Pt 2):H2312–H2324
22. Wu LL, Liu MS (1992) Altered ryanodine receptor of canine cardiac sarcoplasmic reticulum and its underlying mechanism in endotoxin shock. J Surg Res 53(1):82–90
23. Zhu X, Bernecker OY, Manohar NS, Hajjar RJ, Hellman J, Ichinose F, Valdivia HH, Schmidt U (2005) Increased leakage of sarcoplasmic reticulum Ca2+ contributes to abnormal myocyte Ca2+ handling and shortening in sepsis. Crit Care Med 33(3):598–604
24. Ayers L, Nieuwland R, Kohler M, Kraenkel N, Ferry B, Leeson P (2015) Dynamic microvesicle release and clearance within the cardiovascular system: triggers and mechanisms. Clin Sci (Lond) 129(11):915–931
25. Dai D-F, Rabinovitch PS, Ungvari Z (2012) Mitochondria and cardiovascular aging. Circ Res 110(8):1109–1124
26. Marzetti E, Csiszar A, Dutta D, Balagopal G, Calvani R, Leeuwenburgh C (2013) Role of mitochondrial dysfunction and altered autophagy in cardiovascular aging and disease: from mechanisms to therapeutics. Am J Phys Heart Circ Phys 305(4):H459–H476
27. Giustarini D, Dalle-Donne I, Tsikas D, Rossi R (2009) Oxidative stress and human diseases: origin, link, measurement, mechanisms, and biomarkers. Crit Rev Clin Lab Sci 46(5–6):241–281
28. Reis JF, Monteiro VVS, de Souza Gomes R, do Carmo MM, da Costa GV, Ribera PC, Monteiro MC (2016) Action mechanism and cardiovascular effect of anthocyanins: a systematic review of animal and human studies. J Transl Med 14(1):315–315
29. Sawyer DB, Colucci WS (2000) Mitochondrial oxidative stress in heart failure. Circ Res 86(2):119–121
30. Singer M (2014) The role of mitochondrial dysfunction in sepsis-induced multi-organ failure. Virulence 5(1):66–72
31. Cimolai MC, Alvarez S, Bode C, Bugger H (2015) Mitochondrial mechanisms in septic cardiomyopathy. Int J Mol Sci 16(8):17763–17778
32. Wachter SB, Gilbert EM (2012) Beta-adrenergic receptors, from their discovery and characterization through their manipulation to beneficial clinical application. Cardiology 122(2):104–112
33. Johnson M (2006) Molecular mechanisms of beta(2)-adrenergic receptor function, response, and regulation. J Allergy Clin Immunol 117(1):18–24. Quiz 25
34. Strosberg AD (1993) Structure, function, and regulation of adrenergic receptors. Protein Sci 2(8):1198–1209
35. Hahn PY, Wang P, Tait SM, Ba ZF, Reich SS, Chaudry IH (1995) Sustained elevation in circulating catecholamine levels during polymicrobial sepsis. Shock 4(4):269–273
36. Reithmann C, Hallstrom S, Pilz G, Kapsner T, Schlag G, Werdan K (1993) Desensitization of rat cardiomyocyte adenylyl cyclase stimulation by plasma of noradrenaline-treated patients with septic shock. Circ Shock 41(1):48–59
37. Matsuda N, Hattori Y, Akaishi Y, Suzuki Y, Kemmotsu O, Gando S (2000) Impairment of cardiac beta-adrenoceptor cellular signaling by decreased expression of G(s alpha) in septic rabbits. Anesthesiology 93(6):1465–1473
38. Wu L-L, Yang S-L, Yang R-C, Hsu H-K, Hsu C, Dong L-W, Liu M-S (2003) G protein and adenylate cyclase complex-mediated signal transduction in the rat heart during sepsis. Shock 19(6):533–537
39. Gilbert EM, Olsen SL, Renlund DG, Bristow MR (1993) beta-Adrenergic receptor regulation and left ventricular function in idiopathic dilated cardiomyopathy. Am J Cardiol 71(9):23C–29C

40. Boyd JH, Mathur S, Wang Y, Bateman RM, Walley KR (2006) Toll-like receptor stimulation in cardiomyoctes decreases contractility and initiates an NF-kappaB dependent inflammatory response. Cardiovasc Res 72(3):384–393
41. Essandoh K, Fan G-C (2014) Role of extracellular and intracellular microRNAs in sepsis. Biochim Biophys Acta 1842(11):2155–2162
42. Barry OP, Praticò D, Savani RC, FitzGerald GA (1998) Modulation of monocyte-endothelial cell interactions by platelet microparticles. J Clin Investig 102(1):136–144
43. Weber A-A, Köppen HO, Schrör K (2000) Platelet-derived microparticles stimulate coronary artery smooth muscle cell mitogenesis by a PDGF-independent mechanism. Thromb Res 98(5):461–466
44. Nieuwland R, Berckmans RJ, McGregor S, Böing AN, Romijn FP, Westendorp RG, Hack CE, Sturk A (2000) Cellular origin and procoagulant properties of microparticles in meningococcal sepsis. Blood J 95(3):930–935
45. Ogura H, Kawasaki T, Tanaka H, Koh T, Tanaka R, Ozeki Y, Hosotsubo H, Kuwagata Y, Shimazu T, Sugimoto H (2001) Activated platelets enhance microparticle formation and platelet-leukocyte interaction in severe trauma and sepsis. J Trauma Acute Care Surg 50(5):801–809
46. Irani K (2000) Oxidant signaling in vascular cell growth, death, and survival. Circ Res 87:179–183
47. Finazzi-Agrò A, Menichelli A, Persiani M, Biancini G, Del Principe D (1982) Hydrogen peroxide release from human blood platelets. Biochim Biophys Acta Gen Subj 718(1):21–25
48. Leoncini G, Maresca M, Colao C (1991) Oxidative metabolism of human platelets. Biochem Int 25(4):647–655
49. Marcus AJ, Silk ST, Safier LB, Ullman HL (1977) Superoxide production and reducing activity in human platelets. J Clin Invest 59:149–158
50. Janiszewski M, Do Carmo AO, Ma P, Silva E, Knobel E, Laurindo FRM (2004) Platelet-derived exosomes of septic individuals possess proapoptotic NAD(P)H oxidase activity: a novel vascular redox pathway. Crit Care Med 32(3):818–825
51. Caccese D, Praticò D, Ghiselli A, Natoli S, Pignatelli P, Sanguigni V, Iuliano L, Violi F (2000) Superoxide anion and hydroxyl radical release by collagen-induced platelet aggregation—role of arachidonic acid metabolism. Thromb Haemost 83(3):485–490
52. Azevedo LCP, Janiszewski M, Pontieri V, Pedro MA, Bassi E, Tucci PJF, Laurindo FRM (2007) Platelet-derived exosomes from septic shock patients induce myocardial dysfunction. Crit Care 11(6):R120–R120
53. Kumar A, Haery C, Parrillo JE (2001) Myocardial dysfunction in septic shock: part I. Clinical manifestation of cardiovascular dysfunction. J Cardiothorac Vasc Anesth 15(3):364–376
54. Paulus WJ, Bronzwaer JGF (2002) Myocardial contractile effects of nitric oxide. Heart Fail Rev 7(4):371–383
55. Ullrich R, Scherrer-Crosbie M, Bloch KD, Ichinose F, Nakajima H, Picard MH, Zapol WM, Quezado ZM (2000) Congenital deficiency of nitric oxide synthase 2 protects against endotoxin-induced myocardial dysfunction in mice. Circulation 102(12):1440–1446
56. Ferdinandy P, Danial H, Ambrus I, Rothery RA, Schulz R (2000) Peroxynitrite is a major contributor to cytokine-induced myocardial contractile failure. Circ Res 87(3):241–247
57. Gambim MH, do Carmo Ade O, Marti L, Verissimo-Filho S, Lopes LR, Janiszewski M (2007) Platelet-derived exosomes induce endothelial cell apoptosis through peroxynitrite generation: experimental evidence for a novel mechanism of septic vascular dysfunction. Crit Care 11(5):R107–R107
58. Zhu C, Wang X, Qiu L, Peeters-Scholte C, Hagberg H, Blomgren K (2004) Nitrosylation precedes caspase-3 activation and translocation of apoptosis-inducing factor in neonatal rat cerebral hypoxia-ischaemia. J Neurochem 90(2):462–471
59. Albina JE, Cui S, Mateo RB, Reichner JS (1993) Nitric oxide-mediated apoptosis in murine peritoneal macrophages. J Immunol 150(11):5080–5085

60. Rössig L, Fichtlscherer B, Breitschopf K, Haendeler J, Zeiher AM, Mülsch A, Dimmeler S (1999) Nitric oxide inhibits caspase-3 by S-nitrosation in vivo. J Biol Chem 274(11):6823–6826
61. Humphreys DT, Westman BJ, Martin DIK, Preiss T (2005) MicroRNAs control translation initiation by inhibiting eukaryotic initiation factor 4E/cap and poly(A) tail function. Proc Natl Acad Sci U S A 102(47):16961–16966
62. Fabian MR, Sonenberg N (2012) The mechanics of miRNA-mediated gene silencing: a look under the hood of miRISC. Nat Struct Mol Biol 19(6):586–593
63. Jackson RJ, Standart N (2007) How do microRNAs regulate gene expression? Sci STKE 367(367):re1
64. Sayed D, Abdellatif M (2011) MicroRNAs in development and disease. Physiol Rev 91(3):827–887
65. Zhu H, Fan G-C (2011) Extracellular/circulating microRNAs and their potential role in cardiovascular disease. Am J Cardiovasc Dis 1(2):138–149
66. Zhu H, Fan G-C (2012) Role of microRNAs in the reperfused myocardium towards post-infarct remodelling. Cardiovasc Res 94(2):284–292
67. Wang H, Zhang P, Chen W, Feng D, Jia Y, Xie L (2012) Serum microRNA signatures identified by Solexa sequencing predict sepsis patients' mortality: a prospective observational study. PLoS One 7(6):e38885
68. Benz F, Roy S, Trautwein C, Roderburg C, Luedde T (2016) Circulating microRNAs as biomarkers for sepsis. Int J Mol Sci 17(78):1–17
69. Taïbi F, Meuth VM-l, Massy ZA, Metzinger L (2014) miR-223 : an inflammatory oncomiR enters the cardiovascular field. Biochim Biophys Acta 1842:1001–1009
70. Wang X, Huang W, Yang Y, Wang Y, Peng T, Chang J, Caldwell CC, Zingarelli B, Fan GC (2014) Loss of duplexmiR-223 (5p and 3p) aggravates myocardial depression and mortality in polymicrobial sepsis. Biochim Biophys Acta Mol Basis Dis 1842(5):701–711
71. Akira S, Nishio Y, Inoue M, Wang X-J, We S, Matsusaka T, Yoshida K, Sudo T, Naruto M, Kishimoto T (1994) Molecular cloning of APRF, a novel IFN-stimulated gene factor 3 p91-related transcription factor involved in the gp130-mediated signaling pathway. Cell 77(1):63–71
72. Yang XO, Panopoulos AD, Nurieva R, Chang SH, Wang D, Watowich SS, Dong C (2007) STAT3 regulates cytokine-mediated generation of inflammatory helper T cells. J Biol Chem 282(13):9358–9363
73. Ieda M, Kanazawa H, Kimura K, Hattori F, Ieda Y, Taniguchi M, Lee J-K, Matsumura K, Tomita Y, Miyoshi S, Shimoda K, Makino S, Sano M, Kodama I, Ogawa S, Fukuda K (2007) Sema3a maintains normal heart rhythm through sympathetic innervation patterning. Nat Med 13(5):604–612
74. Tabet F, Vickers KC, Cuesta Torres LF, Wiese CB, Shoucri BM, Lambert G, Catherinet C, Prado-Lourenco L, Levin MG, Thacker S, Sethupathy P, Barter PJ, Remaley AT, Rye K-A (2014) HDL-transferred microRNA-223 regulates ICAM-1 expression in endothelial cells. Nat Commun 5:3292–3292
75. Wang X, Gu H, Qin D, Yang L, Huang W, Essandoh K, Wang Y, Caldwell CC, Peng T, Zingarelli B, Fan G-C (2015) Exosomal miR-223 contributes to mesenchymal stem cell-elicited cardioprotection in polymicrobial sepsis. Sci Rep 5:1–16

Chapter 8
Pathological Effects of Exosomes in Mediating Diabetic Cardiomyopathy

Esam S.B. Salem and Guo-Chang Fan

8.1 Introduction

Diabetes mellitus (DM) is a chronic endocrine and metabolic disorder characterized by hyperglycemia due to defective insulin secretion, action, peripheral insulin resistance or all of them [1, 2]. The global prevalence has reached epidemic proportions and approximately 285 million cases were diagnosed with diabetes in 2010, and this figure is estimated to reach 439 million in 2030 [3]. Recent studies suggest that 26 million people in the USA are suffering from diabetes, and by 2030 the estimated economic burden would reach to $490 billion [4]. Most cases of diabetes can be classified into two groups: type 1 (T1DM) and type 2 (T2DM) with annual estimation of 150,000 newly diagnosed patients with type 1 and 1.3 million with type 2 in the USA [5]. T1DM was previously known as "insulin dependent diabetes mellitus" (IDDM) [6]. It occurs in 0.3–0.5% of the population, and is characterized by insidious and sudden onset [7]. T1DM is a chronic autoimmune disorder characterized by an absolute deficiency of insulin caused by selective destruction of pancreatic β-cells [8]. T2DM was previously known as "non-insulin-dependent diabetes mellitus" (NIDDM) [9]. It occurs in 3–5% of the population, and typically is characterized by slow and progressive onset [10]. T2DM is often associated with obesity, lack of exercise and family history [11]. Moreover, the pathogenesis of T2DM involves a combination of insulin resistance and diminished insulin secretion from pancreatic β-cells [12]. Notably, patients with diabetes are at two to three folds higher risk of developing heart disease [13]. Indeed, the main cause of mortality and morbidity among subjects with diabetes is cardiovascular disease [14], accounting for more than 60% of death [15]. Although comorbidity factors such as chronic systemic hypertension and vascular atherosclerosis contribute to development of cardiovascular disease

E.S.B. Salem • G.-C. Fan, Ph.D. (✉)
Department of Pharmacology and Cell Biophysics, University of Cincinnati
College of Medicine, 5872 Care Mail Loc-0575, Cincinnati, OH 45267, USA
e-mail: fangg@ucmail.uc.edu

J. Xiao, S. Cretoiu (eds.), *Exosomes in Cardiovascular Diseases*, Advances in Experimental Medicine and Biology 998, DOI 10.1007/978-981-10-4397-0_8

among diabetic patients, but chronic hyperglycemia itself is considered as an independent risk factor for cardiac damage [16]. In diabetic animal models, cardiac dysfunction could be occurred without systemic hypertension, coronary artery disease, or blood vessel atherosclerosis [17, 18]. Furthermore, the incidence of suffering from heart failure is still high even after pharmacological controlling of hypertension or ischemic heart disease in diabetic patients [19]. High levels of inflammatory cytokines, adrenergic and renin-angiotensin hormones in the blood circulation, augmented sympathetic outflow, hyperlipidemia and hyperglycemia enhance toxic effects on the myocardium and cause cardiac dysfunction and damage, and that pathological disorder together known as diabetic cardiomyopathy (DCM) [20–25].

More than a decade has passed since DCM was first described with a substantial amount of research; nonetheless, how DCM develops is still largely complex and remained unclear [26]. The heart is the most active and life-long beating organ, which continuously keeps pumping the blood to perfuse the whole tissues of the body with sufficient amount of oxygen and nutrient [27]. In fact, coordination, synchronization and communication between different cell types of the heart are essential and required to maintain effective myocardial function and normal structure [28]. Histologically, the heart of an adult murine is composed of two to three billion cardiomyocytes, which accounts one-third of the total cellular mass of the heart [29]. However, it should be emphasized that, not only cardiomyocytes are important to maintain normal function and structure of myocardium, but also smooth muscle cells, endothelial cells, cardiac stem cells, fibroblasts, and immunological cells such as macrophages [30–33]. Therefore, it becomes clear that local or long-distance cell-to-cell communications have a major contribution to maintain normal cardiac homeostasis [34]. These include several pathways that are mediated via paracrine or autocrine growth factors, cell matrix interactions, cell gap junctions or adhesion molecules that control physiological cardiac homeostasis [35–38].

Large amount of evidence has suggested that extracellular vesicles generated and released from different mammalian cells' have a potential role for cardiac cells communication [39, 40]. These extracellular vesicles composed of a phospholipid bilayer and contain a specific code of genetic materials, proteins and lipids [40–43]. Several types of extracellular vesicles can be produced by different cells of the heart, including apoptotic bodies, microvesicles and exosomes [43–45]. These different sorts of vesicles are distinguished from each other on the basis of their site of origin, diameter and contents. For example, apoptotic bodies are originated from the plasma membrane of cells when they undergo the last stage of apoptosis with 1–5 μm in diameter [46]. They harbor different cytoplasmic organelles, and many nuclear DNA fragments. Even though the precise role of apoptotic bodies is not completely emphasized, but it is widely accepted that apoptotic bodies can induce the elimination of other damaged cells [47]. On the other hand, microvesicles are 200–1000 nm in diameter and synthesized from cells by a direct external budding (Exocytosis) of the plasma membrane when they are exposed to physiological or pathological stimuli such as cellular proliferation or differentiation, and cellular apoptosis or necrosis [48–50]. Microvesicles can be derived from non-nucleated cells include platelets and red blood corpuscles, or nucleated endothelial and other viable cells [49]. Microvesicles contain variable amounts of bioactive molecules,

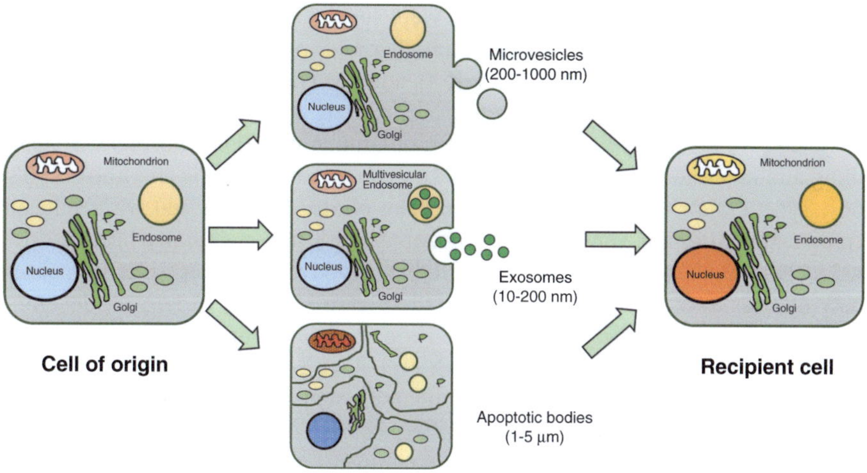

Fig. 8.1 The mechanisms of cell-to-cell communication through several types of extracellular vesicles. Microvesicles are originated by outward budding of plasma membrane. Exosomes are formed and released from the multivesicular endosomal compartment. Apoptotic bodies are generated during the late stage of cell apoptosis

including cytokines and chemokines, cytoplasmic proteins, plasma membrane proteins and lipids, non-coding RNAs (e.g., miRNAs, lncRNAs, circRNAs) and mRNAs [48, 49, 51]. Whereas, exosomes are defined as types of extracellular vesicles whose diameter ranges between 10 and 200 nm. They are characterized by very unique properties that distinct them from apoptotic bodies or microvesicles (Fig. 8.1) [52, 53]. Synthesis and release of exosomes from viable cells was considered as a mechanism through which their waste products can be discard into extracellular environment. However, accumulated reports have suggested that exosomes are appeared to have an important function as biological messengers in mediating intercellular communication both in physiological (e.g. myocardial angiogenesis) and pathological conditions (e.g. metastasis of malignant cancer) [54–56]. This chapter will focus and summarize the available data concerning the detrimental effects of exosomes in diabetic cardiomyopathy.

8.2 Diabetic Cardiomyopathy: What We Have Known and Do Not Know

8.2.1 Concept and Characterizations of Diabetic Cardiomyopathy

In 1972, diabetic cardiomyopathy (DCM) was first identified by Rubler et al. based on postmortem observations of the cardiac autopsy from diabetic patients who were diagnosed with heart failure without other cardiovascular complications [57]. These observations were suggested as specific impairment of the heart muscle itself [57]. DCM has a higher 12% prevalence in type 2 diabetic patients compared to healthy subjects [58], and been described as a long-term and serious complication of sustained toxic effects of hyperglycemia that leads to enhance cardiac oxidative stress [59, 60], inflammation [61], abnormal Ca^{2+} handling and mitochondrial function [62–64], myocardial apoptosis and fibrosis [65, 66]. Type 1 diabetic patients have also been clinically diagnosed with reduced cardiac function, left ventricular hypertrophy [67], and eventually heart failure independent of other comorbidities including blood vessel atherosclerosis, coronary arteries disease or chronic systemic hypertension [68]. Moreover, development of cardiomyopathy has also been reported in animal models which are rodents resistant to atherosclerosis with type 1 or 2 diabetes, providing strong evidence for the occurrence of DCM comparable with that seen in human diabetic subjects [69].

While multiple and different signaling pathways are involved in the onset and development of DCM, it is difficult to distinguish between a pure hyperglycemic toxic effect and other cardiovascular comorbidity factors effect [70]. Consequently, the medical diagnosis or terminology of DCM as an independent cardiac disease has been unpopular among physicians [71], and even the existence of DCM has been disproved among some scholars for a long time [72, 73]. However, large accumulation of data from *in vivo* and *in vitro* studies, both have shown that hyperglycemia itself can cause functional abnormalities and structural changes of the heart. These studies strongly support the existence of DCM [74, 75]. In diabetic heart, cardiomyocytes have to increase β-oxidation of fatty acids in order to generate ATP, which results in an elevation of reactive oxygen species (ROS) production [76]. ROS have harmful effects on cell homeostasis by damaging DNA, proteins, plasma membrane lipids, subcellular organelles, and inhibiting enzymes of the oxidative-phosphorylation mechanism, thereby further reducing efficiency of ATP generation [77–81]. During the early stage, intracellular antioxidant buffers counteract the effects of increased levels of ROS, but these buffering agents are depleted quickly. The cardiomyocytes thereafter undergo an energy deficit, alteration of gene transcription and protein translation, certainly leading to apoptosis [82–85].

DCM clinical presentation can be divided into an initial stage which is preceded by abnormal myocardial metabolism, and the early stage that is characterized by diastolic dysfunction [86]. Then, the late well-established stage which is presented with systolic dysfunction or heart failure [17, 18]. In clinical practice, he initial stage of DCM is asymptomatic and difficult to be diagnosed, but the early stage of DCM is characterized by diastolic dysfunction (including: ventricular hypertrophy and stiffness, high end-diastolic volume and pressure) [87, 88], and cardiac inter-cellular matrix remodeling (including: an accumulation of insoluble type VI collagen and fibrous connective tissue) [89, 90]. Several mechanisms contribute to ventricular stiffness in diabetic heart mainly due to hyperglycemia and formation of advanced glycation end-products (AGEs) [91]. Consequently, AGEs can interact and bind with extracellular matrix proteins including collagen and elastin, leading to a change in the properties of myocardium compliance and eventually impairing ventricular relaxation [92–94]. Also, the diastolic dysfunction prevalence is much higher in type 2 diabetic patients compared to healthy subjects with same age and gender [95, 96]. A diabetic subject with DCM is clinically diagnosed as the impairment of his left ventricle (LV) function in the absence of other concomitant risk factors, including coronary artery disease (CAD), chronic systemic hypertension, or other congenital cardiac anomalies [74]. Additionally, many studies have demonstrated that there is a link between cardiac systolic dysfunction and sustained hyperglycemia [97]. Other studies have indicated that systolic function during the physical rest was not altered in the most patients with type 2 diabetes, as clinically evaluated by measuring the left ventricular ejection fraction (LVEF) [98, 99]. However, some diabetic subjects have cardiac systolic dysfunction during an extensive exercise, which suggests that LVEF is not a sensitive indicator of myocardium contraction to detect early changes in myocardial function during rest [100].

In fact, during the course of hyperglycemia, these processes are exacerbated and led to hypertrophy, apoptosis and fibrosis of myocardium. When the hypertrophy of myocardium can't overcome or compensate for apoptosis and fibrosis, systolic dysfunction appears, indicating the irreversible stage of DCM towards congestive heart failure [101]. At the beginning, DCM is similar and diagnosed as restrictive cardiomyopathy with normal left ventricle ejection fraction [102, 103]. After that, the late stage of DCM is similar and diagnosed as dilated cardiomyopathy with reduced left ventricle ejection fraction [103, 104]. In general, human subjects or animal models with cardiac systolic dysfunction have a worse prognosis and could be exacerbated further in the presence of diabetes [105–107]. Therefore, the main pathophysiological features of DCM include the metabolic changes of cardiomyocytes during the initial stage, followed by increased ROS production and ATP deficiency, altered intracellular calcium transport and recycling, enhanced myocardial hypertrophy, and then cardiac apoptosis and fibrosis, leading to ultimate congestive heart failure [108, 109].

8.2.2 The Mechanisms Underlying the Development of Diabetic Cardiomyopathy

At present, it has been elucidated that diabetes-induced cardiac dysfunction correlates with abnormal changes in Ca^{2+} handling that reduce efficiency of the sarcomere and thereby the electromechanical coupling mechanism, resulting in impairment of myocardium contraction [110]. Furthermore, the cardiac sarco-endo-plasmic reticulum Ca^{2+}-ATPase 2 (*SERCA2*), the Na^+/Ca^{2+} exchanger 1 (*NCX1*), and ryanodine receptor (*RyR*) functions are significantly inhibited in DCM, causing to a reduction in release and slow in recycling of Ca^{2+} from and to endoplasmic reticulum during the relaxation phase [62, 111–115]. Using diabetic animal models to investigate underlying mechanisms, and to understand how hyperglycemia damages the myocardium would be beneficial to develop therapeutic agents to protect or retard onset of heart failure in human diabetic subjects [59, 77]. Hyperglycemia, hyperlipidemia, oxidative stress, apoptosis, autophagy, non-coding RNAs, and epigenetic effects have been investigated as underlying etiologies and mechanisms of DCM [116–118]. For example, there is an evidence that hyperglycemia directly activates caspase-3 resulting in increased cardiac fibrosis due to apoptosis of cardiomyocytes. So enhanced myocardium apoptosis is an essential etiology in the development and progression of heart failure from reversible compensated to irreversible decompensated stage [77, 119]. Recently, studies have been reported that insulin-resistance contributes to the development of DCM. Even though the myocardial insulin signaling is impaired in patients with DCM; however, the underlying mechanisms that could be involved in the insulin resistance are not extensively investigated or deeply understood [17, 120, 121]. Based on previous results, medical therapeutic interventions were developed to treat DCM. For example, administration of antioxidant buffers was failed to protect diabetic heart in the clinical practice [122, 123]. Furthermore, tight control of hyperglycemia also failed to retard the onset or progression of DCM towards heart failure in diabetic patients [124]. To highlight more on the same theme, we will briefly discuss about several sorts of miRNAs and their potential roles in DCM pathogenesis [116]. Structurally, miRNAs are a family of small (~22 nucleotide) and single-strand noncoding RNAs that act as post-transcriptional regulators of target genes via degradation or translation inhibition of their targeted mRNAs [125]. For instance, a recent study has found that miRNA-195 levels were higher in the hearts of STZ-induced diabetic mice, which was negatively correlated with lower expression levels of its targeted molecules including Sirtuin 1 (*SIRT-1*) and B-cell leukemia/lymphoma 2 (*BCL-2*) in comparison to wild type mice. In addition, inhibition of miRNA-195 reduced hypertrophy, oxidative stress, and apoptosis in STZ diabetic myocardium [126]. Moreover, same study has observed that restoration of normal cardiac function, improvement of coronary blood flow, an elevation of expression levels of *SIRT-1* and *BCL-2* and SIRT-1 after miRNA-195 blockade, demonstrating that reduction of miRNA-195 levels can retard the onset or propagation of DCM towards complete heart failure [126]. In the same context, a study has shown that expression levels of miRNA-133a/b were participated in diminishing the protein expression of glucose co-transporter isoform 4 (GLUT4) and then the rate of glucose uptake into neonatal rat cardiomyocytes (NRCMs) [127]. Of interest, another study has emphasized that miRNA-133a expression levels were significantly decreased in diabetic hearts, and simultaneously there was an elevation in the expression levels of cardiac fibrosis markers. Additionally, the same study has found that higher levels of miRNA-133a ameliorate cardiac fibrosis and prevent development of HF in diabetic hearts [128]. In spite of the controversial between these two previous studies that have been conducted on different animal models, more investigations are required to explore the function of miRNAs in DCM pathogenesis [129–131]. Furthermore, higher levels of miRNA-223 in NRCMs were significantly increased the glucose uptake with a similar rate of glucose uptake that measured in cells stimulated with insulin [132].

Overexpression of miRNA-223 in cardiomyocytes also observed to enhance total GLUT4 expression levels and translocation of GLUT4-contained vesicles from the cytoplasm to the cell membrane [132]. By contrast, blockage of myocardial miRNA-223 caused a reduction inexpression levels of insulin-regulated glucose transporter, GLUT4 [132]. Another study has revealed that miRNA-141 levels were significantly higher in a diabetic mouse heart, which was associated with a significant decrease in the phosphate transporter/solute carrier family 25 member 3 (*Slc25a3*) of inner-mitochondrial membrane in the mouse atrial cardiomyocyte cell line (HL-1), demonstrating that miRNA-141 can impact the mitochondrial function of myocardium [133]. In the same context for the importance of miRNAs in the pathogenesis of DCM, miRNA-451 expression levels were highly increased in palmitate-stimulated NRCMs and high fat diet (HFD)-induced obesity mice hearts [134]. Furthermore, cardiac hypertrophy was improved in miRNA-451 knockout mice fed with HFD via inhibition of the LKB1/AMPK signaling cascade. As well as, *in vitro* miRNA-451 knockdown protein expression was protected from toxic and harmful effects of the lipid via the inhibition of LKB1/AMPK pathway [134]. Additionally, induction of miRNA-143 in cardiomyocytes via release activin-(A) from adipose tissue of epicardium membrane can attenuate insulin action due to inhibition the AKT signaling pathway through down-regulation of ORP8. However, further studies are needed to confirm the functional role of miRNA-143/ORP8 in the pathogenesis of DCM [135]. In regard to insulin action, a study has exhibited that miRNA-29 was participated in the development of DCM. In the same context, a recent study has shown that insulin inhibits the expression levels of miRNA-29a, b and c, which was associated with an increase of the mRNA levels of the pro-survival protein myeloid cell leukemia 1 (MCL-1) in (*HL-1*)-cells [136]. Also, the same study has revealed that dis-balance of the (miRNA-29)-(*MCL-1*) levels due to mTORC1 inhibition or lower insulin levels in Zucker Diabetic Fatty (ZDF) rats was involved in the myocardium destruction, and contributed to cardiac malfunction [136]. Additionally, a recent study has found that miRNA-322 modulates the insulin signaling cascade, protects the myocardium against the malfunction and structural damage detected in HFD-fed mice, revealing that overexpression ofmiRNA-322 attenuated insulin-induced AKT phosphorylation [137]. Indeed, miRNA-322 can modulate insulin signaling, protect the cardiac function and structure against pathological effects of hyperinsulinemia as seen in pre-diabetic subjects and patients with type 2 diabetes [137]. On the other hand, a recent study has demonstrated that detrimental effects of cardiac miRNAs induced by sustained hyperglycemia were irreversible even after the normalization of hyperglycemia with anti-diabetic medications because altered levels of miRNAs profile in the diabetic heart were involved in autophagy, oxidative stress, apoptosis, fibrosis and eventually HF. Based on the *in vitro* and *in vivo* data presented above, miRNAs are clearly involved in the pathophysiology mechanisms of DCM. Nonetheless, further studies are needed to investigate the underlying mechanisms of DCM, and thus clear understanding of these mechanisms will control their therapeutic potential by protecting the heart from the toxic effects of hyperlipidemia and hyperglycemia.

8.3 Role of Exosomes in Diabetes: What We Have Known and Do Not Know

8.3.1 Diabetes Affects the Exosome Generation and Exosomal Compositions

Exosomes are generated from inward invagination of the cell plasma membrane to form and produce an early endosome [138]. During the maturation of an early endosome, the inward luminal buddings of the endosomal membrane form and produce multi-vesicular bodies (MVBs) that contain intraluminal vesicles (ILVs) [139]. These newly formed vesicles contain different cytoplasmic components, including proteins, lipids, DNA and several types of RNAs [140]. Then, the endosomal sorting complexes required for cellular transport machinery are essential for the formation of ILVs in MVBs [141]. When MVBs are formed, the ILVs (exosomes) are released to the extracellular environment by docking and fusion of the MVBs with the cell plasma membrane, which is mediated by small GTPase-dependent proteins such as Rab-proteins (e.g., Rab27, Rab35) [142–145]. Alternatively, MVBs can fuse with a lysosome, where the contents of the MVBs are degraded as cell waste products [146, 147]. The functions of secreted exosomes can be exerted directly via interaction of their transmembrane integral proteins or lipid with receptors of recipient cells, or indirectly by delivering their contents, including transcription factors, cytoplasmic proteins, mRNAs, miRNAs into cytoplasm of recipient cells via an endocytosis mechanism (e.g., cell drinking or eating) [148]. Several of the regulatory exosomes are sensitive and regulated by changing of glucose concentration levels [149, 150]. Expression levels of miRNA-9, miRNA-15a, miRNA-30d and miRNA-133a were up-regulated, whereas expression levels of miRNA-375 was down-regulated, by high glucose levels [151, 152]. Also, fatty acids can affect miRNA expression levels; for example, miRNA-34a and miRNA-146 expression levels were up-regulated by increased levels of fatty acids concentrations [153, 154]. Functionally, biologic effects of exosomes can be local or remote by delivering coded messages from the cell of origin to recipient cell via releasing them into intercellular fluid or blood circulation [155].

8.3.2 Detrimental Effects of miRNAs and Exosomes in the Development of Diabetes Mellitus

Currently, exosomes have been implicated as pathological factors in the pathogenesis of several diseases, including malignant tumors, autoimmune and inflammatory diseases; as well as, cardiovascular and chronic metabolic disorders [156–158]. Indeed, recent studies have also demonstrated that exosomes can orchestrate insulin signaling cascade in peripheral tissues and different organs, indicating the pivotal role of exosomes in regulation of glucose metabolism in T2DM [159–161]. Large amount of data has shown that miRNAs regulate pancreatic β-cell activity, but recently some studies reported that these miRNAs are also can be transferred from β-cells to another recipient cells via exosomes [162, 163]. Additionally, exosomes-enriched miRNAs regulate a wide spectrum of genes that are important for pancreatic β-cell homeostasis, and a chronic exposure to high concentrations of glucose and fatty acids impact negatively on their synthesis and secretion of insulin hormone. Therefore, exosomes-enriched with a specific set of miRNAs can participate in progressive β-cell dysfunction or devastation in both types of diabetes [164, 165]. In order to investigate if miRNAs directly involved in mediating pancreatic β-cells dysfunction and apoptosis, miRNAs expression levels have been investigated indifferent cell lines, type 1 and type 2 diabetic animal models, exploring the functional roles of several miRNAs in β-cells dysfunction, and their potential roles in the apoptosis of pancreatic β-cells under hyperglycemic condition [166–168]. For instance, cytokine-treated Min6B1 cells (pancreatic β-cell line) secrete exosomes containing miRNAs that are transferred to neighboring β-cells, leading to apoptosis [169, 170]. Exosomes were also isolated from the culture media of pancreatic islet cells of human and non-obese diabetic (NOD) mice (type 1 diabetes model); as well as, Min6B1-cells treated with pro-inflammatory cytokines (TNFα, IL-1β, IFNγ). Then, by incubating these isolated exosomes with untreated mice pancreatic β-cells or Min6B1-cells leads to apoptosis in the recipient cells [169–171]. The pro-inflammatory cytokines induce β-cells expression of miRNA-21, miRNA-29, miRNA-34a and miRNA-146a, thereby an increase levels of these miRNAs can contribute at least partially in destroying of insulin-secreting pancreatic β-cells and then in the development of diabetes [153, 162, 164, 172]. Additionally, over-expression of miRNA-29, miRNA-34a, miRNA-146a or miRNA199a-5p in Min6B1-cells enhance apoptosis [173], while knockdown of miRNA-34a, miRNA-203, miRNA-210 and miRNA-383 diminishes β-cell apoptosis [173, 174]. High levels of miRNA-34a were also negatively associated with the anti-apoptotic protein Bcl-2 (B-cell lymphoma 2), demonstrating that pro-apoptotic effects of miRNA-34a could be exerted via inhibition of Bcl-2 [175]. Whereas, the pro-apoptotic effect of miRNA-29 was due to its ability to block translation of another pro-survival protein Mcl-1 (Myeloid Cell Leukemia 1) [136]. On the other hand, mRNA translation of the tumor suppressor programmed cell death protein 4 (PDCD4) in pancreatic β-cells was reduced by miRNA-21. Additionally, PDCD4 expression levels were significantly attenuated as result of higher miRNA-21 expression levels in NOD mice [176]. Subsequently, the same study has shown that inhibition of miRNA-21 activity significantly increased β-cell apoptosis. Moreover, β-cells-specificPDCD4Knockoutwas correlated with an elevation in the expression levels of the anti-apoptotic genes (e.g., Bcl-xL), whereas with a reduction in the expression levels of the pro-apoptotic genes (e.g., Bax family), leading to the delay in the onset of diabetes development. Therefore, over-expression of miRNA-21 could be a pivotal protective mechanism of pancreatic β-cells after exposure to pathological stimuli via shifting the balance from pro-apoptotic to anti-apoptotic proteins [176]. In the same route, a recent study has reported that higher expression levels of miRNA-200 in a mouse pancreatic β-cells promote apoptosis ofβ-cells and development of diabetes via inhibition of tumor suppressor (Trp53) and amelioration of pro-apoptotic (Bax) signaling pathway [177]. Mocharla et al., was successfully able to detect that CD34$^+$ peripheral blood mono-nuclear cells (PBMCs) release higher levels of miRNA-126 containing exosomes than CD34$^-$ PBMC, and to observe that these exosomes enriched higher miRNA-126 levels had higher pro-angiogenic effects on endothelial cells (ECs) in comparison to lower miRNA-126 levels enriched exosomes [178]. At the same time, same author also reported that treatment of CD34$^+$ PBMC with anti-miRNA-126 or inhibition of their release diminished their pro-angiogenic effects on ECs. In addition, same study has revealed that treatment of CD34$^+$ PBMCs with high glucose concentration was associated with impairment of their pro-angiogenic properties, which could be rescued by miRNA-mimic-126 treatment [178]. Even though there is still a gap of knowledge and limitation of studies that clearly demonstrate the biological nature of several types extracellular vesicles (EVs), but also there are several studies reported that the levels of EVs are higher in the plasma of type 2 diabetic subjects compared to healthy subjects [179–181]. In specific, the membrane proteins or lipids of EVs which are released from skeletal muscle, platelets, or T lymphocytes increase the risk of metabolic dysfunction and cardiovascular diseases [182, 183]. In addition, an elevation levels of EVs has been reported and positively linked with obesity, diabetic micro or macrovascular complications, and inflammations, suggesting that the exosomes quantity can be clinically used as an early indicator (e.g.; urinary or blood marker) of previous diseases and their complications [184–187]. For example, exosomes released from visceral adipose tissue (VAT) in diabetic mice can fuse with blood monocytes and stimulate their differentiation into mature and active macrophages [160]. Various studies have provided an interesting attention into the way miRNA-containing endothelial micro-particles (EMPs) influence inflammatory effects under hyperglycemic condition [188]. Studies both *in vitro* and *in vivo* have shown that extracellular membrane-bonded micro-vesicles (EMVs) contribute to anti-inflammatory effects by diminishing endothelial ICAM-1 expression since they act as carriers to transfer functional miRNA-222 into recipient cells. Interestingly, their anti-inflammatory effects were reduced under hyperglycaemic condition due to a reduce of miRNA-222 levels of newly generated EMPs [170]. Exosomes released from VAT in type 2 diabetic mice also induce the development of insulin resistance when they were administered to wild type healthy mice. Adipocyte-derived exosomes contain numerous sorts and sets of proteins, which is indicated the important roles of exosomes in the extracellular and intercellular transmission of signals [189]. Further, exosomes released from skeletal muscle, which is another critical tissue regulating blood glucose metabolism, have an auto-regulation effect on skeletal muscle homeostasis as well as on other organs, including the pancreas and liver [159]. Barutta et al., also has provided additional information about the role of exosomes and underlying mechanisms that taking place in diabetes. The author reported that urinary exosomes from micro-albuminuric patients contain higher concentrations of miRNA-130a and miRNA-145, whereas the concentrations of miRNA-155 and miRNA-424 were lower [190]. In a mouse model of diabetic nephropathy, urinary levels miRNA-145-enriched exosomes were elevated while miRNA-145 were overexpressed within the glomeruli, which was simultaneously associated with higher levels of urinary miRNA-145-containing exosomes as detected in the urine specimens from type 2 diabetic patients, reflecting a higher shedding rate of exosomes from the renal parenchyma into the urine [191, 192]. In addition, renal mesangial cells cultured with high glucose concentration showed an increase in miRNA-145 levels in these mesangial cells and their released exosomes [191]. Human β-cells exhibit specific miRNAs sets and profiles which are involved in activation or inhibition specific genes associated with the development of type 2 diabetes [193–196]. Moreover, similar miRNAs setting and profiling of β-cells from diabetic Goto-Kakizaki rat (non-obese type 2 diabetes model) have been identified and detected upregulation levels of miRNA-130a, miRNA-132, miRNA-212 and miRNA-335 [197]. In 2010, Zampetaki et al., investigated and generated an expression profile of miRNAs in plasma of type 2 diabetic subjects. The author observed a significant decrease in plasma levels of miRNA-15a, miRNA-20b, miRNA-21, miRNA-24, miRNA-126, miRNA-191, miRNA-197, miRNA-223, miRNA-320, and miRNA-486, whereas a moderate increase of miRNA-28 and miRNA-3p levels. Additionally, high glucose concentration was attributable to the reduction of miRNA-126 levels in endothelial apoptotic bodies [198]. Furthermore, it has been suggested that type 1 and 2 diabetes are associated with multiple variations of mouse miRNA-encoding genes and miRNA-binding sites in 3′-UTR of mRNA-encoding genes, speculating the importance of preceded alteration in the miRNAs profile that incorporated in diabetes pathogenesis [199, 200]. Fundamentally, the ongoing research are still at the early phase of understanding the ultimate role of exosomes in pathogenesis of both types of diabetes, and demonstrating how we can use them clinically as an effective therapy for the prevention or retardation onset of diabetes, or as an early and sensitive biomarker for its long-term complications including DCM [187, 201].

8.4 Role of Exosomes in Diabetic Cardiomyopathy

The cellular cross-talk mechanisms involving exosomes are often multidirectional and multifunctional rather than unidirectional and unifunctional [55, 202]. Exosomes enclose and carry both numerous species of miRNAs and different types of proteins, of which transfer several and specific signaling codes from the cell of origin to another cells of different tissues and organs at the adjacency, at the proximity or at the distance [203, 204]. In this context, accumulated evidence has emphasized that a major portion of the plasma miRNAs are enveloped in exosomes [205, 206]. On the another hand, a study has suggested that plasma miRNAs can be bound to high density lipoproteins in the blood circulation in addition to exosomes [207]. Recently, based on the multiple properties of cardiac-derived exosomes from all different cell types of the heart, scientific community has speculated that exosomes can be involved in the pathophysiology of cardiovascular disorders including DCM [49]. However, the available data on a such context is still limited due to either gap of knowledge or technical difficulties on exosomes. In fact, we will cover and summarize the present studies that investigated the roles of exosomes in diabetes-induced cardiomyopathy, as reviewed below.

8.4.1 Diabetic Cardiomyocytes Secrete Antiangiogenic Exosomes

During the initial course of diabetes, hyperglycemia can lead to endothelial and microvascular dysfunctions [208, 209]. Interestingly, it has been reported that dysregulation of myocardial angiogenesis is the pivotal cause of diabetic cardiovascular disease [210–212]. Indeed, cardiac endothelial cells play a critical role in cardiomyocyte contraction and structure [213, 214]. However, under hyperglycemic condition, whether cardiomyocytes have an ability to modulate cardiac endothelial cell function remains uncertain [215, 216]. In the context of understanding the mechanism, recent studies have shown that cardiomyocytes-derived exosomes contain variable amounts of mRNAs and miRNAs, proteins and lipids, all of which can transfer to the adjacent cardiac endothelial cells and regulate their functions [217–220]. In prior study by Dr. Fan's team has provided an evidence that cardiomyocyte-derived exosomes can affect cardiac endothelial cells' contents and release (ECs) in the variable ways that are dependent upon the condition to which the cardiomyocytes were exposed to or cultured with. As an example, isolated exosomes from diabetic Goto-Kakizaki (GK) rat's cardiomyocytes diminished the migration and proliferation of cardiac ECs. In contrast, isolated exosomes from control non-diabetic Wistar rat's cardiomyocytes promoted the migration and proliferation of cardiac ECs (Fig. 8.2) [220]. Further, the effects of generated exosomes from both diabetic GK and non-diabetic Wistar rat's cardiomyocyte on cardiac ECs were reversed by blocking their release using an inhibitor of neutral sphingomyelinase/ceramide GW4869. Consequently, previous results suggest that pathological exosomes can transfer and distribute harmful molecules that participating in the development of diabetic cardiomyopathy [220–222].

In addition, the same study has emphasized that cardiomyocytes-derived exosomes from diabetic GK cardiomyocytes enriched with higher levels of miRNA-320, and lower levels of miRNA-126 and heat shock protein 20 (Hsp20) in comparison to non-diabetic Wistar rat's cardiomyocyte-derived exosomes [220]. They also observed that the cardiomyocyte enriched-miRNA-320 exosomes can transfer to cardiac ECs, and subsequently downregulate the expression of IGF-1, Hsp20, and Ets-2, resulting in impairment of angiogenic function of adjacent cardiac ECs (Fig. 8.2). Therefore, these observations provide a strong evidence that exosomes-isolated from diabetic cardiomyocytes reduced the angiogenesis in the diabetic myocardium via transferring detrimental factors that are able to induce, amplify and spread the downstream cascade of anti-angiogenic effects to cardiac ECs [220]. Garcia et al., has been recently shown that exosomes-derived from contractile cardiomyocytes can regulate glucose transport into the cardiac ECs. Same study reported that under hypoglycemic condition, cardiomyocytes produced larger

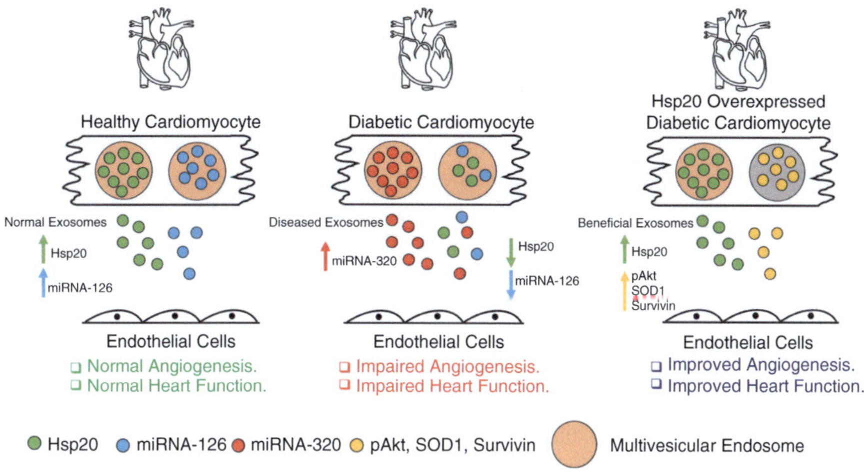

Fig. 8.2 Schematic representation of healthy cardiomyocytes can release Hsp20 and miRNA-126 enriched exosomes, which transfer Hsp20 and miRNA-126 to adjacent endothelial cells, leading to normal angiogenesis and cardiac functions. By contrast, exosomes release from type 2 diabetic cardiomyocytes contain higher levels of miRNA-320 and lower levels of Hsp20, compared with those from non-diabetic healthy cardiomyocytes. Accordingly, miRNA-320 enriched exosomes are transported to endothelial cells, resulting in impairment of angiogenesis and cardiac functions. In Hsp20 overexpressed diabetic cardiomyocytes, Hsp20 promotes quantitative and qualitative changes in cardiomyocyte-derived exosomes, transforming diseased exosomes to beneficial exosomes, and resulting in improvement of angiogenesis and amelioration of cardiac functions

amount of exosomes-enriched with glucose transporters and enzymes involved in glucose metabolism, resulting in increased in the rate of glucose uptake and glycolysis in cardiac ECs under condition of glucose deprivation [223].

8.4.2 Beneficial Effects of Gene-Modified Exosomes in Diabetic Cardiomyopathy

It has recently been elucidated that loaded contents of exosomes could be directly mediated by gene transfection or indirectly modified in parental cells, thereby we can convert diseased exosomes to be beneficial ones [224]. Along this line, it has been reported that Hsp20, an important chaperone protein involves in cellular defense mechanisms against multiple pathological stimuli, can promote the production of exosomes from cardiomyocytes via interacting with tumor susceptibility gene 101 (Tsg101) [225, 226]. Tsg101 is an upstream and integral protein of regulated endosomal membrane transport, thereby it involves in the exosome biogenesis, trafficking, sorting and secretion pathways [139, 222, 227]. An interesting study by Wang et al., has revealed that Hsp20 expression levels were decreased in diabetic cardiomyocytes and inhibited by miRNA-320 post-transcriptionally. In addition, Hsp20 response was sensitive to both acute and chronic course of hyperglycemia in a mouse heart, implying that reduced levels of Hsp20 could be involved in the development of DCM and propagation towards heart failure [226]. Consequently, chemical-induced type 1 diabetes in transgenic mice with cardiac-specific Hsp20 overexpression were generated and used to investigate the underlying mechanisms regard the functional role of Hsp20 in diabetic heart. In fact, this study also observed that deleterious effects of exosomes released from diabetic cardiomyocytes can be transformed to beneficial exosomes after Hsp20 overexpression via altering contents and the secretion pattern of cardiomyocyte-generated exosomes, as result restoring normal cardiac functions under hyperglycemic condition. Importantly, the same author also reported that altered exosomes contain cellular protective proteins, including phosphorylated AKT, SOD1, and Survivin, all of which distributed to adjacent cardiac cells promoting myocardial angiogenesis, alleviating oxidative stress, ameliorating fibrosis and apoptosis in a mouse diabetic heart (Fig. 8.2) [226]. Even though it has been confirmed that beneficial exosomes deliver functionally active Hsp20 to the recipient cardiac cells of type 1 diabetic mouse, but also it is interesting to further investigate in-depth the molecular alterations induced in Hsp20 transgenic cardiomyocytes under hyperglycemia and emphasize what are the pivotal mediators that mediate a such deviation from releasing detrimental exosomes to beneficial exosomes [222]. In the same avenue, a study reported that cardiomyocyte-derived exosomes do not have only local cardiac effects, but also have systemic effects. Interestingly, a study by Pironti et al., revealed that both *in vitro* cellular stretch and *in vivo* pressure overload promote the cardiomyocytes to produce exosomes-enriched with angiotensin II type 1 receptor (AT1R) into the culture media and blood circulation, respectively [228]. Transferring functionally active AT1R to other distinct tissues and organs, principally mesenteric vasculature and skeletal muscles can modulate peripheral vascular resistance and thus blood pressure, when these specific exosomes were injected into the tail vein of AT1 knockout mice [228]. Clearly, the functional role of exosomes in the regulation of different cellular mechanisms can be altered in diabetes and contribute to the onset and development of cardiovascular complications. Finally, the exosomes can deliver and distribute sustained information between distinct cardiovascular cells under hyperglycemic condition has large-scale clinical implications in terms of diabetic cardiomyopathy pathogenesis [222, 226, 229]. Considering all the previous data and taken these findings together, exosomes have potential and functional roles on the pathophysiology of DCM; as well as, promising and possible roles for exosomes to act as therapeutic targets or agents that can be used in the treatment of diabetic cardiomyopathy.

8.5 Conclusion and Future Directions

DCM is still the main cause of morbidity and mortality among diabetic patients. The complete understanding of DCM pathophysiology and the role of exosomes in this pathologic disorder are still under investigations. However, there has been emerging *in vitro, ex-vivo* and *in vivo* experimental data about different underlying mechanisms that might be helpful to answer important questions during the study of DCM. In addition, exosomes are considered and studied as microRNA-carriers that are essential for local tissue-specific or inter-systemic cells' communication. Of interest, miRNAs-enriched exosomes play an important role in the development of DCM based on present evidences such as alteration of miRNAs profile in the diabetic heart, suggesting their sustained detrimental effects on DCM despite normalization of hyperglycemia. As matter of fact, exosomes can be generated and derived from different cell types of the body under different pathological conditions, implicating that molecular pathological messages can be spread throughout the body organs via exosomes. Clearly, exosomes are involved in all stages of DCM, including altered insulin signaling and glucose metabolism, hypertrophy, apoptosis and fibrosis of the myocardium. In this context, the functional and expressional phenotype of exosomes from different sort of cells contributing to the onset and propagation of DCM to HF can be an alternative medical therapy, in which external artificial exosomes will be administered in order to counteract the molecular defects in DCM. Nonetheless, further studies are necessary to emphasize and elucidate the complete molecular mechanisms underlying the development of DCM and the biological nature of exosomes: First, could we be able to isolate normal exosomes from the blood of healthy subjects?; Second, could isolated normal exosomes reverse the phenotype and restore normal cardiac function in patients diagnosed with DCM?; Third, could we be able to use host cells to overexpress beneficial factors that can be loaded and delivered via exosomes?; Fourth, is it clinically possible to intervene the generation or secretion of harmful exosomes?; also, we do not know if blockade of harmful exosomes generation or secretion during DCM pathogenesis is able to have beneficial effects? In general, deep understanding of mechanisms that are contributed to exosome generation, secretion or uptake will be important to modulate the quality and quantity of exosomes in order to develop an effective therapy and improve efficacy of therapeutic intervention for the treatment of DCM. To sum up, there are clear evidence that exosomes play a critical role in the regulation of tissue-specific and/or whole body glucose metabolism. Future studies in the understanding of cell-to-cell communication pathways via exosomes in the setting of diabetic cardiomyopathy will facilitate clinical research towards the identification of effective medical therapy for DCM.

References

1. Eriksson J, Franssila-Kallunki A, Ekstrand A, Saloranta C, Widen E, Schalin C, Groop L (1989) Early metabolic defects in persons at increased risk for non-insulin-dependent diabetes mellitus. N Engl J Med 321(6):337–343
2. Shepherd PR, Kahn BB (1999) Glucose transporters and insulin action—implications for insulin resistance and diabetes mellitus. N Engl J Med 341(4):248–257
3. Egede LE, Ellis C (2010) Diabetes and depression: global perspectives. Diabetes Res Clin Pract 87(3):302–312
4. Farag YM, Gaballa MR (2011) Diabesity: an overview of a rising epidemic. Nephrol Dial Transplant 26(1):28–35

5. Greenberg RA, Sacks DB (2002) Screening for diabetes: is it warranted? Clin Chim Acta 315(1–2):61–69
6. Castano L, Eisenbarth GS (1990) Type-I diabetes: a chronic autoimmune disease of human, mouse, and rat. Annu Rev Immunol 8:647–679
7. American Diabetes Association (2010) Diagnosis and classification of diabetes mellitus. Diabetes Care 33(Suppl 1):S62–S69
8. Hanafusa T, Imagawa A (2007) Fulminant type 1 diabetes: a novel clinical entity requiring special attention by all medical practitioners. Nat Clin Pract Endocrinol Metab 3(1):36–45
9. Reaven GM (1988) Banting lecture 1988. Role of insulin resistance in human disease. Diabetes 37(12):1595–1607
10. Fajans SS (1989) Maturity-onset diabetes of the young (MODY). Diabetes Metab Rev 5(7):579–606
11. American Diabetes Association (2006) Diagnosis and classification of diabetes mellitus. Diabetes Care 29(Suppl 1):S43–S48
12. Sacks DB, McDonald JM (1996) The pathogenesis of type II diabetes mellitus. A polygenic disease. Am J Clin Pathol 105(2):149–156
13. Grundy SM, Benjamin IJ, Burke GL, Chait A, Eckel RH, Howard BV, Mitch W, Smith SC Jr, Sowers JR (1999) Diabetes and cardiovascular disease: a statement for healthcare professionals from the American Heart Association. Circulation 100(10):1134–1146
14. Zimmet P, Alberti KG, Shaw J (2001) Global and societal implications of the diabetes epidemic. Nature 414(6865):782–787
15. Isomaa B, Almgren P, Tuomi T, Forsen B, Lahti K, Nissen M, Taskinen MR, Groop L (2001) Cardiovascular morbidity and mortality associated with the metabolic syndrome. Diabetes Care 24(4):683–689
16. Ryden L, Standl E, Bartnik M, Van den Berghe G, Betteridge J, de Boer MJ, Cosentino F, Jonsson B, Laakso M, Malmberg K, Priori S, Ostergren J, Tuomilehto J, Thrainsdottir I, Vanhorebeek I, Stramba-Badiale M, Lindgren P, Qiao Q, Priori SG, Blanc JJ, Budaj A, Camm J, Dean V, Deckers J, Dickstein K, Lekakis J, McGregor K, Metra M, Morais J, Osterspey A, Tamargo J, Zamorano JL, Deckers JW, Bertrand M, Charbonnel B, Erdmann E, Ferrannini E, Flyvbjerg A, Gohlke H, Juanatey JR, Graham I, Monteiro PF, Parhofer K, Pyorala K, Raz I, Schernthaner G, Volpe M, Wood D, Task Force on D, Cardiovascular Diseases of the European Society of C, European Association for the Study of D (2007) Guidelines on diabetes, pre-diabetes, and cardiovascular diseases: executive summary. The Task Force on Diabetes and Cardiovascular Diseases of the European Society of Cardiology (ESC) and of the European Association for the Study of Diabetes (EASD). Eur Heart J 28(1):88–136
17. Boudina S, Abel ED (2007) Diabetic cardiomyopathy revisited. Circulation 115(25):3213–3223
18. Boudina S, Abel ED (2010) Diabetic cardiomyopathy, causes and effects. Rev Endocr Metab Disord 11(1):31–39
19. Poirier P, Bogaty P, Garneau C, Marois L, Dumesnil JG (2001) Diastolic dysfunction in normotensive men with well-controlled type 2 diabetes: importance of maneuvers in echocardiographic screening for preclinical diabetic cardiomyopathy. Diabetes Care 24(1):5–10
20. Atlas SA (2007) The renin-angiotensin aldosterone system: pathophysiological role and pharmacologic inhibition. J Manag Care Pharm 13(8 Suppl B):9–20
21. Bahrami H, Bluemke DA, Kronmal R, Bertoni AG, Lloyd-Jones DM, Shahar E, Szklo M, Lima JA (2008) Novel metabolic risk factors for incident heart failure and their relationship with obesity: the MESA (Multi-Ethnic Study of Atherosclerosis) study. J Am Coll Cardiol 51(18):1775–1783
22. Floras JS (2009) Sympathetic nervous system activation in human heart failure: clinical implications of an updated model. J Am Coll Cardiol 54(5):375–385
23. Lymperopoulos A, Rengo G, Koch WJ (2013) Adrenergic nervous system in heart failure: pathophysiology and therapy. Circ Res 113(6):739–753
24. Privratsky JR, Wold LE, Sowers JR, Quinn MT, Ren J (2003) AT1 blockade prevents glucose-induced cardiac dysfunction in ventricular myocytes: role of the AT1 receptor and NADPH oxidase. Hypertension 42(2):206–212

25. Wende AR, Abel ED (2010) Lipotoxicity in the heart. Biochim Biophys Acta 1801(3):311–319
26. Schilling JD, Mann DL (2012) Diabetic cardiomyopathy: bench to bedside. Heart Fail Clin 8(4):619–631
27. Skrzypiec-Spring M, Grotthus B, Szelag A, Schulz R (2007) Isolated heart perfusion according to Langendorff—still viable in the new millennium. J Pharmacol Toxicol Methods 55(2):113–126
28. Kakkar R, Lee RT (2010) Intramyocardial fibroblast myocyte communication. Circ Res 106(1):47–57
29. Banerjee I, Fuseler JW, Price RL, Borg TK, Baudino TA (2007) Determination of cell types and numbers during cardiac development in the neonatal and adult rat and mouse. Am J Physiol Heart Circ Physiol 293(3):H1883–H1891
30. Frangogiannis NG (2008) The immune system and cardiac repair. Pharmacol Res 58(2):88–111
31. Orlic D, Hill JM, Arai AE (2002) Stem cells for myocardial regeneration. Circ Res 91(12):1092–1102
32. Souders CA, Bowers SL, Baudino TA (2009) Cardiac fibroblast: the renaissance cell. Circ Res 105(12):1164–1176
33. Williams AR, Hare JM (2011) Mesenchymal stem cells: biology, pathophysiology, translational findings, and therapeutic implications for cardiac disease. Circ Res 109(8):923–940
34. Tirziu D, Giordano FJ, Simons M (2010) Cell communications in the heart. Circulation 122(9):928–937
35. Blankenberg S, Rupprecht HJ, Bickel C, Peetz D, Hafner G, Tiret L, Meyer J (2001) Circulating cell adhesion molecules and death in patients with coronary artery disease. Circulation 104(12):1336–1342
36. Re RN (1989) The cellular biology of angiotensin: paracrine, autocrine and intracrine actions in cardiovascular tissues. J Mol Cell Cardiol 21(Suppl 5):63–69
37. Sapir Y, Kryukov O, Cohen S (2011) Integration of multiple cell-matrix interactions into alginate scaffolds for promoting cardiac tissue regeneration. Biomaterials 32(7):1838–1847
38. van der Velden HM, Wilders R, Jongsma HJ (2002) Atrial fibrillation-induced gap junctional remodeling. J Am Coll Cardiol 39(10):1709
39. Thery C (2011) Exosomes: secreted vesicles and intercellular communications. F1000 Biol Rep 3:15
40. Turturici G, Tinnirello R, Sconzo G, Geraci F (2014) Extracellular membrane vesicles as a mechanism of cell-to-cell communication: advantages and disadvantages. Am J Physiol Cell Physiol 306(7):C621–C633
41. Gupta S, Knowlton AA (2007) HSP60 trafficking in adult cardiac myocytes: role of the exosomal pathway. Am J Physiol Heart Circ Physiol 292(6):H3052–H3056
42. Record M, Carayon K, Poirot M, Silvente-Poirot S (2014) Exosomes as new vesicular lipid transporters involved in cell-cell communication and various pathophysiologies. Biochim Biophys Acta 1841(1):108–120
43. Waldenstrom A, Genneback N, Hellman U, Ronquist G (2012) Cardiomyocyte microvesicles contain DNA/RNA and convey biological messages to target cells. PLoS One 7(4):e34653
44. Ibrahim AG, Cheng K, Marban E (2014) Exosomes as critical agents of cardiac regeneration triggered by cell therapy. Stem Cell Rep 2(5):606–619
45. Sharov VG, Sabbah HN, Shimoyama H, Goussev AV, Lesch M, Goldstein S (1996) Evidence of cardiocyte apoptosis in myocardium of dogs with chronic heart failure. Am J Pathol 148(1):141–149
46. Hristov M, Erl W, Linder S, Weber PC (2004) Apoptotic bodies from endothelial cells enhance the number and initiate the differentiation of human endothelial progenitor cells in vitro. Blood 104(9):2761–2766
47. Elmore S (2007) Apoptosis: a review of programmed cell death. Toxicol Pathol 35(4):495–516
48. Buzas EI, Gyorgy B, Nagy G, Falus A, Gay S (2014) Emerging role of extracellular vesicles in inflammatory diseases. Nat Rev Rheumatol 10(6):356–364
49. Sluijter JP, Verhage V, Deddens JC, van den Akker F, Doevendans PA (2014) Microvesicles and exosomes for intracardiac communication. Cardiovasc Res 102(2):302–311

50. Zhu H, Dardick CD, Beers EP, Callanhan AM, Xia R, Yuan R (2011) Transcriptomics of shading-induced and NAA-induced abscission in apple (Malus domestica) reveals a shared pathway involving reduced photosynthesis, alterations in carbohydrate transport and signaling and hormone crosstalk. BMC Plant Biol 11:138
51. Ling H, Fabbri M, Calin GA (2013) MicroRNAs and other non-coding RNAs as targets for anticancer drug development. Nat Rev Drug Discov 12(11):847–865
52. Johnsen KB, Gudbergsson JM, Skov MN, Pilgaard L, Moos T, Duroux M (2014) A comprehensive overview of exosomes as drug delivery vehicles—endogenous nanocarriers for targeted cancer therapy. Biochim Biophys Acta 1846(1):75–87
53. Vlassov AV, Magdaleno S, Setterquist R, Conrad R (2012) Exosomes: current knowledge of their composition, biological functions, and diagnostic and therapeutic potentials. Biochim Biophys Acta 1820(7):940–948
54. Johnstone RM, Adam M, Hammond JR, Orr L, Turbide C (1987) Vesicle formation during reticulocyte maturation. Association of plasma membrane activities with released vesicles (exosomes). J Biol Chem 262(19):9412–9420
55. Johnstone RM, Bianchini A, Teng K (1989) Reticulocyte maturation and exosome release: transferrin receptor containing exosomes shows multiple plasma membrane functions. Blood 74(5):1844–1851
56. Kharaziha P, Ceder S, Li Q, Panaretakis T (2012) Tumor cell-derived exosomes: a message in a bottle. Biochim Biophys Acta 1826(1):103–111
57. Rubler S, Dlugash J, Yuceoglu YZ, Kumral T, Branwood AW, Grishman A (1972) New type of cardiomyopathy associated with diabetic glomerulosclerosis. Am J Cardiol 30(6):595–602
58. Kiencke S, Handschin R, von Dahlen R, Muser J, Brunner-Larocca HP, Schumann J, Felix B, Berneis K, Rickenbacher P (2010) Pre-clinical diabetic cardiomyopathy: prevalence, screening, and outcome. Eur J Heart Fail 12(9):951–957
59. Cai L, Kang YJ (2001) Oxidative stress and diabetic cardiomyopathy: a brief review. Cardiovasc Toxicol 1(3):181–193
60. Liu Q, Wang S, Cai L (2014) Diabetic cardiomyopathy and its mechanisms: role of oxidative stress and damage. J Diabetes Investig 5(6):623–634
61. Westermann D, Rutschow S, Jager S, Linderer A, Anker S, Riad A, Unger T, Schultheiss HP, Pauschinger M, Tschope C (2007) Contributions of inflammation and cardiac matrix metalloproteinase activity to cardiac failure in diabetic cardiomyopathy: the role of angiotensin type 1 receptor antagonism. Diabetes 56(3):641–646
62. Choi KM, Zhong Y, Hoit BD, Grupp IL, Hahn H, Dilly KW, Guatimosim S, Lederer WJ, Matlib MA (2002) Defective intracellular Ca(2+) signaling contributes to cardiomyopathy in Type 1 diabetic rats. Am J Physiol Heart Circ Physiol 283(4):H1398–H1408
63. Duncan JG (2011) Mitochondrial dysfunction in diabetic cardiomyopathy. Biochim Biophys Acta 1813(7):1351–1359
64. Lebeche D, Davidoff AJ, Hajjar RJ (2008) Interplay between impaired calcium regulation and insulin signaling abnormalities in diabetic cardiomyopathy. Nat Clin Pract Cardiovasc Med 5(11):715–724
65. Asbun J, Villarreal FJ (2006) The pathogenesis of myocardial fibrosis in the setting of diabetic cardiomyopathy. J Am Coll Cardiol 47(4):693–700
66. Singh VP, Le B, Khode R, Baker KM, Kumar R (2008) Intracellular angiotensin II production in diabetic rats is correlated with cardiomyocyte apoptosis, oxidative stress, and cardiac fibrosis. Diabetes 57(12):3297–3306
67. Raev DC (1994) Which left ventricular function is impaired earlier in the evolution of diabetic cardiomyopathy? An echocardiographic study of young type I diabetic patients. Diabetes Care 17(7):633–639
68. Fang ZY, Prins JB, Marwick TH (2004) Diabetic cardiomyopathy: evidence, mechanisms, and therapeutic implications. Endocr Rev 25(4):543–567
69. Bugger H, Abel ED (2009) Rodent models of diabetic cardiomyopathy. Dis Model Mech 2(9–10):454–466

70. Rodrigues B, McNeill JH (1992) The diabetic heart: metabolic causes for the development of a cardiomyopathy. Cardiovasc Res 26(10):913–922
71. Mizamtsidi M, Paschou SA, Grapsa J, Vryonidou A (2016) Diabetic cardiomyopathy: a clinical entity or a cluster of molecular heart changes? Eur J Clin Invest 46(11):947–953
72. Bugger H, Abel ED (2014) Molecular mechanisms of diabetic cardiomyopathy. Diabetologia 57(4):660–671
73. Holscher ME, Bode C, Bugger H (2016) Diabetic cardiomyopathy: does the type of diabetes matter? Int J Mol Sci 17(12):2136
74. Aneja A, Tang WH, Bansilal S, Garcia MJ, Farkouh ME (2008) Diabetic cardiomyopathy: insights into pathogenesis, diagnostic challenges, and therapeutic options. Am J Med 121(9):748–757
75. Galderisi M, Anderson KM, Wilson PW, Levy D (1991) Echocardiographic evidence for the existence of a distinct diabetic cardiomyopathy (the Framingham Heart Study). Am J Cardiol 68(1):85–89
76. Maisch B, Alter P, Pankuweit S (2011) Diabetic cardiomyopathy—fact or fiction? Herz 36(2):102–115
77. Cai L, Li W, Wang G, Guo L, Jiang Y, Kang YJ (2002) Hyperglycemia-induced apoptosis in mouse myocardium: mitochondrial cytochrome C-mediated caspase-3 activation pathway. Diabetes 51(6):1938–1948
78. Huynh K, Kiriazis H, XJ D, Love JE, Gray SP, Jandeleit-Dahm KA, McMullen JR, Ritchie RH (2013) Targeting the upregulation of reactive oxygen species subsequent to hyperglycemia prevents type 1 diabetic cardiomyopathy in mice. Free Radic Biol Med 60:307–317
79. Liu ZW, Zhu HT, Chen KL, Dong X, Wei J, Qiu C, Xue JH (2013) Protein kinase RNA-like endoplasmic reticulum kinase (PERK) signaling pathway plays a major role in reactive oxygen species (ROS)-mediated endoplasmic reticulum stress-induced apoptosis in diabetic cardiomyopathy. Cardiovasc Diabetol 12:158
80. Valko M, Leibfritz D, Moncol J, Cronin MT, Mazur M, Telser J (2007) Free radicals and antioxidants in normal physiological functions and human disease. Int J Biochem Cell Biol 39(1):44–84
81. Ye G, Metreveli NS, Ren J, Epstein PN (2003) Metallothionein prevents diabetes-induced deficits in cardiomyocytes by inhibiting reactive oxygen species production. Diabetes 52(3):777–783
82. Cai L, Wang J, Li Y, Sun X, Wang L, Zhou Z, Kang YJ (2005) Inhibition of superoxide generation and associated nitrosative damage is involved in metallothionein prevention of diabetic cardiomyopathy. Diabetes 54(6):1829–1837
83. Haidara MA, Yassin HZ, Rateb M, Ammar H, Zorkani MA (2006) Role of oxidative stress in development of cardiovascular complications in diabetes mellitus. Curr Vasc Pharmacol 4(3):215–227
84. Hamblin M, Friedman DB, Hill S, Caprioli RM, Smith HM, Hill MF (2007) Alterations in the diabetic myocardial proteome coupled with increased myocardial oxidative stress underlies diabetic cardiomyopathy. J Mol Cell Cardiol 42(4):884–895
85. Li CJ, Lv L, Li H, Yu DM (2012) Cardiac fibrosis and dysfunction in experimental diabetic cardiomyopathy are ameliorated by alpha-lipoic acid. Cardiovasc Diabetol 11:73
86. Schannwell CM, Schneppenheim M, Perings S, Plehn G, Strauer BE (2002) Left ventricular diastolic dysfunction as an early manifestation of diabetic cardiomyopathy. Cardiology 98(1–2):33–39
87. Connelly KA, Gilbert RE, Krum H (2008) Letter by Connelly et al regarding article, "Diastolic stiffness of the failing diabetic heart: importance of fibrosis, advanced glycation end products, and myocyte resting tension". Circulation 117(23):e483
88. van Heerebeek L, Hamdani N, Handoko ML, Falcao-Pires I, Musters RJ, Kupreishvili K, Ijsselmuiden AJ, Schalkwijk CG, Bronzwaer JG, Diamant M, Borbely A, van der Velden J, Stienen GJ, Laarman GJ, Niessen HW, Paulus WJ (2008) Diastolic stiffness of the failing diabetic heart: importance of fibrosis, advanced glycation end products, and myocyte resting tension. Circulation 117(1):43–51

89. Khaidar A, Marx M, Lubec B, Lubec G (1994) L-arginine reduces heart collagen accumulation in the diabetic db/db mouse. Circulation 90(1):479–483
90. Spiro MJ, Crowley TJ (1993) Increased rat myocardial type VI collagen in diabetes mellitus and hypertension. Diabetologia 36(2):93–98
91. Bodiga VL, Eda SR, Bodiga S (2014) Advanced glycation end products: role in pathology of diabetic cardiomyopathy. Heart Fail Rev 19(1):49–63
92. Candido R, Forbes JM, Thomas MC, Thallas V, Dean RG, Burns WC, Tikellis C, Ritchie RH, Twigg SM, Cooper ME, Burrell LM (2003) A breaker of advanced glycation end products attenuates diabetes-induced myocardial structural changes. Circ Res 92(7):785–792
93. Liu J, Masurekar MR, Vatner DE, Jyothirmayi GN, Regan TJ, Vatner SF, Meggs LG, Malhotra A (2003) Glycation end-product cross-link breaker reduces collagen and improves cardiac function in aging diabetic heart. Am J Physiol Heart Circ Physiol 285(6):H2587–H2591
94. Vasan S, Foiles P, Founds H (2003) Therapeutic potential of breakers of advanced glycation end product-protein crosslinks. Arch Biochem Biophys 419(1):89–96
95. From AM, Scott CG, Chen HH (2010) The development of heart failure in patients with diabetes mellitus and pre-clinical diastolic dysfunction a population-based study. J Am Coll Cardiol 55(4):300–305
96. von Bibra H, St John Sutton M (2010) Diastolic dysfunction in diabetes and the metabolic syndrome: promising potential for diagnosis and prognosis. Diabetologia 53(6):1033–1045
97. Devereux RB, Roman MJ, Paranicas M, O'Grady MJ, Lee ET, Welty TK, Fabsitz RR, Robbins D, Rhoades ER, Howard BV (2000) Impact of diabetes on cardiac structure and function: the strong heart study. Circulation 101(19):2271–2276
98. Patil VC, Patil HV, Shah KB, Vasani JD, Shetty P (2011) Diastolic dysfunction in asymptomatic type 2 diabetes mellitus with normal systolic function. J Cardiovasc Dis Res 2(4):213–222
99. Paulus WJ, van Ballegoij JJ (2010) Treatment of heart failure with normal ejection fraction: an inconvenient truth! J Am Coll Cardiol 55(6):526–537
100. Zhang X, Wei X, Liang Y, Liu M, Li C, Tang H (2013) Differential changes of left ventricular myocardial deformation in diabetic patients with controlled and uncontrolled blood glucose: a three-dimensional speckle-tracking echocardiography-based study. J Am Soc Echocardiogr 26(5):499–506
101. MacDonald MR, Petrie MC, Hawkins NM, Petrie JR, Fisher M, McKelvie R, Aguilar D, Krum H, McMurray JJ (2008) Diabetes, left ventricular systolic dysfunction, and chronic heart failure. Eur Heart J 29(10):1224–1240
102. Reisinger J, Dubrey SW, Falk RH (1997) Restrictive cardiomyopathy. N Engl J Med 336(26):1917
103. Seferovic PM, Paulus WJ (2015) Clinical diabetic cardiomyopathy: a two-faced disease with restrictive and dilated phenotypes. Eur Heart J 36(27):1718–1727
104. Kuethe F, Sigusch HH, Bornstein SR, Hilbig K, Kamvissi V, Figulla HR (2007) Apoptosis in patients with dilated cardiomyopathy and diabetes: a feature of diabetic cardiomyopathy? Horm Metab Res 39(9):672–676
105. Dries DL, Sweitzer NK, Drazner MH, Stevenson LW, Gersh BJ (2001) Prognostic impact of diabetes mellitus in patients with heart failure according to the etiology of left ventricular systolic dysfunction. J Am Coll Cardiol 38(2):421–428
106. Shiomi T, Tsutsui H, Ikeuchi M, Matsusaka H, Hayashidani S, Suematsu N, Wen J, Kubota T, Takeshita A (2003) Streptozotocin-induced hyperglycemia exacerbates left ventricular remodeling and failure after experimental myocardial infarction. J Am Coll Cardiol 42(1):165–172
107. Song GY, Wu YJ, Yang YJ, Li JJ, Zhang HL, Pei HJ, Zhao ZY, Zeng ZH, Hui RT (2009) The accelerated post-infarction progression of cardiac remodelling is associated with genetic changes in an untreated streptozotocin-induced diabetic rat model. Eur J Heart Fail 11(10):911–921
108. Miki T, Yuda S, Kouzu H, Miura T (2013) Diabetic cardiomyopathy: pathophysiology and clinical features. Heart Fail Rev 18(2):149–166

109. Voulgari C, Papadogiannis D, Tentolouris N (2010) Diabetic cardiomyopathy: from the pathophysiology of the cardiac myocytes to current diagnosis and management strategies. Vasc Health Risk Manag 6:883–903
110. Morgan JP (1991) Abnormal intracellular modulation of calcium as a major cause of cardiac contractile dysfunction. N Engl J Med 325(9):625–632
111. Hattori Y, Matsuda N, Kimura J, Ishitani T, Tamada A, Gando S, Kemmotsu O, Kanno M (2000) Diminished function and expression of the cardiac Na+-Ca2+ exchanger in diabetic rats: implication in Ca2+ overload. J Physiol 527(Pt 1):85–94
112. Lanner JT (2012) Ryanodine receptor physiology and its role in disease. Adv Exp Med Biol 740:217–234
113. Periasamy M, Bhupathy P, Babu GJ (2008) Regulation of sarcoplasmic reticulum Ca2+ ATPase pump expression and its relevance to cardiac muscle physiology and pathology. Cardiovasc Res 77(2):265–273
114. Teshima Y, Takahashi N, Saikawa T, Hara M, Yasunaga S, Hidaka S, Sakata T (2000) Diminished expression of sarcoplasmic reticulum Ca(2+)-ATPase and ryanodine sensitive Ca(2+)Channel mRNA in streptozotocin-induced diabetic rat heart. J Mol Cell Cardiol 32(4):655–664
115. Trost SU, Belke DD, Bluhm WF, Meyer M, Swanson E, Dillmann WH (2002) Overexpression of the sarcoplasmic reticulum Ca(2+)-ATPase improves myocardial contractility in diabetic cardiomyopathy. Diabetes 51(4):1166–1171
116. Asrih M, Steffens S (2013) Emerging role of epigenetics and miRNA in diabetic cardiomyopathy. Cardiovasc Pathol 22(2):117–125
117. Chavali V, Tyagi SC, Mishra PK (2013) Predictors and prevention of diabetic cardiomyopathy. Diabetes Metab Syndr Obes 6:151–160
118. Finck BN, Han X, Courtois M, Aimond F, Nerbonne JM, Kovacs A, Gross RW, Kelly DP (2003) A critical role for PPARalpha-mediated lipotoxicity in the pathogenesis of diabetic cardiomyopathy: modulation by dietary fat content. Proc Natl Acad Sci U S A 100(3):1226–1231
119. Narula J, Pandey P, Arbustini E, Haider N, Narula N, Kolodgie FD, Dal Bello B, Semigran MJ, Bielsa-Masdeu A, Dec GW, Israels S, Ballester M, Virmani R, Saxena S, Kharbanda S (1999) Apoptosis in heart failure: release of cytochrome c from mitochondria and activation of caspase-3 in human cardiomyopathy. Proc Natl Acad Sci U S A 96(14):8144–8149
120. Rutter MK, Parise H, Benjamin EJ, Levy D, Larson MG, Meigs JB, Nesto RW, Wilson PW, Vasan RS (2003) Impact of glucose intolerance and insulin resistance on cardiac structure and function: sex-related differences in the Framingham Heart Study. Circulation 107(3):448–454
121. Witteles RM, Tang WH, Jamali AH, Chu JW, Reaven GM, Fowler MB (2004) Insulin resistance in idiopathic dilated cardiomyopathy: a possible etiologic link. J Am Coll Cardiol 44(1):78–81
122. Johansen JS, Harris AK, Rychly DJ, Ergul A (2005) Oxidative stress and the use of antioxidants in diabetes: linking basic science to clinical practice. Cardiovasc Diabetol 4:5
123. Vassort G, Turan B (2010) Protective role of antioxidants in diabetes-induced cardiac dysfunction. Cardiovasc Toxicol 10(2):73–86
124. Fein FS (1990) Diabetic cardiomyopathy. Diabetes Care 13(11):1169–1179
125. Bartel DP (2004) MicroRNAs: genomics, biogenesis, mechanism, and function. Cell 116(2):281–297
126. Zheng D, Ma J, Yu Y, Li M, Ni R, Wang G, Chen R, Li J, Fan GC, Lacefield JC, Peng T (2015) Silencing of miR-195 reduces diabetic cardiomyopathy in C57BL/6 mice. Diabetologia 58(8):1949–1958
127. Horie T, Ono K, Nishi H, Iwanaga Y, Nagao K, Kinoshita M, Kuwabara Y, Takanabe R, Hasegawa K, Kita T, Kimura T (2009) MicroRNA-133 regulates the expression of GLUT4 by targeting KLF15 and is involved in metabolic control in cardiac myocytes. Biochem Biophys Res Commun 389(2):315–320
128. Chen S, Puthanveetil P, Feng B, Matkovich SJ, Dorn GW II, Chakrabarti S (2014) Cardiac miR-133a overexpression prevents early cardiac fibrosis in diabetes. J Cell Mol Med 18(3):415–421

129. Chavali V, Tyagi SC, Mishra PK (2012) MicroRNA-133a regulates DNA methylation in diabetic cardiomyocytes. Biochem Biophys Res Commun 425(3):668–672
130. Feng B, Chen S, George B, Feng Q, Chakrabarti S (2010) miR133a regulates cardiomyocyte hypertrophy in diabetes. Diabetes Metab Res Rev 26(1):40–49
131. Yildirim SS, Akman D, Catalucci D, Turan B (2013) Relationship between downregulation of miRNAs and increase of oxidative stress in the development of diabetic cardiac dysfunction: junctin as a target protein of miR-1. Cell Biochem Biophys 67(3):1397–1408
132. Lu H, Buchan RJ, Cook SA (2010) MicroRNA-223 regulates Glut4 expression and cardiomyocyte glucose metabolism. Cardiovasc Res 86(3):410–420
133. Baseler WA, Thapa D, Jagannathan R, Dabkowski ER, Croston TL, Hollander JM (2012) miR-141 as a regulator of the mitochondrial phosphate carrier (Slc25a3) in the type 1 diabetic heart. Am J Physiol Cell Physiol 303(12):C1244–C1251
134. Kuwabara Y, Horie T, Baba O, Watanabe S, Nishiga M, Usami S, Izuhara M, Nakao T, Nishino T, Otsu K, Kita T, Kimura T, Ono K (2015) MicroRNA-451 exacerbates lipotoxicity in cardiac myocytes and high-fat diet-induced cardiac hypertrophy in mice through suppression of the LKB1/AMPK pathway. Circ Res 116(2):279–288
135. Blumensatt M, Greulich S, Herzfeld de Wiza D, Mueller H, Maxhera B, Rabelink MJ, Hoeben RC, Akhyari P, Al-Hasani H, Ruige JB, Ouwens DM (2013) Activin A impairs insulin action in cardiomyocytes via up-regulation of miR-143. Cardiovasc Res 100(2):201–210
136. Arnold N, Koppula PR, Gul R, Luck C, Pulakat L (2014) Regulation of cardiac expression of the diabetic marker microRNA miR-29. PLoS One 9(7):e103284
137. Marchand A, Atassi F, Mougenot N, Clergue M, Codoni V, Berthuin J, Proust C, Tregouet DA, Hulot JS, Lompre AM (2016) miR-322 regulates insulin signaling pathway and protects against metabolic syndrome-induced cardiac dysfunction in mice. Biochim Biophys Acta 1862(4):611–621
138. Thery C, Zitvogel L, Amigorena S (2002) Exosomes: composition, biogenesis and function. Nat Rev Immunol 2(8):569–579
139. Stoorvogel W, Kleijmeer MJ, Geuze HJ, Raposo G (2002) The biogenesis and functions of exosomes. Traffic 3(5):321–330
140. Simons M, Raposo G (2009) Exosomes—vesicular carriers for intercellular communication. Curr Opin Cell Biol 21(4):575–581
141. Fevrier B, Raposo G (2004) Exosomes: endosomal-derived vesicles shipping extracellular messages. Curr Opin Cell Biol 16(4):415–421
142. Babst M (2011) MVB vesicle formation: ESCRT-dependent, ESCRT-independent and everything in between. Curr Opin Cell Biol 23(4):452–457
143. Ostrowski M, Carmo NB, Krumeich S, Fanget I, Raposo G, Savina A, Moita CF, Schauer K, Hume AN, Freitas RP, Goud B, Benaroch P, Hacohen N, Fukuda M, Desnos C, Seabra MC, Darchen F, Amigorena S, Moita LF, Thery C (2010) Rab27a and Rab27b control different steps of the exosome secretion pathway. Nat Cell Biol 12(1):19–30
144. Raiborg C, Stenmark H (2009) The ESCRT machinery in endosomal sorting of ubiquitylated membrane proteins. Nature 458(7237):445–452
145. Slagsvold T, Pattni K, Malerod L, Stenmark H (2006) Endosomal and non-endosomal functions of ESCRT proteins. Trends Cell Biol 16(6):317–326
146. Blott EJ, Griffiths GM (2002) Secretory lysosomes. Nat Rev Mol Cell Biol 3(2):122–131
147. Luzio JP, Rous BA, Bright NA, Pryor PR, Mullock BM, Piper RC (2000) Lysosome-endosome fusion and lysosome biogenesis. J Cell Sci 113(Pt 9):1515–1524
148. Raposo G, Stoorvogel W (2013) Extracellular vesicles: exosomes, microvesicles, and friends. J Cell Biol 200(4):373–383
149. de Jong OG, Verhaar MC, Chen Y, Vader P, Gremmels H, Posthuma G, Schiffelers RM, Gucek M, van Balkom BW (2012) Cellular stress conditions are reflected in the protein and RNA content of endothelial cell-derived exosomes. J Extracell Vesicles 1. doi:10.3402/jev.v1i0.18396
150. Rice GE, Scholz-Romero K, Sweeney E, Peiris H, Kobayashi M, Duncombe G, Mitchell MD, Salomon C (2015) The effect of glucose on the release and bioactivity of exosomes from first trimester trophoblast cells. J Clin Endocrinol Metab 100(10):E1280–E1288

151. Dehwah MA, Xu A, Huang Q (2012) MicroRNAs and type 2 diabetes/obesity. J Genet Genomics 39(1):11–18
152. Williams MD, Mitchell GM (2012) MicroRNAs in insulin resistance and obesity. Exp Diabetes Res 2012:484696
153. Guay C, Roggli E, Nesca V, Jacovetti C, Regazzi R (2011) Diabetes mellitus, a microRNA-related disease? Transl Res 157(4):253–264
154. Kong L, Zhu J, Han W, Jiang X, Xu M, Zhao Y, Dong Q, Pang Z, Guan Q, Gao L, Zhao J, Zhao L (2011) Significance of serum microRNAs in pre-diabetes and newly diagnosed type 2 diabetes: a clinical study. Acta Diabetol 48(1):61–69
155. Salido-Guadarrama I, Romero-Cordoba S, Peralta-Zaragoza O, Hidalgo-Miranda A, Rodriguez-Dorantes M (2014) MicroRNAs transported by exosomes in body fluids as mediators of intercellular communication in cancer. Onco Targets Ther 7:1327–1338
156. Esteller M (2011) Non-coding RNAs in human disease. Nat Rev Genet 12(12):861–874
157. Ha TY (2011) MicroRNAs in human diseases: from cancer to cardiovascular disease. Immune Network 11(3):135–154
158. Iborra M, Bernuzzi F, Invernizzi P, Danese S (2012) MicroRNAs in autoimmunity and inflammatory bowel disease: crucial regulators in immune response. Autoimmun Rev 11(5):305–314
159. Aswad H, Forterre A, Wiklander OP, Vial G, Danty-Berger E, Jalabert A, Lamaziere A, Meugnier E, Pesenti S, Ott C, Chikh K, El-Andaloussi S, Vidal H, Lefai E, Rieusset J, Rome S (2014) Exosomes participate in the alteration of muscle homeostasis during lipid-induced insulin resistance in mice. Diabetologia 57(10):2155–2164
160. Deng ZB, Poliakov A, Hardy RW, Clements R, Liu C, Liu Y, Wang J, Xiang X, Zhang S, Zhuang X, Shah SV, Sun D, Michalek S, Grizzle WE, Garvey T, Mobley J, Zhang HG (2009) Adipose tissue exosome-like vesicles mediate activation of macrophage-induced insulin resistance. Diabetes 58(11):2498–2505
161. Karolina DS, Armugam A, Tavintharan S, Wong MT, Lim SC, Sum CF, Jeyaseelan K (2011) MicroRNA 144 impairs insulin signaling by inhibiting the expression of insulin receptor substrate 1 in type 2 diabetes mellitus. PLoS One 6(8):e22839
162. Guay C, Jacovetti C, Nesca V, Motterle A, Tugay K, Regazzi R (2012) Emerging roles of non-coding RNAs in pancreatic beta-cell function and dysfunction. Diabetes Obes Metab 14(Suppl 3):12–21
163. Ventriglia G, Nigi L, Sebastiani G, Dotta F (2015) MicroRNAs: novel players in the dialogue between pancreatic islets and immune system in autoimmune diabetes. Biomed Res Int 2015:749734
164. Guay C, Regazzi R (2013) Circulating microRNAs as novel biomarkers for diabetes mellitus. Nat Rev Endocrinol 9(9):513–521
165. Rottiers V, Naar AM (2012) MicroRNAs in metabolism and metabolic disorders. Nat Rev Mol Cell Biol 13(4):239–250
166. Lovis P, Roggli E, Laybutt DR, Gattesco S, Yang JY, Widmann C, Abderrahmani A, Regazzi R (2008) Alterations in microRNA expression contribute to fatty acid-induced pancreatic beta-cell dysfunction. Diabetes 57(10):2728–2736
167. Roggli E, Britan A, Gattesco S, Lin-Marq N, Abderrahmani A, Meda P, Regazzi R (2010) Involvement of microRNAs in the cytotoxic effects exerted by proinflammatory cytokines on pancreatic beta-cells. Diabetes 59(4):978–986
168. Roggli E, Gattesco S, Caille D, Briet C, Boitard C, Meda P, Regazzi R (2012) Changes in microRNA expression contribute to pancreatic beta-cell dysfunction in prediabetic NOD mice. Diabetes 61(7):1742–1751
169. Guay C, Menoud V, Rome S, Regazzi R (2015) Horizontal transfer of exosomal microRNAs transduce apoptotic signals between pancreatic beta-cells. Cell Commun Signal 13:17
170. Pfeifer P, Werner N, Jansen F (2015) Role and function of microRNAs in extracellular vesicles in cardiovascular biology. Biomed Res Int 2015:161393
171. Patel D, Ythier D, Brozzi F, Eizirik DL, Thorens B (2015) Clic4, a novel protein that sensitizes beta-cells to apoptosisa. Mol Metab 4(4):253–264

172. Zhao X, Mohan R, Ozcan S, Tang X (2012) MicroRNA-30d induces insulin transcription factor MafA and insulin production by targeting mitogen-activated protein 4 kinase 4 (MAP4K4) in pancreatic beta-cells. J Biol Chem 287(37):31155–31164
173. Osmai M, Osmai Y, Bang-Berthelsen CH, Pallesen EM, Vestergaard AL, Novotny GW, Pociot F, Mandrup-Poulsen T (2016) MicroRNAs as regulators of beta-cell function and dysfunction. Diabetes Metab Res Rev 32(4):334–349
174. Nesca V, Guay C, Jacovetti C, Menoud V, Peyot ML, Laybutt DR, Prentki M, Regazzi R (2013) Identification of particular groups of microRNAs that positively or negatively impact on beta cell function in obese models of type 2 diabetes. Diabetologia 56(10):2203–2212
175. Lin X, Guan H, Huang Z, Liu J, Li H, Wei G, Cao X, Li Y (2014) Downregulation of Bcl-2 expression by miR-34a mediates palmitate-induced Min6 cells apoptosis. J Diabetes Res 2014:258695
176. Ruan Q, Wang T, Kameswaran V, Wei Q, Johnson DS, Matschinsky F, Shi W, Chen YH (2011) The microRNA-21-PDCD4 axis prevents type 1 diabetes by blocking pancreatic beta cell death. Proc Natl Acad Sci U S A 108(29):12030–12035
177. Belgardt BF, Ahmed K, Spranger M, Latreille M, Denzler R, Kondratiuk N, von Meyenn F, Villena FN, Herrmanns K, Bosco D, Kerr-Conte J, Pattou F, Rulicke T, Stoffel M (2015) The microRNA-200 family regulates pancreatic beta cell survival in type 2 diabetes. Nat Med 21(6):619–627
178. Mocharla P, Briand S, Giannotti G, Dorries C, Jakob P, Paneni F, Luscher T, Landmesser U (2013) AngiomiR-126 expression and secretion from circulating CD34(+) and CD14(+) PBMCs: role for proangiogenic effects and alterations in type 2 diabetics. Blood 121(1):226–236
179. Cocucci E, Racchetti G, Meldolesi J (2009) Shedding microvesicles: artefacts no more. Trends Cell Biol 19(2):43–51
180. Gyorgy B, Szabo TG, Pasztoi M, Pal Z, Misjak P, Aradi B, Laszlo V, Pallinger E, Pap E, Kittel A, Nagy G, Falus A, Buzas EI (2011) Membrane vesicles, current state-of-the-art: emerging role of extracellular vesicles. Cell Mol Life Sci 68(16):2667–2688
181. Schatteman GC, Hanlon HD, Jiao C, Dodds SG, Christy BA (2000) Blood-derived angioblasts accelerate blood-flow restoration in diabetic mice. J Clin Investig 106(4):571–578
182. Lawson C, Vicencio JM, Yellon DM, Davidson SM (2016) Microvesicles and exosomes: new players in metabolic and cardiovascular disease. J Endocrinol 228(2):R57–R71
183. Tabit CE, Chung WB, Hamburg NM, Vita JA (2010) Endothelial dysfunction in diabetes mellitus: molecular mechanisms and clinical implications. Rev Endocr Metab Disord 11(1):61–74
184. Burger D, Thibodeau JF, Holterman CE, Burns KD, Touyz RM, Kennedy CR (2014) Urinary podocyte microparticles identify prealbuminuric diabetic glomerular injury. J Am Soc Nephrol 25(7):1401–1407
185. Fais S, O'Driscoll L, Borras FE, Buzas E, Camussi G, Cappello F, Carvalho J, Cordeiro da Silva A, Del Portillo H, El Andaloussi S, Ficko Trcek T, Furlan R, Hendrix A, Gursel I, Kralj-Iglic V, Kaeffer B, Kosanovic M, Lekka ME, Lipps G, Logozzi M, Marcilla A, Sammar M, Llorente A, Nazarenko I, Oliveira C, Pocsfalvi G, Rajendran L, Raposo G, Rohde E, Siljander P, van Niel G, Vasconcelos MH, Yanez-Mo M, Yliperttula ML, Zarovni N, Zavec AB, Giebel B (2016) Evidence-based clinical use of nanoscale extracellular vesicles in nanomedicine. ACS Nano 10(4):3886–3899
186. Kranendonk ME, de Kleijn DP, Kalkhoven E, Kanhai DA, Uiterwaal CS, van der Graaf Y, Pasterkamp G, Visseren FL, Group SS (2014) Extracellular vesicle markers in relation to obesity and metabolic complications in patients with manifest cardiovascular disease. Cardiovasc Diabetol 13:37
187. van der Pol E, Boing AN, Harrison P, Sturk A, Nieuwland R (2012) Classification, functions, and clinical relevance of extracellular vesicles. Pharmacol Rev 64(3):676–705
188. Jansen F, Yang X, Franklin BS, Hoelscher M, Schmitz T, Bedorf J, Nickenig G, Werner N (2013) High glucose condition increases NADPH oxidase activity in endothelial microparticles that promote vascular inflammation. Cardiovasc Res 98(1):94–106

189. Lakhter AJ, Sims EK (2015) Minireview: emerging roles for extracellular vesicles in diabetes and related metabolic disorders. Mol Endocrinol 29(11):1535–1548
190. Barutta F, Tricarico M, Corbelli A, Annaratone L, Pinach S, Grimaldi S, Bruno G, Cimino D, Taverna D, Deregibus MC, Rastaldi MP, Perin PC, Gruden G (2013) Urinary exosomal microRNAs in incipient diabetic nephropathy. PLoS One 8(11):e73798
191. Li R, Chung AC, Yu X, Lan HY (2014) MicroRNAs in diabetic kidney disease. Int J Endocrinol 2014:593956
192. Mann JF, Rossing P, Wiecek A, Rosivall L, Mark P, Mayer G (2015) Diagnosis and treatment of early renal disease in patients with type 2 diabetes mellitus: what are the clinical needs? Nephrol Dial Transplant 30(Suppl 4):iv1–iv5
193. Joglekar MV, Joglekar VM, Hardikar AA (2009) Expression of islet-specific microRNAs during human pancreatic development. Gene Expr Patterns 9(2):109–113
194. Melkman-Zehavi T, Oren R, Kredo-Russo S, Shapira T, Mandelbaum AD, Rivkin N, Nir T, Lennox KA, Behlke MA, Dor Y, Hornstein E (2011) miRNAs control insulin content in pancreatic beta-cells via downregulation of transcriptional repressors. EMBO J 30(5):835–845
195. Tang X, Muniappan L, Tang G, Ozcan S (2009) Identification of glucose-regulated miRNAs from pancreatic {beta} cells reveals a role for miR-30d in insulin transcription. RNA 15(2):287–293
196. Wang Y, Liu J, Liu C, Naji A, Stoffers DA (2013) MicroRNA-7 regulates the mTOR pathway and proliferation in adult pancreatic beta-cells. Diabetes 62(3):887–895
197. Esguerra JL, Bolmeson C, Cilio CM, Eliasson L (2011) Differential glucose-regulation of microRNAs in pancreatic islets of non-obese type 2 diabetes model Goto-Kakizaki rat. PLoS One 6(4):e18613
198. Zampetaki A, Kiechl S, Drozdov I, Willeit P, Mayr U, Prokopi M, Mayr A, Weger S, Oberhollenzer F, Bonora E, Shah A, Willeit J, Mayr M (2010) Plasma microRNA profiling reveals loss of endothelial miR-126 and other microRNAs in type 2 diabetes. Circ Res 107(6):810–817
199. Horikawa Y, Oda N, Cox NJ, Li X, Orho-Melander M, Hara M, Hinokio Y, Lindner TH, Mashima H, Schwarz PE, del Bosque-Plata L, Horikawa Y, Oda Y, Yoshiuchi I, Colilla S, Polonsky KS, Wei S, Concannon P, Iwasaki N, Schulze J, Baier LJ, Bogardus C, Groop L, Boerwinkle E, Hanis CL, Bell GI (2000) Genetic variation in the gene encoding calpain-10 is associated with type 2 diabetes mellitus. Nat Genet 26(2):163–175
200. Wicker LS, Clark J, Fraser HI, Garner VE, Gonzalez-Munoz A, Healy B, Howlett S, Hunter K, Rainbow D, Rosa RL, Smink LJ, Todd JA, Peterson LB (2005) Type 1 diabetes genes and pathways shared by humans and NOD mice. J Autoimmun 25(Suppl):29–33
201. Lai RC, Yeo RW, Tan KH, Lim SK (2013) Exosomes for drug delivery—a novel application for the mesenchymal stem cell. Biotechnol Adv 31(5):543–551
202. Simpson RJ, Lim JW, Moritz RL, Mathivanan S (2009) Exosomes: proteomic insights and diagnostic potential. Expert Rev Proteomics 6(3):267–283
203. Camussi G, Deregibus MC, Bruno S, Cantaluppi V, Biancone L (2010) Exosomes/microvesicles as a mechanism of cell-to-cell communication. Kidney Int 78(9):838–848
204. Lee Y, El Andaloussi S, Wood MJ (2012) Exosomes and microvesicles: extracellular vesicles for genetic information transfer and gene therapy. Hum Mol Genet 21(R1):R125–R134
205. Cortez MA, Calin GA (2009) MicroRNA identification in plasma and serum: a new tool to diagnose and monitor diseases. Expert Opin Biol Ther 9(6):703–711
206. Ge Q, Zhou Y, Lu J, Bai Y, Xie X, Lu Z (2014) miRNA in plasma exosome is stable under different storage conditions. Molecules 19(2):1568–1575
207. Vickers KC, Palmisano BT, Shoucri BM, Shamburek RD, Remaley AT (2011) MicroRNAs are transported in plasma and delivered to recipient cells by high-density lipoproteins. Nat Cell Biol 13(4):423–433
208. Di Carli MF, Janisse J, Grunberger G, Ager J (2003) Role of chronic hyperglycemia in the pathogenesis of coronary microvascular dysfunction in diabetes. J Am Coll Cardiol 41(8):1387–1393

209. Hink U, Li H, Mollnau H, Oelze M, Matheis E, Hartmann M, Skatchkov M, Thaiss F, Stahl RA, Warnholtz A, Meinertz T, Griendling K, Harrison DG, Forstermann U, Munzel T (2001) Mechanisms underlying endothelial dysfunction in diabetes mellitus. Circ Res 88(2):E14–E22
210. Chen JX, Zeng H, Reese J, Aschner JL, Meyrick B (2012) Overexpression of angiopoietin-2 impairs myocardial angiogenesis and exacerbates cardiac fibrosis in the diabetic db/db mouse model. Am J Physiol Heart Circ Physiol 302(4):H1003–H1012
211. Pandey AK, Agarwal P, Kaur K, Datta M (2009) MicroRNAs in diabetes: tiny players in big disease. Cell Physiol Biochem 23(4–6):221–232
212. Wang XH, Qian RZ, Zhang W, Chen SF, Jin HM, Hu RM (2009) MicroRNA-320 expression in myocardial microvascular endothelial cells and its relationship with insulin-like growth factor-1 in type 2 diabetic rats. Clin Exp Pharmacol Physiol 36(2):181–188
213. Brutsaert DL (2003) Cardiac endothelial-myocardial signaling: its role in cardiac growth, contractile performance, and rhythmicity. Physiol Rev 83(1):59–115
214. Narmoneva DA, Vukmirovic R, Davis ME, Kamm RD, Lee RT (2004) Endothelial cells promote cardiac myocyte survival and spatial reorganization: implications for cardiac regeneration. Circulation 110(8):962–968
215. Shantikumar S, Angelini GD, Emanueli C (2014) Diabetes, microRNAs and exosomes: Les liaisons dangereuses. J Mol Cell Cardiol 74:196–198
216. Westermeier F, Riquelme JA, Pavez M, Garrido V, Diaz A, Verdejo HE, Castro PF, Garcia L, Lavandero S (2016) New molecular insights of insulin in diabetic cardiomyopathy. Front Physiol 7:125
217. Das S, Halushka MK (2015) Extracellular vesicle microRNA transfer in cardiovascular disease. Cardiovasc Pathol 24(4):199–206
218. Loyer X, Vion AC, Tedgui A, Boulanger CM (2014) Microvesicles as cell-cell messengers in cardiovascular diseases. Circ Res 114(2):345–353
219. Vrijsen KR, Sluijter JP, Schuchardt MW, van Balkom BW, Noort WA, Chamuleau SA, Doevendans PA (2010) Cardiomyocyte progenitor cell-derived exosomes stimulate migration of endothelial cells. J Cell Mol Med 14(5):1064–1070
220. Wang X, Huang W, Liu G, Cai W, Millard RW, Wang Y, Chang J, Peng T, Fan GC (2014) Cardiomyocytes mediate anti-angiogenesis in type 2 diabetic rats through the exosomal transfer of miR-320 into endothelial cells. J Mol Cell Cardiol 74:139–150
221. Guo BB, Bellingham SA, Hill AF (2015) The neutral sphingomyelinase pathway regulates packaging of the prion protein into exosomes. J Biol Chem 290(6):3455–3467
222. Sahoo S, Emanueli C (2016) Exosomes in diabetic cardiomyopathy: the next-generation therapeutic targets? Diabetes 65(10):2829–2831
223. Garcia NA, Ontoria-Oviedo I, Gonzalez-King H, Diez-Juan A, Sepulveda P (2015) Glucose starvation in cardiomyocytes enhances exosome secretion and promotes angiogenesis in endothelial cells. PLoS One 10(9):e0138849
224. Tan A, Rajadas J, Seifalian AM (2013) Exosomes as nano-theranostic delivery platforms for gene therapy. Adv Drug Deliv Rev 65(3):357–367
225. Ribeiro MF, Zhu H, Millard RW, Fan GC (2013) Exosomes function in pro- and anti-angiogenesis. Current Angiogenes 2(1):54–59
226. Wang X, Gu H, Huang W, Peng J, Li Y, Yang L, Qin D, Essandoh K, Wang Y, Peng T, Fan GC (2016) Hsp20-mediated activation of exosome biogenesis in cardiomyocytes improves cardiac function and angiogenesis in diabetic mice. Diabetes 65(10):3111–3128
227. Kowal J, Tkach M, Thery C (2014) Biogenesis and secretion of exosomes. Curr Opin Cell Biol 29:116–125
228. Pironti G, Strachan RT, Abraham D, Mon-Wei Yu S, Chen M, Chen W, Hanada K, Mao L, Watson LJ, Rockman HA (2015) Circulating exosomes induced by cardiac pressure overload contain functional angiotensin II type 1 receptors. Circulation 131(24):2120–2130
229. Ailawadi S, Wang X, Gu H, Fan GC (2015) Pathologic function and therapeutic potential of exosomes in cardiovascular disease. Biochim Biophys Acta 1852(1):1–11

Chapter 9
Peripartum Cardiomyopathy: Do Exosomes Play a Role?

Huanyu Gu, Qiying Dai, Zhuyuan Liu, Hongbao Wang, Jianhua Yao, and Lei Zhou

9.1 Introduction of PPCM

In the 18th, PPCM has been widely accepted as one of the most dreadful complications during and after pregnancy. It was finally recorded in the medical literature until early in the 19th [1, 2]. It was correlated to heart failure induced by dilated cardiomyopathy initially at that time [3]. The term PPCM was used to name this dangerous disease. Classical definition are: (1) the symptoms of heart failure occur in the last month of pregnancy or within the 5 month after delivery; (2) the absence of pre-existing cardiac disease leading to heat failure before the last month of pregnancy; (3) no other determinable cause of heart failure is found [1]. PPCM is mainly a diagnosis of exclusion. The 6 months' time limit is very important, and all other causes of heart failure must be ruled out. Echocardiogram diagnostic criteria of PPCM includes left ventricular ejection fraction (EF) less than 45%, or fractional shortening less than 30%, or both [2]. Unfortunately, echocardiography is not accessible in some developing countries, left ventricular systolic dysfunction is hard to confirm.

Huanyu Gu, Qiying Dai and Zhuyuan Liu contributed equally to this work.

H. Gu • Z. Liu • L. Zhou (✉)
Department of Cardiology, First Affiliated Hospital of Nanjing Medical University, Nanjing 210029, China
e-mail: zhoulei@njmu.edu.cn

Q. Dai
Department of Cardiology, First Affiliated Hospital of Nanjing Medical University, Nanjing 210029, China

Metrowest Medical Center, Framingham, MA 01702, USA

H. Wang • J. Yao (✉)
Department of Cardiology, Yangpu Hospital, Tongji University School of Medicine, Shanghai 200090, China
e-mail: yaojianhua@tongji.edu.cn

J. Xiao, S. Cretoiu (eds.), *Exosomes in Cardiovascular Diseases*, Advances in Experimental Medicine and Biology 998, DOI 10.1007/978-981-10-4397-0_9

9.2 Epidemiology of PPCM

It is not uncommon that PPCM can occur in healthy young women. This means that all women at reproductive age share the same risk of developing PPCM. However, epidemiologic data on PPCM remains insufficient. Separated diagnostic criteria, different population and individual studies have been blamed for. In addition, lack of systematic reporting, misdiagnosis and under-diagnosis, the true morbidity of PPCM is hard to ascertain. Incidence as high as 1 in 100 to 1 in 300 pregnant women were reported in Haiti and Nigeria, where are two global hotpots in the world map [3, 4]. The reasons for variation in the incidence among countries remains unclear. PPCM seem to be more common in African women. The incidence rate of South Africa is 1 in 1000 live births. What's more, the prognosis is poorest in African women [5]. Even in the same country, the prevalence of PPCM is different based on geographical regions, races and social classes. For example, the incidences range from 1 in 1149 to 1 in 4350 in the United States, where the most studies have been conducted [6]. Hispanics have a comparatively lower morbidity, as opposed to African races. Most of the studies were conducted in the three countries mentioned above. The rest of the studies were done in other parts of the world, including Europe where the cases are relatively rare [7]. Although PPCM is rare, it is associated with a high maternal mortality rate, especially in African females [8]. About 28% of the patients died after 6 months despite conventional treatment [9]. Increasing studies have be done to explore the underlying mechanism of the disease in order to find more specific treatment.

9.3 Etiology of PPCM

Factors like black race, multiple pregnancy, advanced maternal age, pregnancy induced hypertension, multiparity, abuse of tocolysis medications, smoking, malnutrition, breastfeeding are all related to the development of PPCM [10, 11]. Among these risks, black race is considered to be the greatest risk for PPCM. This is often confused by socioeconomic and ethic risk factors [8]. Women of advanced maternal age are associated with increased risk, whereas risk still exists in young group [12]. Multiparity has been considered as risk factors for all the complications for a lot time. However, its role in PPCM is controversial since most of the studied females are at first or second pregnancy [13]. In contrast to traditional thought, breast-feeding is not associated with worsened outcomes in PPCM. In fact, about 67% of PPCM patients choose to breast feed without any adverse effects.

The pathology of PPCM presents as a general cardiomyocyte damage which is provoked by myocarditis, increased myocyte apoptosis, inflammation, abnormal autoimmune responses, viral infection, stress-induced release of cytokines and hormonal abnormality [14].

Myocarditis was the first recognized as a cause for PPCM 30 years ago, but the reported incidence rate of myocarditis is high variable over the years. Sanderson showed that 45% PPCM patients could suffer from myocarditis by histological detection [15]. Another study showed supportive evidence on this association [16]. Moreover, a 62% incidence rate of myocarditis was found in PPCM patients [17]. Even with the results provided, sampling error cannot be excluded since these studies did not group patients by race and had no unified diagnostic criteria.

Apoptosis refers to the programmed cell death. In *vivo* study has proven that apoptosis occurs in PPCM [18]. Fas and Fas ligand, located in cell surface, are pivotal molecules in the process of apoptosis. During a 6 month's longitudinal study with 100 PPCM patients from South Africa, 15 patients died and they had higher levels of Fas/Apo-1 in serum [12]. A caspase inhibitor can rescue the apoptosis in PPCM mouse model, thus provide a novel therapeutic strategies for PPCM and an evidence for the role of apoptosis in PPCM [19].

Inflammation has been found to play an important role in many pathological process of PPCM. Levels of serum markers of inflammation like interferon-gamma (IFN-γ), tumor necrosis factor (TNF), interleukin-6 (IL-6) and C-reactive protein are higher in serum of patients with PPCM than in healthy controls. Besides that, these inflammatory mediators are closely related with echocardiographic outcomes of impaired left ventricular function. Adding anti-inflammatory agent Pentoxifylline to conventional therapy could improve clinical presentation in a trial of 59 PPCM patients [12].

During pregnancy, maternal immune system is suppressed to ensure the safety and normal development of fetus. A maladaptive response with the feature of autoimmunity could occur. It can produce high titers of antibodies recognizing maternal cardiac tissue and initiate inflammatory damage. These antibodies is detectable in patients with PPCM, but not in normal controls [20]. The inflammatory damage to the heart tissue will lead to autoimmune myocarditis. One explanation to the development of these autoantibodies is that rapid degeneration of uterus after delivery triggers the release of myosin, actin and lots of other proteins into maternal circulation. Antibodies against these proteins will make cross-reacting with analogous proteins of myocardium, leading to auto attack.

Viral infection is one of common causes of myocarditis. It could also contribute to the pathological changes of PPCM. Similar to the autoimmune attack theory, autoantibodies against hear tissue can develop after episode of viral infection. EpsteinBarr virus (EBV), human cytomegalovirus (CMV), human herpes virus 6 (HHV6) and parvovirus B19's viral genomes have been found in biopsy specimens in 31% PPCM patients [21]. The ejection fraction of PPCM patients can be improved from 50.2 to 58.1% by anti-viral therapy, while the ejection fractions of those with persisted viral titers are decreased from 54.3 to 51.4% [22]. Among these known viruses, human immunodeficiency virus (HIV) have not been found involved in PPCM [16]. Animal study further confirmed this theory by providing the result that pregnant mice can develop worse myocarditis after experimentally infected with echovirus or coxsachievirus than those non-pregnant mice [17].

It is obvious to suspect hormone may contribute to PPCM since hormonal changes extremely dramatically in pregnancy. It is found that levels of prolactin, progesterone and estrogen are lower in serum of PPCM patients. These hormones are important vasodilatory elements to prevent hypertension from intravascular volume expansion [18]. Prolactin precipitates oxidative cascade through cathepsin. Initial mild oxidative stress can activate cathepsin D which cleaves prolactin into 16-kDa prolactin. The 16-kDa prolactin enhance cardiomyocytes apoptosis, contributing to PPCM [23]. Higher levels of activated cathepsin D, total prolactin and cleaved 16-kDa prolactin fragment are found in PPCM patients' serum. In animal study, this effect can be prevented by inhibiting secretion of prolactin [23]. PGC-1alpha is a transcriptional regulator for angiogenesis and metabolism. It modulates the expression of SOD2, which is able to enhance oxidative stress. It is demonstrated that down regulation of PGC-1alpha in heart make female mice prone to develop PPCM [24].

9.4 Clinical Presentation of PPCM

Pregnancy is a physiological state in which there is a dramatic change in maternal hemodynamic. The most common clinical presentation is heart failure [25]. These symptoms overlap with physiological changes in pregnancy, such as persistent cough, fatigue, dyspnea, peripheral edema and atypical chest pain. Progression of PPCM can be as fast as a couple of days. Classical signs of late PPCM include jugular venous distension, positive hepatojugular reflex, tachycardia, tachypnea, gallop rhythm, hepatomegaly, and ascites. Blood pressure fluctuates in PPCM. Multiple organ failure is the most dangerous complication of PPCM [26].

Approximately 78% of PPAM patients are symptomatic in the first 4 months after delivery. Only 9% of PPCM patients develop symptoms in the last month of pregnancy. The rest of patients have symptoms either after 4 month postpartum or before the last month of pregnancy [27]. The initial presentation might vary from NYHA I to IV. The majority of patients fall into NYHA functional class III or IV [5]. In rare circumstances, cardiac arrest or complex ventricular arrhythmias can happen if treatment is delayed or not appropriate [28].

PPCM is associated more to thromboembolic events than cardiomyopathy from other causes. Left ventricular thrombosis is common in PPCM patients who have a left ventricular ejection fraction less than 35% [29]. Peripheral embolic such as mesenteric embolism, coronary embolism and cerebral embolism can happen occasionally [30].

9.5 Diagnosis of PPCM

As mentioned above, PPCM is possible even in patients has no history of heart disease. The diagnoses usually can be delayed or missed as most symptoms are similar to physiological changes during pregnancy [31]. Early diagnosis and treatment correlates to better outcomes for PPCM [32]. As a diagnosis of exclusion, a thorough investigation is required to rule out other alternatives, such as thyroid disorders, sepsis, myocardial infarction, idiopathic dilated cardiomyopathy, postpartum depression, severe preeclampsia, anemia and pulmonary vasculitis [33]. Routine assessment of PPCM includes history, basic physical examination, blood analysis and imaging.

Although there has no specific biomarkers of PPCM, routine blood work is still helpful for early diagnosis of PPCM. Inflammatory markers including IL-6, TNF-α and CRP are elevated. Markers of apoptosis, Fas/Apo-1 is also elevated. Brain natriuretic peptide (BNP) and N-terminal portion of proBNP (pro-NT BNP) are sensitive biomarkers to heart failure but not specific for PPCM. High level of pro-NT BNP is associated with worse prognosis [34]. In addition, microRNA (miRNA, miR) has recently revealed its potential to detect cardiac disease. miR-146a is a promising specific biomarker for PPCM. It is reported that a higher level of miR-146a can be detected in PPCM patients than healthy control or patients with cardiomyopathy from other causes [35]. Other studies suggested that troponin and endomyocardial biopsy (EMB) could also be useful for testing PPCM. In a study on 106 newly diagnosed PPCM patients, elevated troponin level is related to persistent left ventricular dysfunction for up to 6 months [36]. EMB is seldom used but could be very specific. Given its invasive nature and comparatively high risks, it should be performed at professional hospital [37].

Imaging studies are recommended for PPCM. The combination of blood work and imaging is helpful not only for diagnosis, but also for risk stratification. There is no particular electrocardiogram (EKG) patterns for PPCM. Sixty-six percent of PPCM patients can present left ventricular hypertrophy pattern on EKG, and 96% of them may have abnormal ST-T wave. Other abnormalities such as atrial fibrillation, ventricular tachycardia and bundle branch block could also present [38]. Through EKG, different degrees of left ventricular dilatation can be shown. Suppression of systolic function vary from moderate to severe. Left ventricular thrombus can been found on initial echocardiography in 10–17% of PPCM patients [32, 39]. Cardiomegaly, pleural effusion and pulmonary venous congestion can be found by chest X ray [40, 41]. MRI is more precise in measuring ventricular function and chamber volumes. However, it is less commonly used since it is more time consuming and there is no definite imaging finding for PPCM after all.

9.6 Treatment of PPCM

The basic treatment of PPCM is similar to heart failure, but also needs to be individualized [42]. The most important concept is to relief afterload and increase contractility. As we all know, traditional medication for heart failure involves angiotensin-converting enzyme inhibitor (ACEI) β blocker, spironolactone and digoxin [43]. Medication for PPCM is different depend on whether patient develop the disease in pregnancy or not. ACEI is contraindicated during pregnancy.

Digitalis is an effective inotropic agent which can improve contractility and control rate. More importantly, it is safe both for women and fetus. However, it needs to be closely monitored for that pregnant women are more sensitive to the effect of digitalis. High level of digoxin in serum indicates a worse outcome in PPCM [44].

Diuretics are commonly used to relief preload. Loop diuretics like furosemide can reduce intravascular volume rapidly. Though it can be excreted by breast milk, no adverse effects has been reported [45].

ACEI has been recognized as one of the fundamental medication for heart failure for many years. It decrease mortality and is recommended for any tolerated heart failure patients. Despite of all the benefits, numerous adverse effects on the fetus has been reported, including prematurity, bony malformation, oligohydramnios, limb contractures, intrauterine growth retardation, neonatal death and pulmonary hypoplasia [46]. For sake of safety, ACEI is usually replaced by hydralazine.

B blockeris another significant medication for PPCM. It suppresses arrhythmia, decrease risk for sudden death and is associated with decreased mortality [47]. Metoprolol is detectable in breast milk, but the drug level in milk can be tolerated well for infants [48]. Among all the different types of β blockers, β1 selective blocker is more favorable. Because other non-selective β blocker can interfere with the uterus activity [47].

It is necessary to add anticoagulation considering the high risk of thromboembolism in PPCM. Anticoagulation is advised to be used at the time of diagnosis until the EF recovers to more than 35% [49]. Common anticoagulation medication is warfarin. PPCM patients with EF less than 35% requires warfarin. Heparin is also often used and is considered to be safe since it can't cross placenta [50]. Effective and safety issues regarding to other newer anticoagulation agents remains unaddressed. More data are necessary before putting them on recommendation.

Prolactin levels contributes to the specificity of pathologic progression of PPCM. The mechanism has been previously discussed. This raise the question whether prolactin inhibitor could be a targeted therapy for this disease. It has been used in post-partum mothers to stop lactation during the past 20 years. Study have provided evidence that there is beneficial effects by adding bromocriptine to standard treatment [29]. The safety of bromocriptine was assured in 1400 mothers who had been taking bromocriptine during the first few weeks of pregnancy. No significant increase in congenital malformation or abortion has been reported [51]. It provided a new prospect in treating PPCM, but more trials regarding to the mortality and side effects are needed.

Modulating immune response is another target for exploring treatment for PPCM. Intravenous immunoglobulin has showed its potential to ameliorate cardiac damage and accelerate recovery of ventricular function [52]. However, its beneficial effect is nonrepeatable in other studies [53]. Further researches are obligatory to clarify its use in PPCM.

Apart from drug therapy, monitoring before and 24 h after delivery is also of great significance. Cardiovascular monitoring is necessary according to the severity of PPCM. It includes continuous EKG and full invasive haemodynamic monitoring. ICU level of care is required for severe clinical cases. Continuous monitoring can also be recommended as an early detection of heart decomposition or a guideline in the treatment for PPCM [54].

In rare conditions, cardiac transplantation is warranted in patients with worsening symptoms despite maximal drug therap. Unfortunately, only fewer than 3000 patients have access to the procedure per year in the world. In this case, ventricular assist device (VAD) has emerged as an important bridge to transplantation [55]. Defibrillator implantation is useful for PPCM patient who has symptomatic ventricular arrhythmias [56].

9.7 Prognosis of PPCM

In general, most of PPCM patients have higher rate of spontaneous recovery than those cardiomyopathy caused by other reasons. Twenty-three to fifty-four percent of PPCM patients' left ventricular function is able to fully recover to baseline [57]. It had been found that mortality of PPCM in the US was the lowest through the world. Report suggested that survival rates of PPCM are similar in the USA, South Africa and Haiti [58].

Available studies in the USA, South Africa, Haiti and Turkey indicate that prognosis of PPCM varies depend on geography [59]. They have reported that mortality rates range from 1.4 to 30% [60].

Factors that can predict mortality of PPCM are: African descent, older age, increased left ventricular size and multiparity.

9.8 Exosomes and PPCM

Exosome is a nanosized extracellular vesicles, secreted by different types of cells. It has proven itself as a crucial mediator for cell-cell communications. Exosome carries molecules like proteins, nucleic acids and lipids. The role of exosomes in tumor has been well established. Its function in heart has become a research hotpot in recent years. Exosomes are found to be secreted from cardiomyocytes, cardiac fibroblasts, endothelial cells and cardiac progenitor cells. They are also proven to have cardioprotective and proangiogenic effect, help to improve cardiac function [52].

The detailed function of exosome in cardiac tissue and the relationship between exosome and some other cardiac diseases have been described in other chapters. As a matter of fact, very few publications have investigated the role of exosome in PPCM.

As discussed above, 16-kDa N-terminal prolactin can trigger PPCM. 16-kDa prolactin is reported to enhance the expression of miR-146a in endothelial cells, thus inhibiting angiogenesis. miR-146a is a key factor in inflammatory diseases, sepsis and innate immunity [53]. It can regulate the activation of NF-κB through tumor necrosis factor receptor associated factor 6 (TRAF6) and interleukin-1 associated Kinase 1(IRAK1) [61]. 16-kDa prolactin can also promote miR-146a-enriched exosomes release from endothelial cells [35]. These exosomes will be taken up by targeted cardiomyocytes, resulting in upregulating of expression of miR-146a. Downstream factors of miR-146a, Notch1, Irak1 and Erbb4 are decreased then. These changes can trigger cellular mitogenesis and differentiation. Besides endothelial cells, fibroblasts are also capable of secreting miR-146a-enriched exosomal in hearts after being stimulated by 16-kDa prolactin. Significant elevated level of exosomal miR-146a are detected in 38 acute PPCM patients. In contrast, no change are observed in 30 dilated cardiomyopathy patients or 18 healthy postpartum mothers. Based on these evidence, exosomal miR-146a can be developed as a distinct biomarker for diagnosis and risk stratification for PPCM patients [35].

9.9 Summary

PPCM is a rare complications but associated with high mortality. Although influenced by various risk factors, it still can affect both previously healthy patients and patients with history of cardiac disease. As most of the early symptoms resemble the physiologic changes in pregnancy, it is not uncommon to miss the diagnosis. Delay in treatment may result in increased mortality. Current treatment of PPCM are similar to treatment for dilated heart failure. Extra caution should be taken given the vulnerability of fetus. Progress has been made in recent years both for the lab test and treatment strategy. Increasing evidence support that exosomes and miRNA have great potential in the field. However, more trials are needed to testify the availability and safety of these new regimen.

Acknowledgements This work was supported by the grants from National Natural Science Foundation of China (81370280 and 81570332), the grant from Yangpu Commission of Science and Technology Commission, and Yangpu Commission of Health and Family Planning (YP15M07), and the "Chenguang Program" supported by Yangpu Hospital, Tongji University School of Medicine (Ye1201409).

Competing Financial Interests The authors declare no competing financial interests.

References

1. Demakis JG, Rahimtoola SH, Sutton GC, Meadows WR, Szanto PB, Tobin JR, Gunnar RM (1971) Natural course of peripartum cardiomyopathy. Circulation 44(6):1053–1061
2. Hibbard JU, Lindheimer M, Lang RM (1999) A modified definition for peripartum cardiomyopathy and prognosis based on echocardiography. Obstet Gynecol 94(2):311–316
3. Deneux-Tharaux C, Berg C, Bouvier-Colle MH, Gissler M, Harper M, Nannini A, Alexander S, Wildman K, Breart G, Buekens P (2005) Underreporting of pregnancy-related mortality in the United States and Europe. Obstet Gynecol 106(4):684–692

4. Brar SS, Khan SS, Sandhu GK, Jorgensen MB, Parikh N, Hsu JW, Shen AY (2007) Incidence, mortality, and racial differences in peripartum cardiomyopathy. Am J Cardiol 100(2):302–304
5. Desai D, Moodley J, Naidoo D (1995) Peripartum cardiomyopathy: experiences at King Edward VIII Hospital, Durban, South Africa and a review of the literature. Trop Dr 25(3):118–123
6. Elkayam U (2011) Clinical characteristics of peripartum cardiomyopathy in the United States: diagnosis, prognosis, and management. J Am Coll Cardiol 58(7):659–670
7. Ferriere M, Sacrez A, Bouhour JB, Cassagnes J, Geslin P, Dubourg O, Komajda M, Degeorges M (1990) Cardiomyopathy in the peripartum period: current aspects. A multicenter study. 11 cases. Arch Mal Coeur Vaiss 83(10):1563–1569
8. Gentry MB, Dias JK, Luis A, Patel R, Thornton J, Reed GL (2010) African-American women have a higher risk for developing peripartum cardiomyopathy. J am Coll Cardiol 55(7):654–659
9. Sliwa K, Skudicky D, Candy G, Bergemann A, Hopley M, Sareli P (2002) The addition of pentoxifylline to conventional therapy improves outcome in patients with peripartum cardiomyopathy. Eur J Heart Fail 4(3):305–309
10. Pearson GD, Veille JC, Rahimtoola S, Hsia J, Oakley CM, Hosenpud JD, Ansari A, Baughman KL (2000) Peripartum cardiomyopathy: National Heart, Lung, and Blood Institute and Office of Rare Diseases (National Institutes of Health) workshop recommendations and review. J Am Med Assoc 283(9):1183–1188
11. Sliwa K, Skudicky D, Bergemann A, Candy G, Puren A, Sareli P (2000) Peripartum cardiomyopathy: analysis of clinical outcome, left ventricular function, plasma levels of cytokines and Fas/APO-1. J Am Coll Cardiol 35(3):701–705
12. Sliwa K, Forster O, Libhaber E, Fett JD, Sundstrom JB, Hilfiker-Kleiner D, Ansari AA (2006) Peripartum cardiomyopathy: inflammatory markers as predictors of outcome in 100 prospectively studied patients. Eur Heart J 27(4):441–446
13. Chapa JB, Heiberger HB, Weinert L, Decara J, Lang RM, Hibbard JU (2005) Prognostic value of echocardiography in peripartum cardiomyopathy. Obstet Gynecol 105(6):1303–1308
14. Ntusi NB, Mayosi BM (2009) Aetiology and risk factors of peripartum cardiomyopathy: a systematic review. Int J Cardiol 131(2):168–179
15. Sanderson JE, Olsen EG, Gatei D (1986) Peripartum heart disease: an endomyocardial biopsy study. Br Heart J 56(3):285–291
16. Sliwa K, Forster O, Tibazarwa K, Libhaber E, Becker A, Yip A, Hilfiker-Kleiner D (2011) Long-term outcome of peripartum cardiomyopathy in a population with high seropositivity for human immunodeficiency virus. Int J Cardiol 147(2):202–208
17. Lyden DC, Huber SA (1984) Aggravation of coxsackievirus, group B, type 3-induced myocarditis and increase in cellular immunity to myocyte antigens in pregnant Balb/c mice and animals treated with progesterone. Cell Immunol 87(2):462–472
18. Ansari AA, Fett JD, Carraway RE, Mayne AE, Onlamoon N, Sundstrom JB (2002) Autoimmune mechanisms as the basis for human peripartum cardiomyopathy. Clin Rev Allergy Immunol 23(3):301–324
19. Hayakawa Y, Chandra M, Miao W, Shirani J, Brown JH, Dorn GW II, Armstrong RC, Kitsis RN (2003) Inhibition of cardiac myocyte apoptosis improves cardiac function and abolishes mortality in the peripartum cardiomyopathy of Galpha(q) transgenic mice. Circulation 108(24):3036–3041
20. Selle T, Renger I, Labidi S, Bultmann I, Hilfiker-Kleiner D (2009) Reviewing peripartum cardiomyopathy: current state of knowledge. Futur Cardiol 5(2):175–189
21. Bültmann BD, Klingel K, Näbauer M, Wallwiener D, Kandolf R (2005) High prevalence of viral genomes and inflammation in peripartum cardiomyopathy. Am J Obstet Gynecol 193(2):363–365
22. Kuhl U, Pauschinger M, Seeberg B, Lassner D, Noutsias M, Poller W, Schultheiss HP (2005) Viral persistence in the myocardium is associated with progressive cardiac dysfunction. Circulation 112(13):1965–1970
23. Hilfiker-Kleiner D, Kaminski K, Podewski E, Bonda T, Schaefer A, Sliwa K, Forster O, Quint A, Landmesser U, Doerries C, Luchtefeld M, Poli V, Schneider MD, Balligand JL, Desjardins

F, Ansari A, Struman I, Nguyen NQ, Zschemisch NH, Klein G, Heusch G, Schulz R, Hilfiker A, Drexler H (2007) A cathepsin D-cleaved 16 kDa form of prolactin mediates postpartum cardiomyopathy. Cell 128(3):589–600
24. Patten IS, Rana S, Shahul S, Rowe GC, Jang C, Liu L, Hacker MR, Rhee JS, Mitchell J, Mahmood F, Hess P, Farrell C, Koulisis N, Khankin EV, Burke SD, Tudorache I, Bauersachs J, del Monte F, Hilfiker-Kleiner D, Karumanchi SA, Arany Z (2012) Cardiac angiogenic imbalance leads to peripartum cardiomyopathy. Nature 485(7398):333–338
25. Fett JD, Carraway RD, Dowell DL, King ME, Pierre R (2002) Peripartum cardiomyopathy in the Hospital Albert Schweitzer District of Haiti. Am J Obstet Gynecol 186(5):1005–1010
26. Keyser EA, Reed BG, Gonzalez-Brown V, Fausett B, Staat BC, Leath CA (2012) Peripartum cardiomyopathy and acute fatty liver of pregnancy: one patient with two zebras. Mil Med 177(4):470–473
27. Shah T, Ather S, Bavishi C, Bambhroliya A, Ma T, Bozkurt B (2013) Peripartum cardiomyopathy: a contemporary review. Methodist Debakey Cardiovasc J 9(1):38–43
28. Diao M, Diop IB, Kane A, Camara S, Kane A, Sarr M, Ba SA, Diouf SM (2004) Electrocardiographic recording of long duration (Holter) of 24 hours during idiopathic cardiomyopathy of the peripartum. Arch Mal Coeur Vaiss 97(1):25–30
29. Meyer GP, Labidi S, Podewski E, Sliwa K, Drexler H, Hilfiker-Kleiner D (2010) Bromocriptine treatment associated with recovery from peripartum cardiomyopathy in siblings: two case reports. J Med Case Rep 4(1):80–84
30. Helms AK, Kittner SJ (2005) Pregnancy and stroke. CNS Spectr 10(7):580–587
31. Mayosi BM (2007) Contemporary trends in the epidemiology and management of cardiomyopathy and pericarditis in sub-Saharan Africa. Heart 93(10):1176–1183
32. Napporn A, Kane A, Damorou J, Dia A, Diop I, Sarr M, Ba S, Diouf S (2000) Intraventricular thrombosis complicating peri-partum idiopathic myocardiopathy. Annales de Cardiologie et D'angeiologie 49(5):309–314
33. Bhakta P, Biswas BK, Banerjee B (2007) Peripartum cardiomyopathy: review of the literature. Yonsei Med J 48(5):731–747
34. Forster O, Hilfiker-Kleiner D, Ansari AA, Sundstrom JB, Libhaber E, Tshani W, Becker A, Yip A, Klein G, Sliwa K (2008) Reversal of IFN-gamma, oxLDL and prolactin serum levels correlate with clinical improvement in patients with peripartum cardiomyopathy. Eur J Heart Fail 10(9):861–868
35. Halkein J, Tabruyn SP, Ricke-Hoch M, Haghikia A, Nguyen NQ, Scherr M, Castermans K, Malvaux L, Lambert V, Thiry M, Sliwa K, Noel A, Martial JA, Hilfiker-Kleiner D, Struman I (2013) MicroRNA-146a is a therapeutic target and biomarker for peripartum cardiomyopathy. J Clin Investig 123(5):2143–2154
36. Hu CL, Li YB, Zou YG, Zhang JM, Chen JB, Liu J, Tang YH, Tang QZ, Huang CX (2007) Troponin T measurement can predict persistent left ventricular dysfunction in peripartum cardiomyopathy. Heart 93(4):488–490
37. Holzmann M, Nicko A, Kuhl U, Noutsias M, Poller W, Hoffmann W, Morguet A, Witzenbichler B, Tschope C, Schultheiss HP, Pauschinger M (2008) Complication rate of right ventricular endomyocardial biopsy via the femoral approach: a retrospective and prospective study analyzing 3048 diagnostic procedures over an 11-year period. Circulation 118(17):1722–1728
38. Bhattacharyya A, Basra SS, Sen P, Kar B (2012) Peripartum cardiomyopathy: a review. Tex Heart Inst J 39(1):8–16
39. Amos AM, Jaber WA, Russell SD (2006) Improved outcomes in peripartum cardiomyopathy with contemporary. Am Heart J 152(3):509–513
40. Witlin AG, Mabie WC, Sibai BM (1997) Peripartum cardiomyopathy: an ominous diagnosis. Am J Obstet Gynecol 176(1):182–188
41. Witlin AG, Mabie WC, Sibai BM (1997) Peripartum cardiomyopathy: a longitudinal echocardiographic study. Am J Obstet Gynecol 177(5):1129–1132
42. Murali S, Baldisseri MR (2005) Peripartum cardiomyopathy. Crit Care Med 33(10):340–346

43. Heart Failure Society of A, Lindenfeld J, Albert NM, Boehmer JP, Collins SP, Ezekowitz JA, Givertz MM, Katz SD, Klapholz M, Moser DK, Rogers JG, Starling RC, Stevenson WG, Tang WH, Teerlink JR, Walsh MN (2010) HFSA 2010 comprehensive heart failure practice guideline. J Card Fail 16(6):1–194
44. Adams KF Jr, Patterson JH, Gattis WA, O'Connor CM, Lee CR, Schwartz TA, Gheorghiade M (2005) Relationship of serum digoxin concentration to mortality and morbidity in women in the digitalis investigation group trial: a retrospective analysis. J Am Coll Cardiol 46(3):497–504
45. Briggs GG, Ambrose PJ, Nageotte MP, Padilla G (2008) High-dose carisoprodol during pregnancy and lactation. Ann Pharmacother 42(6):898–901
46. Shotan A, Widerhorn J, Hurst A, Elkayam U (1994) Risks of angiotensin-converting enzyme inhibition during pregnancy: experimental and clinical evidence, potential mechanisms, and recommendations for use. Am J Med 96(5):451–456
47. Hurst A, Hoffman K, Frishman W, Elkayam U (1998) The use of beta adrenergic blocking agents in pregnancy and lactation. Wiley Liss 3(3):357–372
48. American Academy of Pediatrics Committee on D (2001) Transfer of drugs and other chemicals into human milk. Pediatrics 108(3):776–789
49. Shimamoto T, Marui A, Oda M, Tomita S, Nakajima H, Takeuchi T, Komeda M (2008) A case of peripartum cardiomyopathy with recurrent left ventricular apical thrombus. Circ J 72(5):853–854
50. Briggs GG, Freeman RK, Yaffe SJ (2012) Drugs in pregnancy and lactation: a reference guide to fetal and neonatal risk. Lippincott Williams & Wilkins 10:1–1600
51. Turkalj I, Braun P, Krupp P (1982) Surveillance of bromocriptine in pregnancy. J Am Med Assoc 247(11):1589–1591
52. Barile L, Moccetti T, Marban E, Vassalli G (2016) Roles of exosomes in cardioprotection. Eur Heart J 7(21):304–312
53. Hajivalili M, Pourgholi F, Majidi J, Aghebati-Maleki L, Movassaghpour AA, Samadi Kafil H, Mirshafiey A, Yousefi M (2016) G2013 modulates TLR4 signaling pathway in IRAK-1 and TARF-6 dependent and miR-146a independent manner. Cell Mol Biol 62(4):1–5
54. Lorello G, Cubillos J, McDonald M, Balki M (2014) Peripartum cardiomyopathy: postpartum decompensation and use of non-invasive cardiac output monitoring. Int J Obstet Anesth 23(1):66–70
55. Rose EA, Moskowitz AJ, Packer M, Sollano JA, Williams DL, Tierney AR, Heitjan DF, Meier P, Ascheim DD, Levitan RG, Weinberg AD, Stevenson LW, Shapiro PA, Lazar RM, Watson JT, Goldstein DJ, Gelijns AC (1999) The rematch trial: rationale, design, and end points. Randomized evaluation of mechanical assistance for the treatment of congestive heart failure. Ann Thorac Surg 67(3):723–730
56. Jessup M, Brozena S (2003) Heart failure. N Engl J Med 348(20):2007–2018
57. Duran N, Gunes H, Duran I, Biteker M, Ozkan M (2008) Predictors of prognosis in patients with peripartum cardiomyopathy. Int J Obstet Gynaecol 101(2):137–140
58. Modi KA, Illum S, Jariatul K, Caldito G, Reddy PC (2009) Poor outcome of indigent patients with peripartum cardiomyopathy in the United States. Am J Obstet Gynecol 201(2):171–175
59. Biteker M (2012) Peripartum cardiomyopathy in Turkey. Int J Cardiol 158(3):60–61
60. Blauwet LA, Sliwa K (2011) Peripartum cardiomyopathy. Obstet Med 4(2):44–52
61. Gao M, Wang X, Zhang X, Ha T, Ma H, Liu L, Kalbfleisch JH, Gao X, Kao RL, Williams DL, Li C (2015) Attenuation of cardiac dysfunction in polymicrobial sepsis by microRNA-146a is mediated via targeting of IRAK1 and TRAF6 expression. J Immunol 195(2):672–682

Chapter 10
Vascular Calcification Regulation by Exosomes in the Vascular Wall

Marcel Liberman and Luciana Cavalheiro Marti

10.1 Introduction

Exosomes were described initially as microvesicles (MVs) containing 5′-nucleotidase activity that were released from neoplastic cell lines [1]. A few years later, in 1983 and 1985, other groups reported in electron micrographic studies the secretion of vesicles from endocytic origin by cultured sheep reticulocytes [2, 3]. They suggested that after endocytosis occurs the formation of MVs (~50 nm in diameter) containing transferrin receptors. These MVs were present inside large multivesicular endosomes eventually fused with the plasma membrane and releasing these MVs 50 nm buds in the extracellular milieu (Fig. 10.1) [3].

Later, Johnstone et al., reported to have found two major peptide bands in these released vesicles, the 94-kDa monomer of the transferrin receptor and a 70-kDa peptide identified as a clathrin uncoating ATPase [4]. Next, they found that microvesicles (exosomes) released during sheep reticulocyte maturation exhibited several plasma membrane functions. Using an antibody coated with magnetic core bead, they were able to demonstrate that the vesicles containing transferrin receptor also contained other plasma membrane activities, such as the nucleoside transporter and acetylcholinesterase. Lysosomal activities, normally found in the same pellet, were excluded from the transferrin receptor-containing exosomes, suggesting that there was a common mechanism to isolate and externalize specific plasma membrane proteins. In addition to the sheep, electron micrographic studies show that exosomes can be recovered from the circulation of anemic pigs, rats and rabbits [5].

Several electron-microscopy studies, however, have established the existence of fusion profile between multivesicular late endosomes and the plasma membrane in living cells of hematopoietic origin, such as cytotoxic T lymphocytes (CTLs) [6],

M. Liberman, MD, Ph. D. • L.C. Marti, Ph.D. (✉)
Experimental Research Department, Hospital Israelita Albert Einstein,
Av. Albert Einstein, 627, Bloco A, 2oSS Morumbi, São Paulo 05652 000, Brazil
e-mail: luciana.marti@einstein.br

J. Xiao, S. Cretoiu (eds.), *Exosomes in Cardiovascular Diseases*, Advances in Experimental Medicine and Biology 998, DOI 10.1007/978-981-10-4397-0_10

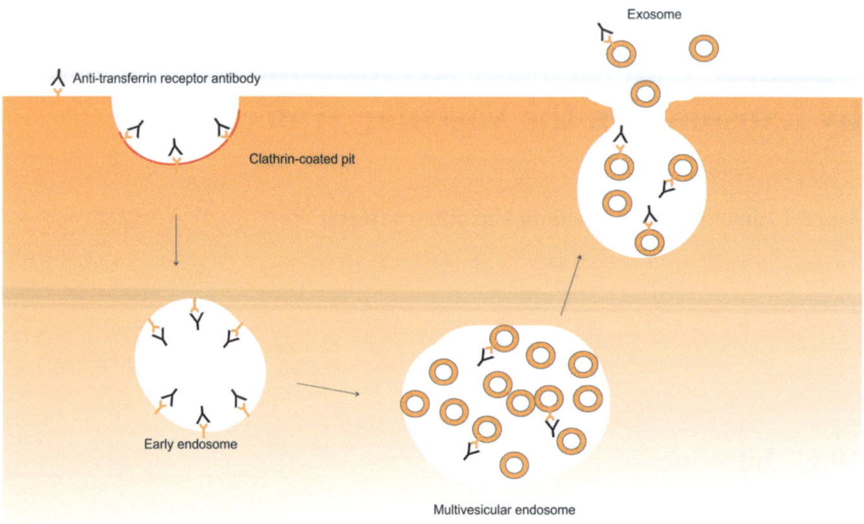

Fig. 10.1 Summary of the observations that led to the first description of exosomes. The secretion of vesicles present inside multivesicular endocytic compartments was reported in 1983 by Harding et al., and confirmed in 1985 by Pan et al.

Epstein–Barr virus (EBV)-transformed B cells [7], mast cells [8], dendritic cells [9] and platelets [10]. These cells types, and perhaps every cell that contains multivesicular endocytic compartments, could potentially secrete exosomes. Indeed, lipid vesicles purified from the culture supernatant of various hematopoietic cells [11–14], tumors of hematopoietic or non-hematopoietic origin [15], or epithelial cells [16] have been described in several studies and recently in vascular smooth muscle cell—VSMC [17]. Assembling the data from these studies they were able to define the characteristic properties of exosomes that distinguish them from vesicles that originate from other cellular sites, such as the plasma membrane.

10.2 Exosomes Physical Properties and Purification

The most common procedure to purify exosomes from cell culture supernatants includes a series of centrifugations to remove dead cells and large debris, followed by a final high-speed ultracentrifugation to pellet exosomes [7, 18]. However, this technique is not sufficient to discriminate between exosomes and other small vesicular structures, or large protein aggregates. Additional criteria must be used to confirm exosomes identity. Since, exosomes float on sucrose gradients, as all lipid vesicles, and their density ranges from 1.13 g mL^{-1} (for B-cell-derived exosomes) to 1.19 g mL^{-1} (for intestinal cell-derived exosomes) [7, 9, 10, 15, 16]. Contaminating material such as protein aggregates or nucleosomal fragments released by apoptotic cells are separated from exosomes by a different density and gradient of flotation on sucrose.

When analyzed by electron microscopy, exosomes have a characteristic 'dish-like' morphology, a compacted sphere that is limited by a lipid bi-layer. Generally, they are between 30 and 100 nm in diameter, with B-cell-derived exosomes being the most homogeneous in size (60–80 nm). These characteristics are consistent with the observed size and morphology of inner vesicles in multivesicular endosomes [7].

Since, other membrane vesicles can be secreted by cells, it is essential to purify exosomes generated in cell-culture supernatants. Filtration of the cell-culture supernatant through 0.22 μm filters, followed by direct high-speed ultracentrifugation reduces the contamination of exosome preparations with larger vesicles that are detached from the plasma membrane [11].

In addition, because exosomes are present in serum [19], it is crucial to avoid contamination with bovine exosomes from the fetal calf serum that is used to culture the exosome-producing cells. For this reason, these cells are cultured in medium in which fetal calf serum is substituted with insulin–transferrin–sodium-selenite supplement [12] or with bovine/human serum albumin [16]. Alternatively, culture medium that contains up to 20% fetal calf serum can be depleted from endogenous exosomes by overnight highspeed ultracentrifugation [14]. In previous studies, the purification of exosomes from metabolically radiolabeled cells showed that exosomes are originated from the cells, rather than from the fetal calf serum, due to the presence of radioactive labeled proteins inside the exosomes.

10.2.1 Molecular Composition of Exosomes

The presence of known cellular proteins in exosome preparations from various cellular sources has been analyzed by western blotting [14, 15, 20] and by fluorescence-activated cell sorting (FACS) analysis of exosome-coated beads [19–21]. The available proteomic studies define a subset of cellular proteins that are targeted specifically to exosomes. The function of most of these proteins in exosomes is unknown at present. Importantly, these studies also showed that exosomes are clearly distinct from the vesicles that are produced by apoptotic cells and they are only secreted by living cells.

Both general and cell-specific proteins might be targeted selectively to exosomes. These proteins are probably involved in exosome biogenesis and, possibly, in some unknown exosome functions. They include cytosolic proteins—such as tubulin, actin and actin-binding proteins as well as Annexins and RAB proteins. They also include molecules that are involved in signal transduction such as protein kinases and heterotrimeric G proteins. Various metabolic enzymes are found in exosomes from enterocytes and human dendritic cells. Exosomes also contain heatshock proteins, such as constitutive isoforms of HSP70 and HSP90. MHC class I molecules are also present in exosomes from most cell types. Finally, one of the most abundant protein families that are found in exosomes comprises the tetraspanins. Several members of this family—including CD9, CD63, CD81 and CD82—are highly enriched in exosomes from virtually any cell type. Tetraspanins interact with many protein partners—including MHC molecules and integrins, which indicates that they are involved in the organization of large molecular complexes and membrane subdomains.

10.3 Vascular Calcification and Vascular Smooth Muscle Cells

Recent advances have shown that vascular calcification is a tightly regulated process resembling bone mineralization [22] driven by VSMC osteogenic conversion and commonly observed in the aged population and patients with chronic kidney disease and type 2 diabetes. Deposition of insoluble calcium phosphate crystals reduces vascular homeostasis, promotes inflammation and stimulates VSMC death resulting in plaques with biomechanical instability [23–26].

The phenotypic plasticity of vascular smooth muscle cells (VSMC) is associated with expression of multiple antigens expression including those present in macrophages, mesenchymal stem cells, myofibroblasts and osteoblasts [27–29] and this is especially observed within the atherosclerotic plaque. Importantly, the functionality of newly acquired phenotypes is limited and often maladaptive; take for example osteogenic VSMCs, which mediate vascular calcification [27, 29, 30]. A breakthrough in calcification studies came from the identification of small membrane-enclosed extracellular vesicles found within the matrix and secreted by VSMCs [31, 32] as well as infiltrating macrophages in atherosclerotic plaques [33].

10.3.1 Exosome Fusion in Vascular Calcification

Changes in cell phenotype, specifically the epithelial-to-mesenchymal transition, were previously linked to elevated exosome-like vesicle secretion, although specific mechanisms between these phenotypic changes and exosome production are unknown [34]. In order to counterbalance calcifying milieu, MVs secreted by VSMCs are physiologically loaded with calcification inhibitors, such an endogenously expressed matrix Gla protein (MGP) and circulating fetuin-A (α2-Heremens–Schmid glycoprotein) [35] in physiological conditions. Fetuin A is a glycoprotein secreted mainly by the liver. On contrary, in patients with exaggerated vascular calcification, circulating fetuin-A is reduced [36].

In parallel, investigators established a pivotal role of MGP, which is able to attenuate vascular calcification in murine models of vascular calcification and diabetes

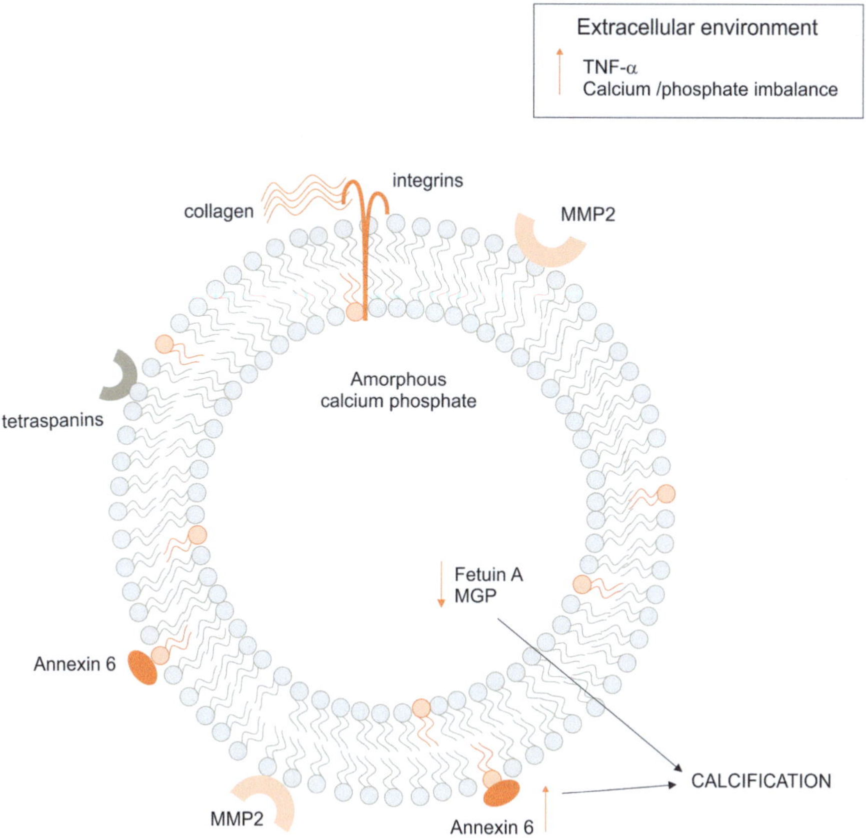

Fig. 10.2 Schematic representation of exosome and proteins related to calcification (Adapted from Shanahan CM et al.). *PS* phosphatidylserine, *MMP2* matrix metallopeptidase 2, *MGP* matrix Gla protein, *TNF-α* tumor necrosis factor-α

(db/db) and in human islet amyloid polypeptide transgenic (HIP) rats, by specifically inhibiting BMP-2 signaling [37]. However, a prolonged mineral imbalance and/or inflammation induces reduction of MGP and fetuin-A inside MVs and enrichment with a protein–lipid complex consisting of phosphatidylserine (PS) and Annexin A6, which converts MVs into a nest for calcification (Fig. 10.2) [35, 36, 38, 39].

Conversely, until recently it was unclear whether MV were derived from apoptotic cells or formed in an intracellular compartment and budded from live cells. Using fetuin-A as a tracer, Kapustin et al. performed an elegant study to identify the origin of calcifying VSMC MVs. They demonstrated that Alexa488-labelled fetuin-A is rapidly taken up by human VSMCs and delivered to early and late endosomal compartments [40].

From here a subset of late endosomal compartment, multivesicular bodies (MVBs), is involved in the production of small (100 nm), extracellular microvesicles or exosomes which are generated by the inverted budding of the MVBs limiting membrane into their own lumen. MVBs are transported to the cell periphery where the fusion of the MVB limiting membrane and plasma membrane releases these intraluminal exosomes to the extracellular matrix [2, 41]. Moreover, they showed co-localization of fetuin-A positive intracellular organelles and a MVB marker in VSMCs indicating that fetuin-A is delivered to MVBs and recycled via the exosome pathway. Interestingly, proteomic comparison revealed high levels of similarity between VSMC-derived MVs and exosomes secreted by other cells.

Furthermore, inhibition of Sphingomyelin phosphodiesterase 3 (SMPD3), which is a crucial protein for exosome biogenesis, will block exosome secretion and consequently VSMC calcification [40, 42]. Conversely, elevated extracellular calcium, a known cause of calcification, induces SMPD3 expression and exosome production [40]. In osteoblast-derived MVs calcium phosphate crystal formation is triggered by nucleation sites consisting of phosphatidylserine and annexin A5 [43]. Osteogenic conditions *in vitro* results in elevated cytosolic calcium levels followed by accumulation of Annexins A2, A5 and A6 in calcifying MVs from condrocytes [44, 45]. Accumulation of alkaline phosphatase in calcifying chondrocyte-derived MVs, which are mediated by SMPD3, has been observed in a number of studies and these data directly implicate osteogenic transcription factors in MV composition regulation [44–46]. Moreover oxLDLs, TNF-α and oxidative stress stimulate SMPD3 activity and proliferation of VSMCs and fibroblasts [47–49]. Given the role of the SMPD3 pathway in exosome biogenesis, inhibition of SMPD3 reduced exosome production [40, 42]. Secretion of calcifying exosomes by VSMCs is also driven by pathological changes in cytosolic calcium homeostasis that triggers dramatic changes in exosome composition including enrichment with nucleating phosphatidylserine/Annexin A6 complexes, loss of MGP and appearance of amorphous calcium phosphate [39, 40].

Furthermore, alterations in cytosolic calcium also stimulate exosome secretion, probably by regulating calcium-dependent fusion events or by activation of calcium-dependent calpains, proteases involved in the remodeling of the cortical cytoskeleton required for plasma membrane dynamics [50, 51]. Interestingly, MVs obtained from the media of coronary artery VSMCs were enriched with alkaline phosphatase [26, 52] after long-term treatment in osteogenic media [52, 53]. Consequently, osteogenic conditions may affect the VSMC extracellular vesicle composition in a cell and related osteogenic transcription factors in the production of calcifying extracellular vesicles by VSMCs, but this data is yet to be determined.

Additionally, recent studies showed that BMP2-activated Runx2 up-regulates SMPD3 expression in C2C12 myoblasts and chondrocytes which directly links the exosome biogenesis machinery with osteogenic master genes [54, 55].

10.4 Final Remarks

In conclusion, although there are missing data about the relationship between pathological factors and activation of calcifying exosomes, the modulation of exosome biogenesis may be a novel therapeutic approach to help improvement in vascular repair.

References

1. Trams EG, Lauter CJ, Salem N Jr, Heine U (1981) Exfoliation of membrane ecto-enzymes in the form of micro-vesicles. Biochim Biophys Acta 645(1):63–70
2. Harding C, Heuser J, Stahl P (1983) Receptor-mediated endocytosis of transferrin and recycling of the transferrin receptor in rat reticulocytes. J Cell Biol 97(2):329–339
3. Pan BT, Teng K, Wu C, Adam M, Johnstone RM (1985) Electron microscopic evidence for externalization of the transferrin receptor in vesicular form in sheep reticulocytes. J Cell Biol 101(3):942–948
4. Johnstone RM, Adam M, Hammond JR, Orr L, Turbide C (1987) Vesicle formation during reticulocyte maturation. Association of plasma membrane activities with released vesicles (exosomes). J Biol Chem 262(19):9412–9420
5. Johnstone RM, Bianchini A, Teng K (1989) Reticulocyte maturation and exosome release: transferrin receptor containing exosomes shows multiple plasma membrane functions. Blood 74(5):1844–1851
6. Peters PJ, Geuze HJ, Van der Donk HA, Slot JW, Griffith JM, Stam NJ, Clevers HC, Borst J (1989) Molecules relevant for T cell-target cell interaction are present in cytolytic granules of human T lymphocytes. Eur J Immunol 19(8):1469–1475
7. Raposo G, Nijman HW, Stoorvogel W, Liejendekker R, Harding CV, Melief CJ, Geuze HJ (1996) B lymphocytes secrete antigen-presenting vesicles. J Exp Med 183(3):1161–1172
8. Raposo G, Tenza D, Mecheri S, Peronet R, Bonnerot C, Desaymard C (1997) Accumulation of major histocompatibility complex class II molecules in mast cell secretory granules and their release upon degranulation. Mol Biol Cell 8(12):2631–2645
9. Thery C, Regnault A, Garin J, Wolfers J, Zitvogel L, Ricciardi-Castagnoli P, Raposo G, Amigorena S (1999) Molecular characterization of dendritic cell-derived exosomes. Selective accumulation of the heat shock protein hsc73. J Cell Biol 147(3):599–610
10. Heijnen HF, Schiel AE, Fijnheer R, Geuze HJ, Sixma JJ (1999) Activated platelets release two types of membrane vesicles: microvesicles by surface shedding and exosomes derived from exocytosis of multivesicular bodies and alpha-granules. Blood 94(11):3791–3799
11. Thery C, Boussac M, Veron P, Ricciardi-Castagnoli P, Raposo G, Garin J, Amigorena S (2001) Proteomic analysis of dendritic cell-derived exosomes: a secreted subcellular compartment distinct from apoptotic vesicles. J Immunol 166(12):7309–7318
12. Skokos D, Le Panse S, Villa I, Rousselle JC, Peronet R, David B, Namane A, Mecheri S (2001) Mast cell-dependent B and T lymphocyte activation is mediated by the secretion of immunologically active exosomes. J Immunol 166(2):868–876
13. Martinez-Lorenzo MJ, Anel A, Gamen S, Monle n I, Lasierra P, Larrad L, Pineiro A, Alava MA, Naval J (1999) Activated human T cells release bioactive Fas ligand and APO2 ligand in microvesicles. J Immunol 163(3):1274–1281

14. Blanchard N, Lankar D, Faure F, Regnault A, Dumont C, Raposo G, Hivroz C (2002) TCR activation of human T cells induces the production of exosomes bearing the TCR/CD3/zeta complex. J Immunol 168(7):3235–3241
15. Wolfers J, Lozier A, Raposo G, Regnault A, Thery C, Masurier C, Flament C, Pouzieux S, Faure F, Tursz T, Angevin E, Amigorena S, Zitvogel L (2001) Tumor-derived exosomes are a source of shared tumor rejection antigens for CTL cross-priming. Nat Med 7(3):297–303
16. van Niel G, Raposo G, Candalh C, Boussac M, Hershberg R, Cerf-Bensussan N, Heyman M (2001) Intestinal epithelial cells secrete exosome-like vesicles. Gastroenterology 121(2):337–349
17. Kapustin AN, Shanahan CM (2016) Emerging roles for vascular smooth muscle cell exosomes in calcification and coagulation. J Physiol 594(11):2905–2914
18. Davis JQ, Dansereau D, Johnstone RM, Bennett V (1986) Selective externalization of an ATP-binding protein structurally related to the clathrin-uncoating ATPase/heat shock protein in vesicles containing terminal transferrin receptors during reticulocyte maturation. J Biol Chem 261(33):15368–15371
19. Gambim MH, do Carmo Ade O, Marti L, Verissimo-Filho S, Lopes LR, Janiszewski M (2007) Platelet-derived exosomes induce endothelial cell apoptosis through peroxynitrite generation: experimental evidence for a novel mechanism of septic vascular dysfunction. Crit Care 11(5):R107
20. Escola JM, Kleijmeer MJ, Stoorvogel W, Griffith JM, Yoshie O, Geuze HJ (1998) Selective enrichment of tetraspan proteins on the internal vesicles of multivesicular endosomes and on exosomes secreted by human B-lymphocytes. J Biol Chem 273(32):20121–20127
21. Rabesandratana H, Toutant JP, Reggio H, Vidal M (1998) Decay-accelerating factor (CD55) and membrane inhibitor of reactive lysis (CD59) are released within exosomes during In vitro maturation of reticulocytes. Blood 91(7):2573–2580
22. Shanahan CM, Crouthamel MH, Kapustin A, Giachelli CM (2011) Arterial calcification in chronic kidney disease: key roles for calcium and phosphate. Circ Res 109(6):697–711
23. Nadra I, Mason JC, Philippidis P, Florey O, Smythe CD, McCarthy GM, Landis RC, Haskard DO (2005) Proinflammatory activation of macrophages by basic calcium phosphate crystals via protein kinase C and MAP kinase pathways: a vicious cycle of inflammation and arterial calcification? Circ Res 96(12):1248–1256
24. Ewence AE, Bootman M, Roderick HL, Skepper JN, McCarthy G, Epple M, Neumann M, Shanahan CM, Proudfoot D (2008) Calcium phosphate crystals induce cell death in human vascular smooth muscle cells: a potential mechanism in atherosclerotic plaque destabilization. Circ Res 103(5):e28–e34
25. Sage AP, Lu J, Tintut Y, Demer LL (2011) Hyperphosphatemia-induced nanocrystals upregulate the expression of bone morphogenetic protein-2 and osteopontin genes in mouse smooth muscle cells in vitro. Kidney Int 79(4):414–422
26. Hutcheson JD, Maldonado N, Aikawa E (2014) Small entities with large impact: microcalcifications and atherosclerotic plaque vulnerability. Curr Opin Lipidol 25(5):327–332
27. Shanahan CM, Cary NR, Metcalfe JC, Weissberg PL (1994) High expression of genes for calcification-regulating proteins in human atherosclerotic plaques. J Clin Invest 93(6):2393–2402
28. Rong JX, Shapiro M, Trogan E, Fisher EA (2003) Transdifferentiation of mouse aortic smooth muscle cells to a macrophage-like state after cholesterol loading. Proc Natl Acad Sci U S A 100(23):13531–13536
29. Shankman LS, Gomez D, Cherepanova OA, Salmon M, Alencar GF, Haskins RM, Swiatlowska P, Newman AA, Greene ES, Straub AC, Isakson B, Randolph GJ, Owens GK (2015) KLF4-dependent phenotypic modulation of smooth muscle cells has a key role in atherosclerotic plaque pathogenesis. Nat Med 21(6):628–637
30. Vengrenyuk Y, Nishi H, Long X, Ouimet M, Savji N, Martinez FO, Cassella CP, Moore KJ, Ramsey SA, Miano JM, Fisher EA (2015) Cholesterol loading reprograms the microRNA-

143/145-myocardin axis to convert aortic smooth muscle cells to a dysfunctional macrophage-like phenotype. Arterioscler Thromb Vasc Biol 35(3):535–546
31. Tanimura A, McGregor DH, Anderson HC (1983) Matrix vesicles in atherosclerotic calcification. Proc Soc Exp Biol Med 172(2):173–177
32. Reynolds JL, Joannides AJ, Skepper JN, McNair R, Schurgers LJ, Proudfoot D, Jahnen-Dechent W, Weissberg PL, Shanahan CM (2004) Human vascular smooth muscle cells undergo vesicle-mediated calcification in response to changes in extracellular calcium and phosphate concentrations: a potential mechanism for accelerated vascular calcification in ESRD. J Am Soc Nephrol 15(11):2857–2867
33. New SE, Goettsch C, Aikawa M, Marchini JF, Shibasaki M, Yabusaki K, Libby P, Shanahan CM, Croce K, Aikawa E (2013) Macrophage-derived matrix vesicles: an alternative novel mechanism for microcalcification in atherosclerotic plaques. Circ Res 113(1):72–77
34. Garnier D, Magnus N, Lee TH, Bentley V, Meehan B, Milsom C, Montermini L, Kislinger T, Rak J (2012) Cancer cells induced to express mesenchymal phenotype release exosome-like extracellular vesicles carrying tissue factor. J Biol Chem 287(52):43565–43572
35. Reynolds JL, Skepper JN, McNair R, Kasama T, Gupta K, Weissberg PL, Jahnen-Dechent W, Shanahan CM (2005) Multifunctional roles for serum protein fetuin-a in inhibition of human vascular smooth muscle cell calcification. J Am Soc Nephrol 16(10):2920–2930
36. Shroff RC, Shah V, Hiorns MP, Schoppet M, Hofbauer LC, Hawa G, Schurgers LJ, Singhal A, Merryweather I, Brogan P, Shanahan C, Deanfield J, Rees L (2008) The circulating calcification inhibitors, fetuin-A and osteoprotegerin, but not matrix Gla protein, are associated with vascular stiffness and calcification in children on dialysis. Nephrol Dial Transplant 23(10):3263–3271
37. Bostrom KI, Rajamannan NM, Towler DA (2011) The regulation of valvular and vascular sclerosis by osteogenic morphogens. Circ Res 109(5):564–577
38. Bobryshev YV, Killingsworth MC, Huynh TG, Lord RS, Grabs AJ, Valenzuela SM (2007) Are calcifying matrix vesicles in atherosclerotic lesions of cellular origin? Basic Res Cardiol 102(2):133–143
39. Kapustin AN, Davies JD, Reynolds JL, McNair R, Jones GT, Sidibe A, Schurgers LJ, Skepper JN, Proudfoot D, Mayr M, Shanahan CM (2011) Calcium regulates key components of vascular smooth muscle cell-derived matrix vesicles to enhance mineralization. Circ Res 109(1):e1–e12
40. Kapustin AN, Chatrou ML, Drozdov I, Zheng Y, Davidson SM, Soong D, Furmanik M, Sanchis P, De Rosales RT, Alvarez-Hernandez D, Shroff R, Yin X, Muller K, Skepper JN, Mayr M, Reutelingsperger CP, Chester A, Bertazzo S, Schurgers LJ, Shanahan CM (2015) Vascular smooth muscle cell calcification is mediated by regulated exosome secretion. Circ Res 116(8):1312–1323
41. Raposo G, Stoorvogel W (2013) Extracellular vesicles: exosomes, microvesicles, and friends. J Cell Biol 200(4):373–383
42. Trajkovic K, Hsu C, Chiantia S, Rajendran L, Wenzel D, Wieland F, Schwille P, Brugger B, Simons M (2008) Ceramide triggers budding of exosome vesicles into multivesicular endosomes. Science 319(5867):1244–1247
43. Genge BR, Wu LN, Wuthier RE (2007) In vitro modeling of matrix vesicle nucleation: synergistic stimulation of mineral formation by annexin A5 and phosphatidylserine. J Biol Chem 282(36):26035–26045
44. Kirsch T, Nah HD, Shapiro IM, Pacifici M (1997) Regulated production of mineralization-competent matrix vesicles in hypertrophic chondrocytes. J Cell Biol 137(5):1149–1160
45. Wang W, Kirsch T (2002) Retinoic acid stimulates annexin-mediated growth plate chondrocyte mineralization. J Cell Biol 157(6):1061–1069
46. Anderson HC (1995) Molecular biology of matrix vesicles. Clin Orthop Relat Res (314):266–280
47. Auge N, Nikolova-Karakashian M, Carpentier S, Parthasarathy S, Negre-Salvayre A, Salvayre R, Merrill AH Jr, Levade T (1999) Role of sphingosine 1-phosphate in the mitogenesis induced

by oxidized low density lipoprotein in smooth muscle cells via activation of sphingomyelinase, ceramidase, and sphingosine kinase. J Biol Chem 274(31):21533–21538
48. Tellier E, Negre-Salvayre A, Bocquet B, Itohara S, Hannun YA, Salvayre R, Auge N (2007) Role for furin in tumor necrosis factor alpha-induced activation of the matrix metalloproteinase/sphingolipid mitogenic pathway. Mol Cell Biochem 27(8):2997–3007
49. Cinq-Frais C, Coatrieux C, Savary A, D'Angelo R, Bernis C, Salvayre R, Negre-Salvayre A, Auge N (2015) Annexin II-dependent actin remodelling evoked by hydrogen peroxide requires the metalloproteinase/sphingolipid pathway. Redox Biol 4:169–179
50. Savina A, Fader CM, Damiani MT, Colombo MI (2005) Rab11 promotes docking and fusion of multivesicular bodies in a calcium-dependent manner. Traffic 6(2):131–143
51. Mellgren RL, Zhang W, Miyake K, McNeil PL (2007) Calpain is required for the rapid, calcium-dependent repair of wounded plasma membrane. J Biol Chem 282(4):2567–2575
52. Chen NX, O'Neill KD, Chen X, Moe SM (2008) Annexin-mediated matrix vesicle calcification in vascular smooth muscle cells. J Bone Miner Res 23(11):1798–1805
53. Hutcheson JD, Goettsch C, Pham T, Iwashita M, Aikawa M, Singh SA, Aikawa E (2014) Enrichment of calcifying extracellular vesicles using density-based ultracentrifugation protocol. J Extracell Vesicles 3:25129
54. Chae YM, Heo SH, Kim JY, Lee JM, Ryoo HM, Cho JY (2009) Upregulation of smpd3 via BMP2 stimulation and Runx2. BMB Rep 42(2):86–90
55. Kakoi H, Maeda S, Shinohara N, Matsuyama K, Imamura K, Kawamura I, Nagano S, Setoguchi T, Yokouchi M, Ishidou Y, Komiya S (2014) Bone morphogenic protein (BMP) signaling up-regulates neutral sphingomyelinase 2 to suppress chondrocyte maturation via the Akt protein signaling pathway as a negative feedback mechanism. J Biol Chem 289(12):8135–8150

Part V
Therapeutic Effects of Exosomes

Chapter 11
Cardioprotective Effects of Exosomes and Their Potential Therapeutic Use

Shengguang Ding, Jingying Zhang, Qiying Dai, Mengfei Zhao, Haitao Huang, Yiming Xu, and Chongjun Zhong

11.1 Introduction

Cardiovascular disease is the leading cause of mortality and morbidity in human beings. It is estimated that approximately one third of American adults have at least one type of cardiovascular disease, particularly among older Americans. Of these estimated 85.6 million American adults, 80 million have high blood pressure and 15.5 million have chronic heart disease. In 2013, about one in three people died from cardiovascular disease in the United States. More specifically, more than 2200 Americans died from cardiovascular disease each day, an average of one death every 40 s [1]. The total medical cost of cardiovascular disease and stroke in the United States increased from $315.4 billion in 2010 to $316.6 billion in 2013 [1, 2]. Consequently, the huge medical expenses will put a burden on the whole society. Therefore, it is imperative to identify novel therapeutics for cardiovascular diseases.

Shengguang Ding, Jingying Zhang and Qiying Dai contributed equally to this work.

S. Ding • H. Huang • Y. Xu • C. Zhong (✉)
Department of Thoracic and Cardiovascular Surgery, The Second Affiliated Hospital of Nantong University, Nantong 226001, China
e-mail: chongjunzhong@hotmail.com

J. Zhang
Department of Cardiology, Shanghai Tenth People's Hospital, Tongji University School of Medicine, Shanghai 200072, China

Q. Dai
Metrowest Medical Center, Framingham, MA 01702, USA

Department of Cardiology, First Affiliated Hospital of Nanjing Medical University, Nanjing 210029, China

M. Zhao
Department of Rehabilitation Medicine, Second Medical School of Nanjing University of Chinese Medicine, Nanjing 210001, China

J. Xiao, S. Cretoiu (eds.), *Exosomes in Cardiovascular Diseases*, Advances in Experimental Medicine and Biology 998, DOI 10.1007/978-981-10-4397-0_11

11.2 Mammalian Heart

Mammalian heart is composed of various types of cells, including cardiac fibroblasts, myocardial cells, vascular cells, and neural cells [3]. In the past few years, cardiac stem cells have been identified in postnatal hearts by using a variety of approaches, such as expression of surface markers like c-Kit, or physiological properties like the ability to efflux fluorescent dye or come into being multicellular spheroids [4–6]. Due to its limited capacity to replace itself, loss of terminally differentiated cardiomyocytes will result in irreversible heart damage [3, 6, 7]. Cardiac fibroblasts play an important role in heart structural coordination and development [8, 9]. Cardiac myocytes serve as a source of diverse paracrine signals that can adjust the vascular tone to meet myocardial requirements for oxygen and nutrients. Apart from that, cardiomyocyte also have an effect on the long-term growth and development of coronary arterial, venous, and lymphatic trees [10]. Cardiac microvascular endothelial cell protects cardiomyocytes against acute ischemic-reperfusion injury (IR injury) through a peptide called intermedin. Intermedin could reduce oxidative damage to cellular proteins, preserve the cellular cytoskeleton, attenuate the apoptotic signaling cascade and maintain cellular viability by acting in a paracrine manner on cardiomyocytes [11]. Cardiac progenitor cell is able to differentiate into three cardiovascular lineages both in *vitro* and in *vivo* [7]. Furthermore, a recent study shows that cardiac stem cell regulates cardiac responses to injury and physiological turnover during aging [12]. Consequently, well-organized and efficient communication among different cell types within the heart is warranted to make the heart function properly. The process includes paracrine/autocrine and endocrine signals, cell-cell contacts, and cell-extracellular matrix interactions [13–16]. The balance in between relies on a wide array of cell types, such as smooth muscle cells, endothelial cells, other connective tissue cells, mast cells, immune system-related cells, and pluripotent cardiac stem cells [10].

11.3 Extracellular Vesicles

Extracellular vesicles (EVs) are membrane-contained vesicles which are highly abundant in cytoskeletal-, cytosolic-, heat shock- and plasma membrane proteins, as well as in proteins involved in vesicles trafficking. They could be released by different types of cells [17]. Accumulating evidence suggests that extracellular membrane vesicles released by cells into extracellular space have been recognized as a novel mode for cardiac cellular crosstalk [12, 18–20]. This is attributed to their capacity to transfer multiple messenger molecules to other cells thereby affecting various physiological and pathological processes of both recipient and parent cells, such as inflammation, immune dysfunction, neurological diseases, and cancer [21]. EVs have been extracted from many body fluids, including blood, milk, saliva, malignant ascites, amniotic fluid and urine [22]. However, they vary in origins, content and size. Based on their biogenesis, EVs can be classified into three main subgroups: (1) exosomes derived from exocytic fusion of multivesicular bodies (MVBs), (2) microvesicles generated by budding of vesicles directly from the plasma membranes, and (3) apoptotic bodies released from the plasma membrane as blebs when cells undergo apoptosis [23–25]. Apoptosis is a biological process in which cells undergo a series of programmed changes involving chromatin condensation, internucleosomal DNA fragmentation, cytochrome c release, proteolytic cleavage of the cytoskeleton, externalization of phosphatidylserine, plasma membrane blebbing, cell shrinkage, and finally formation of apoptotic bodies. Apoptotic bodies are the largest extracellular vesicles, with a diameter range of 1–4 μm [26]. They are able to float on a sucrose substance with a density between 1.16 and 1.28 g/mL. The release of apoptotic bodies is dependent on Rho-associated kinase I (ROCK) and myosin ATPase activity. Apoptotic bodies consist of multiple cellular organelles, intercellular fragments, histones and fragmented DNA, and accompanied by externalization of phosphatidylserine [18]. It is believed that apoptotic bodies could be served as an active signal to promote the elimination of other damaged cells. But the exact role of apoptotic bodies is not fully elucidated. Recently, it was found that apoptotic bodies are responsible for transferring oncogenes from cancer to recipient cells, precipitating metastasis. Membrane permeability increased during early stage of apoptosis, which allow release of death signals to neighbor cells, causing subsequent cell damage [21].

Microvesicles, another type of extracellular vesicles, often referred as microparticles, ectosomes, and membrane particles. Microvesicles are small membrane-enclosed vesicles which appear to be rather heterogeneous in size, ranging from 100 to 1000 nm in diameter. Generally, microvesicles are secreted from cell by outward budding and fission of the plasma membrane when they respond to various physiological or psychological response including prothrombotic and pro-inflammatory stimulations as well as cellular differentiation and senescence [27, 28]. Microvesicles have been isolated from peripheral blood, urine and ascitic fluids at 100,000 g [29]. As the general features of EVs, these microvesicles contains various factors which regulate cell-cell communication. The functions of these vesicles differ depends on the cell type from which they originate [29]. For example, microvesicles originated from skeletal cells contribute to bone mineralization [30], whereas those originated from normal endothelial cells participate in angiogenesis [31]. Once shed, these microvesicles affect both surrounding environment and distant organs. These microvesicles are able to horizontally transfer biological molecules and deposit packaged biological effectors at distal sites [29]. For instance, microvesicles derived from glioma cells expressing epidermal growth factor receptor variant III (EGFRvIII) play a role in inducing morphological transformation and accelerating cancer growth by transferring EGFRvIII to EGFRvIII-negative cancer cells in the same primary tumor [32]. Glioblastoma-derived microvesicles also promote primary tumor growth as well as endothelial cell proliferation by transporting miRNA and angiogenic proteins to recipient cells [33]. Consequently, microvesicles are deemed to have essential roles in multiple biological process and diseases, such as tumor invasion and metastases, inflammation, coagulation, and angiogenesis [29, 34].

This, in turn, may make microvesicles potential diagnostic and prognostic biomarkers as well as therapeutic targets.

Exosomes are the smallest and a more uniform population of these extracellular vesicles, with a diameter range of 30–100 nm [35–38]. The density of exosomes ranges from 1.10 to 1.20 g mL^{-1} [39], which enables these vesicles to float on a sucrose gradient at a density of 1.13–1.19 g/mL [18]. Exosomes are initially released intracellularly into a structure known as multivesicular bodies by invagination of the endosomal membrane. The multivesicular body then fuses with the plasma membrane and release its cargo of exosomes into the extracellular space [40]. Exosomes have membranes rich of lipid proteins [41]. Exosomes also contain a wide range of functional nucleic acids, including mRNAs, microRNAs (miRNAs, miRs), and other non-coding RNAs (ncRNAs) [12, 42]. Carrying these information, exosomes work as messengers between different cells. Their potential to serve as biomarkers in the diagnosis, prognosis and surveillance of a variety of health conditions, has heightened the level of interest in these structures [43].

11.4 Exosomes

Exosomes have been widely detected in eukaryotes, such as amoeboid protists, fungi, plants, and animals [44]. They are characterized by size (30–100 nm), density (1.10–1.20 g mL^{-1}), protein and lipid content [40], which separate it from the other extracellular vesicles. By electron microscopy, people could observe the morphology, structure, positional relationship with neighboring cells, and most importantly, to explore the rules of proper sampling. Murine cardiomyocytes-derived exosomes are small and rounded vesicles enclosed by a bilayered membrane, which revealed by transmission electron microscopy. However, they appear to be rather heterogeneous in electron density [45]. Exosomes are generated by a series of biological process. It starts with the inward budding of the cell membrane to form early endosomes, followed by second inward budding of the endosomal membrane. The second inward budding of the endosomal membrane results in the formation of various intraluminal vesicles (ILVs) (late endosomes). The late endosomes containing ILVs are also known as multivesicular bodies (MVBs) [18, 43, 44]. Some of the multivesicular bodies may move towards the perinuclear space of the cell where they can directly merge with lysosomes for degradation, resulting in hybrid and degradative organelles [23]. Surface proteins lied on the plasma membrane may be sequestered into the inner membrane of these endocytic vesicles during the process [46]. Invaginations of the plasma membrane fuse to these sequestered cargoes sorted into the endoplasmic reticulum and processed in the Golgi complex [44]. Finally, intraluminal vesicles are released into the extracellular space in an exocytotic way when the multivesicular bodies fuse with the plasma membrane. These intraluminal vesicles are thought to be exosomes. The endosomal network has been implicated in the intake and processing of various molecules from the extracellular space, and the transfer of proteins and lipids to the trans-Golgi network. Thus, the endosomal network plays a vital role in regulating the dynamic state of various receptors recycling between the cell surface, endosomes, and the trans-Golgi network to maintain cell homeostasis [47].

Despite the fact that exosome has been discovered for decades, the mechanism for multivesicular bodies biogenesis and exosome release remains unclear. Several models have been suggested as a mechanism for exosome formation. One of the well-studied theory is the endosomal sorting complexes required for transport (ESCRT) system, which regulates protein sequestration and budding [48–51]. Endosomal sorting complexes is required for transport (ESCRT) not only arranges for multi-vesical formation [52], but also recruits exosomal cargo content [17]. The endosomal sorting complex consists of four soluble multi-protein complexes: ESCRT-0, ESCRT-I, ESCRTII and ESCRT-III, and their associated proteins [43]. ESCRT-0 is responsible for cargo clustering and transferring via its ubiquitin-interacting domains [53]. Furthermore, it recruits other soluble multi-protein complexes of ESCRT to the endosomal membrane, including ESCRT-I, ESCRT-II and ESCRT-III [54, 55]. ESCRT-I and ESCRT-II participate in the bud formation. ESCRT-III promotes vesicle fission and regulates the formation of polymeric filaments, resulting in membrane invagination and intraluminal vesicles formation [52, 56]. The associated proteins (especially the VPS4 ATPase) could affect the dissociation and recycle of the ESCRT machinery [44]. Accumulating evidence supports that the ESCRT machinery plays a role in exosome formation [57, 58]. However, a number of studies oppose this theory by reporting ESCRT-independent mechanisms for intraluminal vesicles formation and exosome release [59–62]. One of the studies mentioned that cytoskeletal adapter molecules, one of the intraluminal components of the EV membrane, may also edit and maintain the morphology of the vesicles. Other studies argue that the post-synaptic density protein, disc-large, zonulin I (PDZ) protein syntenin is also involved in the formation of intraluminal vesicles, multivesicular bodies and exosomes [49, 63].

Exosomes are characterized with the typical membrane proteins tetraspanins, such as CD81, CD82 CD63, and CD9. The membrane proteins also include GPI-anchored proteins and receptors like Tumor Necrosis Factor Receptor 1 (TNFR1) [43]. Other types of proteins contained in exosomes include proteins involved in exosome biogenesis (ESCRT complex, i.e.Tsg101, Alix), proteins that direct trafficking and membrane fusion (Annexins, Rab protein family, ARF), and heat shock proteins (Hsp90, Hsp70, Hsp60) [64, 65]. Lipid specific to exosomes are ceramide, cholesterol, PS, and sphingolipids [60, 61]. In addition, emerging evidences suggest that exosomes also contain a wide range of nucleic acids, including double-stranded DNA, mRNA and noncoding RNA (microRNA and lncRNA) [66–69]. These exosomal RNAs can be recognized by neighboring cells or distant cells when exosomes are released [70–72].

The process that protein and RNA sorting into exosomes is changed when cell are subjected to different patho-physiological stress [73]. This, in turn, allows cells to produce specific type of exosomes. Thus, these exosomes could reflect the status of their parent cells. By studying the circulating exosomes, one could obtain information of the status of the original organ status. These suggested the potential of exosomes as biomarkers for diagnosis and prognosis of disease [12, 66, 74]. As a consequence, the discovery of their novel function and properties has brought increasing attention to exosomes.

11.5 Exosomal Therapy in Cardiovascular Disease

11.5.1 Myocardial Infarction

Myocardial infarction can result in myocardial ischemia, and chronic heart failure and death eventually if blood flow is not restored [75, 76]. Myocardial infarction is regarded as the most common cause of death worldwide [77, 78]. In 2012, the Global Myocardial Infarction Task Force has issued the third universal definition of myocardial infarction (MI) expert consensus document. Myocardial infarction (MI) is characterized by cardiac myocyte necrosis and content change of cardiac troponin (cTn) in plasma [79]. Myocardial necrosis due to ischemia triggers an intense inflammatory response by releasing factors like box-1, heat shock proteins, adenosine, extracellular RNA, matrix fragments, and Interleukin (IL)-1α [80–83]. This, in turn, contributes to the formation of collagen-based scar tissue [84]. Due to the absence of contractile function, scar tissue leads to global left ventricular (LV) dysfunction and finally to heart failure [75, 76].

Nowadays, coronary artery by-pass grafting (CABG) surgery is thought to be one of the most commonly performed operations in the world [85]. Coronary artery by-pass grafting (CABG) has become established as the optimal strategy in achieving complete revascularization in patients with moderately severe and severe stable ischemic heart disease [86–88]. By grafting arteries or veins from the patient's body or synthetic conduits to the stenotic coronary arteries, CABG surgery is able to improve and rebuild the blood supply to the coronary circulation and bypass the atherosclerotic narrowing [89]. In general, CABG is indicated for high-risk patients with other severe complications, such as those with severe coronary artery diseases, severe ventricular dysfunction or diabetes mellitus [90]. Exosomes are found to be increased at 24 and 48 h after the procedure. Plasma concentrations of cardiac-enriched, ischemia responsive miRs (miR-1, miR-133a and miR-133b) and high sensitive cardiac troponin I (hs-cTn-I) were all found to be upregulated in the similar period of time. There was a positive correlation between the concentration of total exosomes and the concentration of these substances [85]. This evidence suggests that the circulating exosomes secreted from heart may serve as novel biomarker of cardiac damage though their roles in the treatment for cardiac damage is unclear.

Chronic heart failure (CHF) is commonly caused by myocardial infarction [91–93]. New evidence proved that repeated remote ischemic conditioning could attenuate left ventricular remodeling and oxidative stress on chronic heart failure. Exosome has been related closely to the protective effect of repeated remote ischemic conditioning treatment [94]. Remote ischemic conditioning (RIC) is a clinically applicable method which has been successfully testified on animals and humans [95, 96]. Rats were first induced myocardial infarction for 4 weeks, and followed by RIC treatment for another 4 weeks [94]. Exosomes were found to be increased after RIC treatment compared to controlled group. The expression of or exosomal miR-29a and miR-30a, which are recognized as anti-fibrosis factor were also found to be increased in the RIC group. However, the level of miR-21, known as positive regulator for fibrosis, has no difference between the two groups [94].

In a study investigating exosome's cardioprotective effects, clusterin is detected by shotgun proteomics analysis. It is a heterodimeric secreted glycoprotein detectable in exosomes. It improves cardiac function after ischemic injury by reducing cardiomyocyte apoptosis and inducing vascular regeneration. It is also an important mediator in the regulation of TGF-β-induced epithelial-to-mesenchymal transition [97]. Exosomes isolated from mouse embryonic stem cells were found to be positively related to the improvement of ejection fraction, fractional shortening and decreased left ventricular end-systolic diameter [98]. Similarly exosomes derived from bone marrow mesenchymal stem cells could reduce infarct size, and preserve cardiac systolic and diastolic performance, thereby improving heart function in rat myocardial infarction model [99]. Another example is the GATA-4 expressed in marrow mesenchymal stem cells. Overexpression of GATA-4 down-regulates apoptosis and increases cardiomyocyte survival in neonatal rats [100]. It promotes angiogenesis in the ischemic myocardium and consequently protects against ischemic injury [101]. Apart from these direct benefits from GATA-4, a paracrine effect is also identified. After direct transplanting stem cells at the border of an ischemic region for 4 weeks, it was discovered that the expression of miR-19a was significantly increased in released exosomes. Theses exosomes originated from stems cells that overexpress GATA-4 are enriched in miR-19a. miR-19a preserve the cardiac function by decreasing apoptosis through Akt and ERK pathway [102]. Unlike mesenchymal stem cell-derived exosomes, exosomes released by dendritic cells preserve cardiac function by a modified immune system activation [103].

11.5.2 Myocardial Ischemia/Reperfusion Injury

Ischemia refers to deficient blood supply to cardiac tissues and causes an imbalance between oxygen/nutrients demand and supply, leading to damage or dysfunction of the cardiac tissue [104].

Generally, early and fast restoration of blood supply is regarded as the preferred treatment. However, significant cardiac injury could still happen after restoration of blood flow. This type of damage is known as myocardial ischemia/reperfusion injury (IR injury) [104, 105], which has been associated with high morbidity and mortality [75]. It is a process involves a series of biological response, like reduction in cellular adenosine triphosphate (ATP) levels, accumulation of hydrogen ions, calcium overload, and production of reactive oxygen species (ROS).

An emerging new concept of ischemic preconditioning (IPC) has been coming up because of its protective effect against ischemia. IPC consists of repeated short episodes of ischemia and the subsequent brief phases of reperfusion, leading to tolerance to ischemia [106]. The tolerance is brought up by underlying biological defense reactions [107]. In a myocardial ischemia/reperfusion injury rat model, hearts were exposed to remote ischemic preconditioning including 3 × 5–5 min ischemia and reperfusion before the surgery. Results demonstrated that an increase level of exosomes and microvesicles from the heart after preconditioning. Further evaluation suggested that cardioprotection induced by remote ischemic preconditioning was mediated by these exosomes and microvesicles [108]. As mentioned above, exosomes carry various miRs, which serves as modifiers for target cells. miR-144 is one of these miRs carried by exosomes. It was found that miR-144 levels decreased in mouse myocardium during IR injury, and it increased after IPC treatment. miR-144 precipitates functional recovery and reduces infarct size by involving P-Akt,P-GSK3β and P-p44/42 MAPK, decrease in p-mTOR level and initiation of autophagy signaling, Importantly, miR-144 precursor was found to be increased in the exosome, while no changes was observed in plasma microparticle. This, in turn, means that the role of miR-144 was mediated by exosome [109]. Several other studies also show cardioprotective effects' of exosome, implying its therapeutic role [110–112].

Cardiac progenitor cells (CPC) derived from adult heart have the capacity for cardiac repair as well [113] and the effect has been found to be mediated by exosomes. Both in vitro and in vivo study showed that CPC derived exosomes inhibit apoptosis [111]. Exosomes secreted by mesenchymal stem cell also protect against IR injury by reducing the infarct size in a mouse model [112].

11.6 Conclusion

Exosomes are considered as novel therapeutic approaches for cardiovascular diseases. First, exosomes play a therapeutic role by serving as carriers of biological therapeutics to target cells. Second, exosomes seem a direct source to exploit for cardiac regeneration therapy. This provides a new therapeutic perspective with intercellular mediation of tissue injury and repair. With the development of experimental techniques and molecular technologies, exosome therapy may become a clinically available and practical treatment in cardiovascular diseases including coronary heart disease, heart failure, myocardial infarction, and other related diseases in the future. However, we should aware that the therapeutic role of exosomes has only been identified in animal experiments and in multiple cardiac-related cells. It is imperative that such experiments on other mammals and human should be performed in the future.

Competing Financial Interests The authors declare no competing financial interests.

References

1. Writing Group Members, Mozaffarian D, Benjamin EJ, Go AS, Arnett DK, Blaha MJ, Cushman M, Das SR, de Ferranti S, Despres JP, Fullerton HJ, Howard VJ, Huffman MD, Isasi CR, Jimenez MC, Judd SE, Kissela BM, Lichtman JH, Lisabeth LD, Liu S, Mackey RH, Magid DJ, McGuire DK, Mohler ER 3rd, Moy CS, Muntner P, Mussolino ME, Nasir K, Neumar RW, Nichol G, Palaniappan L, Pandey DK, Reeves MJ, Rodriguez CJ, Rosamond W, Sorlie PD, Stein J, Towfighi A, Turan TN, Virani SS, Woo D, Yeh RW, Turner MB, American Heart Association Statistics Committee, Stroke Statistics Subcommittee (2016) Heart disease and stroke statistics-2016 update: a report from the American Heart Association. Circulation 133(4):e38–360
2. Go AS, Mozaffarian D, Roger VL, Benjamin EJ, Berry JD, Blaha MJ, Dai S, Ford ES, Fox CS, Franco S, Fullerton HJ, Gillespie C, Hailpern SM, Heit JA, Howard VJ, Huffman MD, Judd SE, Kissela BM, Kittner SJ, Lackland DT, Lichtman JH, Lisabeth LD, Mackey RH, Magid DJ, Marcus GM, Marelli A, Matchar DB, McGuire DK, Mohler ER 3rd, Moy CS, Mussolino ME, Neumar RW, Nichol G, Pandey DK, Paynter NP, Reeves MJ, Sorlie PD, Stein J, Towfighi A, Turan TN, Virani SS, Wong ND, Woo D, Turner MB, American Heart Association Statistics Committee and Stroke Statistics Subcommittee (2014) Executive summary: heart disease and stroke statistics—2014 update: a report from the American Heart Association. Circulation 129(3):399–410
3. Ieda M (2013) Heart development and regeneration via cellular interaction and reprogramming. Keio J Med 62(4):99–106
4. Segers VF, Lee RT (2008) Stem-cell therapy for cardiac disease. Nature 451(7181):937–942
5. Passier R, van Laake LW, Mummery CL (2008) Stem-cell-based therapy and lessons from the heart. Nature 453(7193):322–329
6. Laflamme MA, Murry CE (2011) Heart regeneration. Nature 473(7347):326–335
7. Nsair A, Schenke-Layland K, Van Handel B, Evseenko D, Kahn M, Zhao P, Mendelis J, Heydarkhan S, Awaji O, Vottler M, Geist S, Chyu J, Gago-Lopez N, Crooks GM, Plath K, Goldhaber J, Mikkola HK, MacLellan WR (2012) Characterization and therapeutic potential of induced pluripotent stem cell-derived cardiovascular progenitor cells. PLoS One 7(10):e45603
8. Miragoli M, Gaudesius G, Rohr S (2006) Electrotonic modulation of cardiac impulse conduction by myofibroblasts. Circ Res 98(6):801–810
9. Civitarese RA, Kapus A, McCulloch CA, Connelly KA (2017) Role of integrins in mediating cardiac fibroblast-cardiomyocyte cross talk: a dynamic relationship in cardiac biology and pathophysiology. Basic Res Cardiol 112(1):6
10. Tirziu D, Giordano FJ, Simons M (2010) Cell communications in the heart. Circulation 122(9):928–937
11. Bell D, Campbell M, McAleer SF, Ferguson M, Donaghy L, Harbinson MT (2016) Endothelium-derived intermedin/adrenomedullin-2 protects human ventricular cardiomyocytes from ischaemia-reoxygenation injury predominantly via the AM(1) receptor. Peptides 76:1–13
12. Cervio E, Barile L, Moccetti T, Vassalli G (2015) Exosomes for intramyocardial intercellular communication. Stem Cells Int 2015:482171
13. Zhang P, Su J, Mende U (2012) Cross talk between cardiac myocytes and fibroblasts: from multiscale investigative approaches to mechanisms and functional consequences. Am J Physiol Heart Circ Physiol 303(12):H1385–H1396
14. Cherian S, Lopaschuk GD, Carvalho E (2012) Cellular cross-talk between epicardial adipose tissue and myocardium in relation to the pathogenesis of cardiovascular disease. Am J Physiol Endocrinol Metab 303(8):937–949
15. Tian Y, Morrisey EE (2012) Importance of myocyte-nonmyocyte interactions in cardiac development and disease. Circ Res 110(7):1023–1034
16. Doroudgar S, Glembotski CC (2011) The cardiokine story unfolds: ischemic stress-induced protein secretion in the heart. Trends Mol Med 17(4):207–214

17. Yanez-Mo M, Siljander PR, Andreu Z, Zavec AB, Borras FE, Buzas EI, Buzas K, Casal E, Cappello F, Carvalho J, Colas E, Cordeiro-da Silva A, Fais S, Falcon-Perez JM, Ghobrial IM, Giebel B, Gimona M, Graner M, Gursel I, Gursel M, Heegaard NH, Hendrix A, Kierulf P, Kokubun K, Kosanovic M, Kralj-Iglic V, Kramer-Albers EM, Laitinen S, Lasser C, Lener T, Ligeti E, Line A, Lipps G, Llorente A, Lotvall J, Mancek-Keber M, Marcilla A, Mittelbrunn M, Nazarenko I, Nolte-'t Hoen EN, Nyman TA, O'Driscoll L, Olivan M, Oliveira C, Pallinger E, Del Portillo HA, Reventos J, Rigau M, Rohde E, Sammar M, Sanchez-Madrid F, Santarem N, Schallmoser K, Ostenfeld MS, Stoorvogel W, Stukelj R, Van der Grein SG, Vasconcelos MH, Wauben MH, De Wever O (2015) Biological properties of extracellular vesicles and their physiological functions. J Extracell Vesicles 4:27066
18. Sluijter JP, Verhage V, Deddens JC, van den Akker F, Doevendans PA (2014) Microvesicles and exosomes for intracardiac communication. Cardiovasc Res 102(2):302–311
19. Chistiakov DA, Orekhov AN, Bobryshev YV (2016) Cardiac extracellular vesicles in normal and infarcted heart. Int J Mol Sci 17(1):63
20. Oliveira-Rodriguez M, Serrano-Pertierra E, Garcia AC, Lopez-Martin S, Yanez-Mo M, Cernuda-Morollon E, Blanco-Lopez MC (2017) Point-of-care detection of extracellular vesicles: sensitivity optimization and multiple-target detection. Biosens Bioelectron 87:38–45
21. Ciardiello C, Cavallini L, Spinelli C, Yang J, Reis-Sobreiro M, de Candia P, Minciacchi VR, Di Vizio D (2016) Focus on extracellular vesicles: new frontiers of cell-to-cell communication in cancer. Int J Mol Sci 17(2):175
22. Kalra H, Drummen GP, Mathivanan S (2016) Focus on extracellular vesicles: introducing the next small big thing. Int J Mol Sci 17(2):170
23. Schorey JS, Cheng Y, Singh PP, Smith VL (2015) Exosomes and other extracellular vesicles in host-pathogen interactions. EMBO Rep 16(1):24–43
24. Crescitelli R, Lasser C, Szabo TG, Kittel A, Eldh M, Dianzani I, Buzas EI, Lotvall J (2013) Distinct RNA profiles in subpopulations of extracellular vesicles: apoptotic bodies, microvesicles and exosomes. J Extracell Vesicles 2:20677
25. Taylor DD, Gercel-Taylor C (2013) The origin, function, and diagnostic potential of RNA within extracellular vesicles present in human biological fluids. Front Genet 4:142
26. Lazaro-Ibanez E, Sanz-Garcia A, Visakorpi T, Escobedo-Lucea C, Siljander P, Ayuso-Sacido A, Yliperttula M (2014) Different gDNA content in the subpopulations of prostate cancer extracellular vesicles: apoptotic bodies, microvesicles, and exosomes. Prostate 74(14):1379–1390
27. Zhu H, Fan GC (2011) Extracellular/circulating microRNAs and their potential role in cardiovascular disease. Am J Cardiovasc Dis 1(2):138–149
28. Buzas EI, Gyorgy B, Nagy G, Falus A, Gay S (2014) Emerging role of extracellular vesicles in inflammatory diseases. Nat Rev Rheumatol 10(6):356–364
29. Muralidharan-Chari V, Clancy JW, Sedgwick A, D'Souza-Schorey C (2010) Microvesicles: mediators of extracellular communication during cancer progression. J Cell Sci 123(Pt 10):1603–1611
30. Anderson HC, Garimella R, Tague SE (2005) The role of matrix vesicles in growth plate development and biomineralization. Front Biosci 10:822–837
31. Morel O, Toti F, Hugel B, Freyssinet JM (2004) Cellular microparticles: a disseminated storage pool of bioactive vascular effectors. Curr Opin Hematol 11(3):156–164
32. Al-Nedawi K, Meehan B, Micallef J, Lhotak V, May L, Guha A, Rak J (2008) Intercellular transfer of the oncogenic receptor EGFRvIII by microvesicles derived from tumour cells. Nat Cell Biol 10(5):619–624
33. Skog J, Wurdinger T, van Rijn S, Meijer DH, Gainche L, Sena-Esteves M, Curry WT Jr, Carter BS, Krichevsky AM, Breakefield XO (2008) Glioblastoma microvesicles transport RNA and proteins that promote tumour growth and provide diagnostic biomarkers. Nat Cell Biol 10(12):1470–1476
34. Becker A, Thakur BK, Weiss JM, Kim HS, Peinado H, Lyden D (2016) Extracellular vesicles in cancer: cell-to-cell mediators of metastasis. Cancer Cell 30(6):836–848

35. Zhang Z, Yang J, Yan W, Li Y, Shen Z, Asahara T (2016) Pretreatment of cardiac stem cells with exosomes derived from mesenchymal stem cells enhances myocardial repair. J Am Heart Assoc 5(1):e002856
36. Johnsen KB, Gudbergsson JM, Skov MN, Pilgaard L, Moos T, Duroux M (2014) A comprehensive overview of exosomes as drug delivery vehicles—endogenous nanocarriers for targeted cancer therapy. Biochim Biophys Acta 1846(1):75–87
37. Witwer KW, Buzas EI, Bemis LT, Bora A, Lasser C, Lotvall J, Nolte-'t Hoen EN, Piper MG, Sivaraman S, Skog J, Thery C, Wauben MH, Hochberg F (2013) Standardization of sample collection, isolation and analysis methods in extracellular vesicle research. J Extracell Vesicles 2. doi: 10.3402/jev.v2i0.20360
38. Vlassov AV, Magdaleno S, Setterquist R, Conrad R (2012) Exosomes: current knowledge of their composition, biological functions, and diagnostic and therapeutic potentials. Biochim Biophys Acta 1820(7):940–948
39. Vrijsen KR, Maring JA, Chamuleau SA, Verhage V, Mol EA, Deddens JC, Metz CH, Lodder K, van Eeuwijk EC, van Dommelen SM, Doevendans PA, Smits AM, Goumans MJ, Sluijter JP (2016) Exosomes from cardiomyocyte progenitor cells and mesenchymal stem cells stimulate angiogenesis via EMMPRIN. Adv Healthc Mater 5(19):2555–2565
40. Batrakova EV, Kim MS (2015) Using exosomes, naturally-equipped nanocarriers, for drug delivery. J Control Release 219:396–405
41. Gross JC, Chaudhary V, Bartscherer K, Boutros M (2012) Active Wnt proteins are secreted on exosomes. Nat Cell Biol 14(10):1036–1045
42. Sato-Kuwabara Y, Melo SA, Soares FA, Calin GA (2015) The fusion of two worlds: non-coding RNAs and extracellular vesicles—diagnostic and therapeutic implications (Review). Int J Oncol 46(1):17–27
43. Ailawadi S, Wang X, Gu H, Fan GC (2015) Pathologic function and therapeutic potential of exosomes in cardiovascular disease. Biochim Biophys Acta 1852(1):1–11
44. Barile L, Moccetti T, Marban E, Vassalli G (2017) Roles of exosomes in cardioprotection. Eur Heart J 38(18):1372–1379. doi:10.1093/eurheartj/ehw304
45. Waldenstrom A, Genneback N, Hellman U, Ronquist G (2012) Cardiomyocyte microvesicles contain DNA/RNA and convey biological messages to target cells. PLoS One 7(4):e34653
46. Villarroya-Beltri C, Baixauli F, Gutierrez-Vazquez C, Sanchez-Madrid F, Mittelbrunn M (2014) Sorting it out: regulation of exosome loading. Semin Cancer Biol 28:3–13
47. Anitei M, Wassmer T, Stange C, Hoflack B (2010) Bidirectional transport between the trans-Golgi network and the endosomal system. Mol Membr Biol 27(8):443–456
48. Akers JC, Gonda D, Kim R, Carter BS, Chen CC (2013) Biogenesis of extracellular vesicles (EV): exosomes, microvesicles, retrovirus-like vesicles, and apoptotic bodies. J Neuro-Oncol 113(1):1–11
49. Baietti MF, Zhang Z, Mortier E, Melchior A, Degeest G, Geeraerts A, Ivarsson Y, Depoortere F, Coomans C, Vermeiren E, Zimmermann P, David G (2012) Syndecan-syntenin-ALIX regulates the biogenesis of exosomes. Nat Cell Biol 14(7):677–685
50. Colombo M, Moita C, van Niel G, Kowal J, Vigneron J, Benaroch P, Manel N, Moita LF, Thery C, Raposo G (2013) Analysis of ESCRT functions in exosome biogenesis, composition and secretion highlights the heterogeneity of extracellular vesicles. J Cell Sci 126(Pt 24):5553–5565
51. Pant S, Hilton H, Burczynski ME (2012) The multifaceted exosome: biogenesis, role in normal and aberrant cellular function, and frontiers for pharmacological and biomarker opportunities. Biochem Pharmacol 83(11):1484–1494
52. Hanson PI, Cashikar A (2012) Multivesicular body morphogenesis. Annu Rev Cell Dev Biol 28:337–362
53. Raiborg C, Stenmark H (2002) Hrs and endocytic sorting of ubiquitinated membrane proteins. Cell Struct Funct 27(6):403–408
54. Babst M, Katzmann DJ, Snyder WB, Wendland B, Emr SD (2002) Endosome-associated complex, ESCRT-II, recruits transport machinery for protein sorting at the multivesicular body. Dev Cell 3(2):283–289

55. Katzmann DJ, Babst M, Emr SD (2001) Ubiquitin-dependent sorting into the multivesicular body pathway requires the function of a conserved endosomal protein sorting complex, ESCRT-I. Cell 106(2):145–155
56. Wollert T, Wunder C, Lippincott-Schwartz J, Hurley JH (2009) Membrane scission by the ESCRT-III complex. Nature 458(7235):172–177
57. Kowal J, Tkach M, Thery C (2014) Biogenesis and secretion of exosomes. Curr Opin Cell Biol 29:116–125
58. Metcalf D, Isaacs AM (2010) The role of ESCRT proteins in fusion events involving lysosomes, endosomes and autophagosomes. Biochem Soc Trans 38(6):1469–1473
59. de Gassart A, Geminard C, Fevrier B, Raposo G, Vidal M (2003) Lipid raft-associated protein sorting in exosomes. Blood 102(13):4336–4344
60. Wubbolts R, Leckie RS, Veenhuizen PT, Schwarzmann G, Mobius W, Hoernschemeyer J, Slot JW, Geuze HJ, Stoorvogel W (2003) Proteomic and biochemical analyses of human B cell-derived exosomes. Potential implications for their function and multivesicular body formation. J Biol Chem 278(13):10963–10972
61. Trajkovic K, Hsu C, Chiantia S, Rajendran L, Wenzel D, Wieland F, Schwille P, Brugger B, Simons M (2008) Ceramide triggers budding of exosome vesicles into multivesicular endosomes. Science 319(5867):1244–1247
62. Kajimoto T, Okada T, Miya S, Zhang L, Nakamura S (2013) Ongoing activation of sphingosine 1-phosphate receptors mediates maturation of exosomal multivesicular endosomes. Nat Commun 4:2712
63. Ghossoub R, Lembo F, Rubio A, Gaillard CB, Bouchet J, Vitale N, Slavik J, Machala M, Zimmermann P (2014) Syntenin-ALIX exosome biogenesis and budding into multivesicular bodies are controlled by ARF6 and PLD2. Nat Commun 5:3477
64. Ostrowski M, Carmo NB, Krumeich S, Fanget I, Raposo G, Savina A, Moita CF, Schauer K, Hume AN, Freitas RP, Goud B, Benaroch P, Hacohen N, Fukuda M, Desnos C, Seabra MC, Darchen F, Amigorena S, Moita LF, Thery C (2010) Rab27a and Rab27b control different steps of the exosome secretion pathway. Nat Cell Biol 12(1):19–30
65. Subra C, Grand D, Laulagnier K, Stella A, Lambeau G, Paillasse M, De Medina P, Monsarrat B, Perret B, Silvente-Poirot S, Poirot M, Record M (2010) Exosomes account for vesicle-mediated transcellular transport of activatable phospholipases and prostaglandins. J Lipid Res 51(8):2105–2120
66. Emanueli C, Shearn AI, Angelini GD, Sahoo S (2015) Exosomes and exosomal miRNAs in cardiovascular protection and repair. Vasc Pharmacol 71:24–30
67. Gezer U, Ozgur E, Cetinkaya M, Isin M, Dalay N (2014) Long non-coding RNAs with low expression levels in cells are enriched in secreted exosomes. Cell Biol Int 38(9):1076–1079
68. Mathivanan S, Simpson RJ (2009) ExoCarta: a compendium of exosomal proteins and RNA. Proteomics 9(21):4997–5000
69. Valadi H, Ekstrom K, Bossios A, Sjostrand M, Lee JJ, Lotvall JO (2007) Exosome-mediated transfer of mRNAs and microRNAs is a novel mechanism of genetic exchange between cells. Nat Cell Biol 9(6):654–659
70. Zhang J, Li S, Li L, Li M, Guo C, Yao J, Mi S (2015) Exosome and exosomal microRNA: trafficking, sorting, and function. Genomics Proteomics Bioinformatics 13(1):17–24
71. Wang X, Huang W, Liu G, Cai W, Millard RW, Wang Y, Chang J, Peng T, Fan GC (2014) Cardiomyocytes mediate anti-angiogenesis in type 2 diabetic rats through the exosomal transfer of miR-320 into endothelial cells. J Mol Cell Cardiol 74:139–150
72. Pegtel DM, Cosmopoulos K, Thorley-Lawson DA, van Eijndhoven MA, Hopmans ES, Lindenberg JL, de Gruijl TD, Wurdinger T, Middeldorp JM (2010) Functional delivery of viral miRNAs via exosomes. Proc Natl Acad Sci U S A 107(14):6328–6333
73. de Jong OG, Verhaar MC, Chen Y, Vader P, Gremmels H, Posthuma G, Schiffelers RM, Gucek M, van Balkom BW (2012) Cellular stress conditions are reflected in the protein and RNA content of endothelial cell-derived exosomes. J Extracell Vesicles 1. doi:10.3402/jev.v1i0.18396
74. Khalyfa A, Gozal D (2014) Exosomal miRNAs as potential biomarkers of cardiovascular risk in children. J Transl Med 12:162

75. Li X, Liu M, Sun R, Zeng Y, Chen S, Zhang P (2016) Protective approaches against myocardial ischemia reperfusion injury. Exp Ther Med 12(6):3823–3829
76. Callender T, Woodward M, Roth G, Farzadfar F, Lemarie JC, Gicquel S, Atherton J, Rahimzadeh S, Ghaziani M, Shaikh M, Bennett D, Patel A, Lam CS, Sliwa K, Barretto A, Siswanto BB, Diaz A, Herpin D, Krum H, Eliasz T, Forbes A, Kiszely A, Khosla R, Petrinic T, Praveen D, Shrivastava R, Xin D, MacMahon S, McMurray J, Rahimi K (2014) Heart failure care in low- and middle-income countries: a systematic review and meta-analysis. PLoS Med 11(8):e1001699
77. Mozaffarian D, Benjamin EJ, Go AS, Arnett DK, Blaha MJ, Cushman M, de Ferranti S, Despres JP, Fullerton HJ, Howard VJ, Huffman MD, Judd SE, Kissela BM, Lackland DT, Lichtman JH, Lisabeth LD, Liu S, Mackey RH, Matchar DB, DK MG, Mohler ER 3rd, Moy CS, Muntner P, Mussolino ME, Nasir K, Neumar RW, Nichol G, Palaniappan L, Pandey DK, Reeves MJ, Rodriguez CJ, Sorlie PD, Stein J, Towfighi A, Turan TN, Virani SS, Willey JZ, Woo D, Yeh RW, Turner MB, American Heart Association Statistics Committee and Stroke Statistics Subcommittee (2015) Heart disease and stroke statistics—2015 update: a report from the American Heart Association. Circulation 131(4):e29–322
78. Yuan MJ, Maghsoudi T, Wang T (2016) Exosomes mediate the intercellular communication after myocardial infarction. Int J Med Sci 13(2):113–116
79. Reddy K, Khaliq A, Henning RJ (2015) Recent advances in the diagnosis and treatment of acute myocardial infarction. World J Cardiol 7(5):243–276
80. Huebener P, Abou-Khamis T, Zymek P, Bujak M, Ying X, Chatila K, Haudek S, Thakker G, Frangogiannis NG (2008) CD44 is critically involved in infarct healing by regulating the inflammatory and fibrotic response. J Immunol 180(4):2625–2633
81. Andrassy M, Volz HC, Igwe JC, Funke B, Eichberger SN, Kaya Z, Buss S, Autschbach F, Pleger ST, Lukic IK, Bea F, Hardt SE, Humpert PM, Bianchi ME, Mairbaurl H, Nawroth PP, Remppis A, Katus HA, Bierhaus A (2008) High-mobility group box-1 in ischemia-reperfusion injury of the heart. Circulation 117(25):3216–3226
82. Lugrin J, Parapanov R, Rosenblatt-Velin N, Rignault-Clerc S, Feihl F, Waeber B, Muller O, Vergely C, Zeller M, Tardivel A, Schneider P, Pacher P, Liaudet L (2015) Cutting edge: IL-1alpha is a crucial danger signal triggering acute myocardial inflammation during myocardial infarction. J Immunol 194(2):499–503
83. Chen C, Feng Y, Zou L, Wang L, Chen HH, Cai JY, Xu JM, Sosnovik DE, Chao W (2014) Role of extracellular RNA and TLR3-Trif signaling in myocardial ischemia-reperfusion injury. J Am Heart Assoc 3(1):e000683
84. Frangogiannis NG (2014) The immune system and the remodeling infarcted heart: cell biological insights and therapeutic opportunities. J Cardiovasc Pharmacol 63(3):185–195
85. Emanueli C, Shearn AI, Laftah A, Fiorentino F, Reeves BC, Beltrami C, Mumford A, Clayton A, Gurney M, Shantikumar S, Angelini GD (2016) Coronary artery-bypass-graft surgery increases the plasma concentration of exosomes carrying a cargo of cardiac microRNAs: an example of exosome trafficking out of the human heart with potential for cardiac biomarker discovery. PLoS One 11(4):e0154274
86. Lindblom RP, Lytsy B, Sandstrom C, Ligata N, Larsson B, Ransjo U, Swenne CL (2015) Outcomes following the implementation of a quality control campaign to decrease sternal wound infections after coronary artery by-pass grafting. BMC Cardiovasc Disord 15:154
87. Fortuna D, Nicolini F, Guastaroba P, De Palma R, Di Bartolomeo S, Saia F, Pacini D, Grilli R, RERIC (Regional Registry of Cardiac Surgery), REAL (Regional Registry of Coronary Angioplasties) Investigators (2013) Coronary artery bypass grafting vs percutaneous coronary intervention in a 'real-world' setting: a comparative effectiveness study based on propensity score-matched cohorts. Eur J Cardiothorac Surg 44(1):e16–e24
88. Ferguson TB Jr (2016) Physiology of in-situ arterial revascularization in coronary artery bypass grafting: preoperative, intraoperative and postoperative factors and influences. World J Cardiol 8(11):623–637
89. Ghista DN, Kabinejadian F (2013) Coronary artery bypass grafting hemodynamics and anastomosis design: a biomedical engineering review. Biomed Eng Online 12:129

90. Rihal CS, Raco DL, Gersh BJ, Yusuf S (2003) Indications for coronary artery bypass surgery and percutaneous coronary intervention in chronic stable angina: review of the evidence and methodological considerations. Circulation 108(20):2439–2445
91. Ma TK, Kam KK, Yan BP, Lam YY (2010) Renin-angiotensin-aldosterone system blockade for cardiovascular diseases: current status. Br J Pharmacol 160(6):1273–1292
92. Rywik TM (2011) [Summary of the article: Zannad F, McMurray JJV, Krum H et al.; for the EMPHASIS-HF Study Group. Eplerenone in patients with systolic heart failure and mild symptoms. N Engl J Med, 2011; 364: 11-21]. Kardiol Pol 69(6):631–632
93. Gring CN, Francis GS (2004) A hard look at angiotensin receptor blockers in heart failure. J Am Coll Cardiol 44(9):1841–1846
94. Yamaguchi T, Izumi Y, Nakamura Y, Yamazaki T, Shiota M, Sano S, Tanaka M, Osada-Oka M, Shimada K, Miura K, Yoshiyama M, Iwao H (2015) Repeated remote ischemic conditioning attenuates left ventricular remodeling via exosome-mediated intercellular communication on chronic heart failure after myocardial infarction. Int J Cardiol 178:239–246
95. Kleinbongard P, Skyschally A, Heusch G (2017) Cardioprotection by remote ischemic conditioning and its signal transduction. Pflugers Arch 469(2):159–181
96. Elbadawi A, Ha LD, Abuzaid AS, Crimi G, Azzouz MS (2017) Meta-analysis of randomized trials on remote ischemic conditioning during primary percutaneous coronary intervention in patients with ST-segment elevation myocardial infarction. Am J Cardiol 119(6):832–838
97. Foglio E, Puddighinu G, Fasanaro P, D'Arcangelo D, Perrone GA, Mocini D, Campanella C, Coppola L, Logozzi M, Azzarito T, Marzoli F, Fais S, Pieroni L, Marzano V, Germani A, Capogrossi MC, Russo MA, Limana F (2015) Exosomal clusterin, identified in the pericardial fluid, improves myocardial performance following MI through epicardial activation, enhanced arteriogenesis and reduced apoptosis. Int J Cardiol 197:333–347
98. Khan M, Nickoloff E, Abramova T, Johnson J, Verma SK, Krishnamurthy P, Mackie AR, Vaughan E, Garikipati VN, Benedict C, Ramirez V, Lambers E, Ito A, Gao E, Misener S, Luongo T, Elrod J, Qin G, Houser SR, Koch WJ, Kishore R (2015) Embryonic stem cell-derived exosomes promote endogenous repair mechanisms and enhance cardiac function following myocardial infarction. Circ Res 117(1):52–64
99. Teng X, Chen L, Chen W, Yang J, Yang Z, Shen Z (2015) Mesenchymal stem cell-derived exosomes improve the microenvironment of infarcted myocardium contributing to angiogenesis and anti-inflammation. Cell Physiol Biochem 37(6):2415–2424
100. Yu B, Gong M, Wang Y, Millard RW, Pasha Z, Yang Y, Ashraf M, Xu M (2013) Cardiomyocyte protection by GATA-4 gene engineered mesenchymal stem cells is partially mediated by translocation of miR-221 in microvesicles. PLoS One 8(8):e73304
101. Li H, Zuo S, He Z, Yang Y, Pasha Z, Wang Y, Xu M (2010) Paracrine factors released by GATA-4 overexpressed mesenchymal stem cells increase angiogenesis and cell survival. Am J Physiol Heart Circ Physiol 299(6):H1772–H1781
102. Yu B, Kim HW, Gong M, Wang J, Millard RW, Wang Y, Ashraf M, Xu M (2015) Exosomes secreted from GATA-4 overexpressing mesenchymal stem cells serve as a reservoir of anti-apoptotic microRNAs for cardioprotection. Int J Cardiol 182:349–360
103. Liu H, Gao W, Yuan J, Wu C, Yao K, Zhang L, Ma L, Zhu J, Zou Y, Ge J (2016) Exosomes derived from dendritic cells improve cardiac function via activation of CD4(+) T lymphocytes after myocardial infarction. J Mol Cell Cardiol 91:123–133
104. Frank A, Bonney M, Bonney S, Weitzel L, Koeppen M, Eckle T (2012) Myocardial ischemia reperfusion injury: from basic science to clinical bedside. Semin Cardiothorac Vasc Anesth 16(3):123–132
105. Kalogeris T, Baines CP, Krenz M, Korthuis RJ (2012) Cell biology of ischemia/reperfusion injury. Int Rev Cell Mol Biol 298:229–317
106. Murry CE, Jennings RB, Reimer KA (1986) Preconditioning with ischemia: a delay of lethal cell injury in ischemic myocardium. Circulation 74(5):1124–1136
107. Ueno K, Samura M, Nakamura T, Tanaka Y, Takeuchi Y, Kawamura D, Takahashi M, Hosoyama T, Morikage N, Hamano K (2016) Increased plasma VEGF levels following isch-

emic preconditioning are associated with downregulation of miRNA-762 and miR-3072-5p. Sci Rep 6:36758

108. Giricz Z, Varga ZV, Baranyai T, Sipos P, Paloczi K, Kittel A, Buzas EI, Ferdinandy P (2014) Cardioprotection by remote ischemic preconditioning of the rat heart is mediated by extracellular vesicles. J Mol Cell Cardiol 68:75–78
109. Li J, Rohailla S, Gelber N, Rutka J, Sabah N, Gladstone RA, Wei C, Hu P, Kharbanda RK, Redington AN (2014) MicroRNA-144 is a circulating effector of remote ischemic preconditioning. Basic Res Cardiol 109(5):423
110. Vicencio JM, Yellon DM, Sivaraman V, Das D, Boi-Doku C, Arjun S, Zheng Y, Riquelme JA, Kearney J, Sharma V, Multhoff G, Hall AR, Davidson SM (2015) Plasma exosomes protect the myocardium from ischemia-reperfusion injury. J Am Coll Cardiol 65(15):1525–1536
111. Chen L, Wang Y, Pan Y, Zhang L, Shen C, Qin G, Ashraf M, Weintraub N, Ma G, Tang Y (2013) Cardiac progenitor-derived exosomes protect ischemic myocardium from acute ischemia/reperfusion injury. Biochem Biophys Res Commun 431(3):566–571
112. Lai RC, Arslan F, Lee MM, Sze NS, Choo A, Chen TS, Salto-Tellez M, Timmers L, Lee CN, El Oakley RM, Pasterkamp G, de Kleijn DP, Lim SK (2010) Exosome secreted by MSC reduces myocardial ischemia/reperfusion injury. Stem Cell Res 4(3):214–222
113. Tang YL, Zhu W, Cheng M, Chen L, Zhang J, Sun T, Kishore R, Phillips MI, Losordo DW, Qin G (2009) Hypoxic preconditioning enhances the benefit of cardiac progenitor cell therapy for treatment of myocardial infarction by inducing CXCR4 expression. Circ Res 104(10):1209–1216

Chapter 12
Therapeutic Effects of Mesenchymal Stem Cell-Derived Exosomes in Cardiovascular Disease

Etsu Suzuki, Daishi Fujita, Masao Takahashi, Shigeyoshi Oba, and Hiroaki Nishimatsu

12.1 Introduction

Mesenchymal stem cells (MSCs) are multipotent stem cells that reside in various organs, such as bone marrow, subcutaneous adipose tissue, skeletal muscles, lungs and dental pulp [1–4]. They have the capacity to differentiate into various cell types, such as bone, cartilage, cardiac muscle, skeletal muscle, vascular endothelial cells (VECs), and vascular smooth muscle cells (VSMCs) [5, 6]. Among of the various MSCs, bone marrow-derived MSCs (BMMSCs) have been widely used for treating acute myocardial infarction (AMI) and ischemic heart failure (IHF). The efficacy of BMMSCs for treating AMI and IHF has been demonstrated in both preclinical [7–10] and clinical studies [11–16]. Although the earliest preclinical studies suggested that BMMSCs have the potential to differentiate into multiple cardiac cell types, including cardiomyocytes, VECs, and VSMCs [9, 7, 8], subsequent studies did not demonstrate the potent differentiation capacity of MSCs. Most intravenously injected cells are trapped in the lungs rather than engrafted in the heart [17, 18]. Even when MSCs are injected to the swine heart via the coronary artery following AMI induction, only 6% of the injected cells remain in the infarct zones 14 days after AMI induction [17]. Furthermore, the supernatant of MSC cultures improves cardiac function to the same extent as MSCs *per se* [19–21]. These results suggest

E. Suzuki, M.D., Ph.D. (✉)
Institute of Medical Science, St. Marianna University School of Medicine,
2-16-1 Sugao, Miyamae-ku, Kawasaki 216-8512, Japan
e-mail: esuzuki-tky@umin.ac.jp

D. Fujita • M. Takahashi • S. Oba
Faculty of Medicine, The Department of Internal Medicine, University of Tokyo,
7-3-1 Hongo, Bunkyo-ku, Tokyo 113-8655, Japan

H. Nishimatsu
Faculty of Medicine, The Department of Urology, University of Tokyo,
7-3-1 Hongo, Bunkyo-ku, Tokyo 113-8655, Japan

J. Xiao, S. Cretoiu (eds.), *Exosomes in Cardiovascular Diseases*, Advances in Experimental Medicine and Biology 998, DOI 10.1007/978-981-10-4397-0_12

that MSCs improve cardiac function via the secretion of paracrine factors rather than via the direct differentiation of MSCs to cardiac cell types. In this context, MSC-derived exosomes have recently gained much attention as a vehicle to transport cardioprotective molecules to the heart. Although clinical studies for treating cardiovascular diseases (CVDs) using MSC-derived exosomes have not been initiated, we review preclinical studies in which MSC-derived exosomes are being used for treating CVDs (summarized in Table 12.1).

Table 12.1 A summary of the effects of MSC-derived exosome administration on CVD therapy

Origin of exosomes	Experimental model	Findings	References
Human ESC-derived MSCs	Ischemia/reperfusion injury	Reduction in infarct size	Lai et al. [22]
		Recovery of cardiac function	
		Decrease in oxidative stress	
		Activation of Akt and GSK3	
		Inhibition of c-JNK	
Human MSCs	AMI	Reduction in infarct size	Bian et al. [24]
		Recovery of cardiac function	
		Increase in angiogenesis	
Mouse MSCs	AMI	Exosomes were enriched in miR-22	Feng et al. [25]
		miR-22 was implicated in the anti-apoptotic effect of exosomes	
Rat MSCs overexpressing GATA-4	AMI	Reduction in infarct size	Yu et al. [26]
		Recovery of cardiac function	
		Exosomes were enriched in miR-19a	
Human MSCs	AMI	Reduction in infarct size	Wang et al. [27]
		Recovery of cardiac function	
		Stimulation of angiogenesis	
		miR-21 mediates cardioprotective effect	
Rat MSCs	Stroke	Recovery of neurological function	Xin et al. [28]
		Stimulation of neurogenesis and angiogenesis	
Rat MSCs overexpressing miR-133b and those whose expression of miR-133b was knocked down	Stroke	Recovery of neurological function was mediated by miR-133b expressed in exosomes	Xin et al. [29]
Human MSCs	Stroke	Both MSC and MSC-derived exosome administration restored neurological function to the same extent	Doeppner et al. [30]
Mouse MSCs	Pulmonary hypertension	Reduction in the progression of pulmonary hypertension and right ventricular hypertrophy	Lee et al. [31]
Mouse MSCs	Sepsis	Recovery of cardiac function was mediated by miR-223	Wang et al. [32]

MSCs mesenchymal stem cells, *AMI* acute myocardial infarction, *GSK3* glycogen synthase kinase 3, *c-JNK* c-jun N-terminal kinase

12.2 Application of MSC-Derived Exosomes in the Treatment of CVDs

12.2.1 AMI and Ischemia/Reperfusion Injury

Lai et al. fractionated the supernatant of human embryonic stem cell (ESC)-derived MSCs and found that a fraction containing small particles (50–100 nm in diameter) had cardioprotective effects [22]. This fraction included particles that corresponded to exosomes. When administered to a mouse myocardial ischemia/reperfusion injury model, these exosomes remarkably reduced the infarct size. The same group also administered exosomes secreted from human ESC-derived MSCs to a mouse myocardial ischemia/reperfusion injury model and demonstrated a reduced infarct size and improved cardiac function [23]. In addition, they found that tissue levels of ATP and nicotinamide adenine dinucleotide were significantly increased, whereas those of reactive oxygen species were significantly decreased after exosome administration. Furthermore, they demonstrated that following exosome administration, phosphorylation of Akt and glycogen synthase kinase 3 (which has anti-apoptotic effects) significantly increased and that of c-jun N-terminal kinase (which has proapoptotic effects) significantly decreased in cardiac tissue. Finally, they showed that exosome administration significantly reduced neutrophil and macrophage infiltration in the heart after reperfusion, suggesting that exosome treatment has an anti-inflammatory effect. Bian et al. collected extracellular vesicles (EVs) from hypoxic human BMMSCs [24]. The mean diameter of EVs was 100 nm, suggesting that EVs are a mixture of microvesicles and exosomes. They administered these EVs to human umbilical vein endothelial cells (HUVECs) and demonstrated that EVs were internalized by HUVECs and that the proliferation and migration of HUVECs significantly increased. They also administered EVs to a rat AMI model and showed that EV administration significantly reduced infarct size, restored cardiac function, and stimulated angiogenesis in the ischemic zone. Feng et al. demonstrated that exosomes secreted from mouse BMMSCs after ischemic preconditioning contained a large amount of miR-22 and that exosome-derived miR-22 was internalized into cultured cardiomyocytes [25]. When administered to mice with AMI, these miR-22-enriched exosomes significantly reduced infarct size and cardiac fibrosis probably via the downregulation of methyl-CpG-binding protein 2. Yu et al. used rat BMMSCs overexpressing the transcription factor GATA-4 (MSC_GATA-4) and demonstrated that administering MSC_GATA-4-derived exosomes had an anti-apoptotic effect under hypoxic conditions *in vitro* [26]. They also showed that MSC_GATA-4-derived exosome administration restored cardiac function and reduced the infarct size in a rat AMI model. Furthermore, the authors showed that MSC_GATA-4-derived exosomes expressed a greater amount of miRs, particularly miR-19a, than control MSCs and that miR-19a appeared to be involved in the cardioprotective effect of MSC_GATA-4-derived exosomes via the downregulation of phosphatase and tensin homolog (PTEN) and subsequent activation of anti-apoptotic Akt and extracellular signal-regulated kinase. Wang et al. compared the cardioprotective effects of human BMMSCs, adipose tissue-derived MSCs, and endometrium-derived

MSCs (EnMSCs) [27]. They administered these MSCs to a rat AMI model and demonstrated EnMSCs were the most effective in reducing infarct size, restoring cardiac function, and stimulating angiogenesis in the ischemic zone. They also demonstrated that miR-21 contained in EnMSC-derived exosomes mediated cardioprotective effects via the downregulation of PTEN and subsequent activation of Akt, resulting in the upregulation of Bcl-2 and vascular endothelial growth factor.

12.2.2 Stroke

Preclinical studies have also reported favorable effects of exosome administration on neurological recovery following stroke induction. Xin et al. found that the systemic administration of rat BMMSC-derived exosomes after inducing stroke by ligating the middle cerebral artery significantly accelerated neurological recovery and stimulated neurogenesis and angiogenesis in the ischemic boundary zone [28]. The same group prepared BMMSCs overexpressing miR-133b (MSCs_miR-133b+) and BMMSCs with miR-133b knockdown (MSCs_miR-133b−), in addition to wild type BMMSCs (MSCs_wt) [29]. They administered these MSCs to a rat stroke model and demonstrated that compared with MSC_wt administration, MSCs_miR-133b+ enhanced the recovery of neurological function, whereas MSCs_miR-133b− decreased the recovery of neurological function. Furthermore, they showed that compared with the group administered MSCs_wt, miR-133b levels in exosomes isolated from cerebrospinal fluid were higher in the group that received MSCs_miR-133b+ and lower in the group that received MSCs_miR-133b−. They also demonstrated that MSC-derived exosomes could be transferred to neighboring cells. Finally, they showed that the expression of connective tissue growth factor (CTGF), a target for miR-133b, was significantly reduced in the ischemic boundary zone after MSCs_miR-133b+ administration compared with after MSCs_wt administration, whereas CTGF expression remained unchanged after MSCs_miR-133b− administration compared with after MSCs_wt administration. They concluded that exosome-derived miR-133b was implicated in the MSC-mediated recovery of neurological function in this model. Doeppner et al. compared the effect of BMMSC administration and BMMSC-derived EV administration on the neurological function after inducing stroke by ligating the middle cerebral artery [30]. They demonstrated that both treatments improved neurological function and stimulated neurogenesis and angiogenesis at the ischemic boundary zone to the same extent.

12.2.3 Pulmonary Hypertension

The beneficial effects of MSC-derived exosome administration have also been reported in a mouse hypoxic pulmonary hypertension model. Lee et al. demonstrated that administering BMMSC-derived exosomes significantly ameliorated the progression of pulmonary hypertension and right ventricular hypertrophy, possibly via the suppression of signal transducer and activator of transcription 3 (STAT3) and the subsequent modulation of the expression of miR-17 and miR-204 (i.e., suppression of proproliferative miR-17 expression and an increase in miR-204 expression that inhibits STAT3 activation in a feed-forward loop) [31].

12.2.4 Sepsis

Wang et al. prepared exosomes from wild-type murine BMMSCs and miR-223 null BMMSCs (miR-223-KO BMMSCs) [32]. They administered these exosomes to a murine sepsis model that was induced by cecal ligation and puncture, and examined their cardioprotective effect in sepsis. They demonstrated that restoration of cardiac function was observed when wild-type BMMSC-derived exosomes were administered, whereas diminished effect was observed when miR-223-KO BMMSC-derived exosomes were administered. In addition, they demonstrated that administering wild type BMMSC-derived exosomes suppressed the expression of the miR-223 targets semaphorin-3A and STAT3, which might be implicated in the cardioprotective effect of wild type BMMSC-derived exosomes.

12.3 Conclusion

MSC-derived exosomes appear to protect the heart and brain via their anti-apoptotic, anti-inflammatory, proangiogenic, and immunomodulatory effects in animal models. Clinical trials will be required in the future to confirm the beneficial effects of MSC-derived exosomes in treating CVDs.

References

1. Hipp J, Atala A (2008) Sources of stem cells for regenerative medicine. Stem Cell Rev 4(1):3–11
2. Jiang Y, Vaessen B, Lenvik T, Blackstad M, Reyes M, Verfaillie CM (2002) Multipotent progenitor cells can be isolated from postnatal murine bone marrow, muscle, and brain. Exp Hematol 30(8):896–904
3. Presnell SC, Petersen B, Heidaran M (2002) Stem cells in adult tissues. Semin Cell Dev Biol 13(5):369–376
4. Schaffler A, Buchler C (2007) Concise review: adipose tissue-derived stromal cells—basic and clinical implications for novel cell-based therapies. Stem Cells 25(4):818–827
5. Herzog EL, Chai L, Krause DS (2003) Plasticity of marrow-derived stem cells. Blood 102(10):3483–3493
6. Grove JE, Bruscia E, Krause DS (2004) Plasticity of bone marrow-derived stem cells. Stem Cells 22(4):487–500

7. Shake JG, Gruber PJ, Baumgartner WA, Senechal G, Meyers J, Redmond JM, Pittenger MF, Martin BJ (2002) Mesenchymal stem cell implantation in a swine myocardial infarct model: engraftment and functional effects. Ann Thorac Surg 73(6):1919–1925
8. Toma C, Pittenger MF, Cahill KS, Byrne BJ, Kessler PD (2002) Human mesenchymal stem cells differentiate to a cardiomyocyte phenotype in the adult murine heart. Circulation 105(1):93–98
9. Amado LC, Saliaris AP, Schuleri KH, St John M, Xie JS, Cattaneo S, Durand DJ, Fitton T, Kuang JQ, Stewart G, Lehrke S, Baumgartner WW, Martin BJ, Heldman AW, Hare JM (2005) Cardiac repair with intramyocardial injection of allogeneic mesenchymal stem cells after myocardial infarction. Proc Natl Acad Sci U S A 102(32):11474–11479
10. Schuleri KH, Feigenbaum GS, Centola M, Weiss ES, Zimmet JM, Turney J, Kellner J, Zviman MM, Hatzistergos KE, Detrick B, Conte JV, McNiece I, Steenbergen C, Lardo AC, Hare JM (2009) Autologous mesenchymal stem cells produce reverse remodelling in chronic ischaemic cardiomyopathy. Eur Heart J 30(22):2722–2732
11. Chen SL, Fang WW, Ye F, Liu YH, Qian J, Shan SJ, Zhang JJ, Chunhua RZ, Liao LM, Lin S, Sun JP (2004) Effect on left ventricular function of intracoronary transplantation of autologous bone marrow mesenchymal stem cell in patients with acute myocardial infarction. Am J Cardiol 94(1):92–95
12. Hare JM, Traverse JH, Henry TD, Dib N, Strumpf RK, Schulman SP, Gerstenblith G, DeMaria AN, Denktas AE, Gammon RS, Hermiller JB Jr, Reisman MA, Schaer GL, Sherman W (2009) A randomized, double-blind, placebo-controlled, dose-escalation study of intravenous adult human mesenchymal stem cells (prochymal) after acute myocardial infarction. J Am Coll Cardiol 54(24):2277–2286
13. Perin EC, Silva GV, Henry TD, Cabreira-Hansen MG, Moore WH, Coulter SA, Herlihy JP, Fernandes MR, Cheong BY, Flamm SD, Traverse JH, Zheng Y, Smith D, Shaw S, Westbrook L, Olson R, Patel D, Gahremanpour A, Canales J, Vaughn WK, Willerson JT (2011) A randomized study of transendocardial injection of autologous bone marrow mononuclear cells and cell function analysis in ischemic heart failure (FOCUS-HF). Am Heart J 161(6):1078–1087
14. Hare JM, Fishman JE, Gerstenblith G, DiFede Velazquez DL, Zambrano JP, Suncion VY, Tracy M, Ghersin E, Johnston PV, Brinker JA, Breton E, Davis-Sproul J, Schulman IH, Byrnes J, Mendizabal AM, Lowery MH, Rouy D, Altman P, Wong Po Foo C, Ruiz P, Amador A, Da Silva J, McNiece IK, Heldman AW, George R, Lardo A (2012) Comparison of allogeneic vs autologous bone marrow-derived mesenchymal stem cells delivered by transendocardial injection in patients with ischemic cardiomyopathy: the POSEIDON randomized trial. JAMA 308(22):2369–2379
15. Bartunek J, Behfar A, Dolatabadi D, Vanderheyden M, Ostojic M, Dens J, El Nakadi B, Banovic M, Beleslin B, Vrolix M, Legrand V, Vrints C, Vanoverschelde JL, Crespo-Diaz R, Homsy C, Tendera M, Waldman S, Wijns W, Terzic A (2013) Cardiopoietic stem cell therapy in heart failure: the C-CURE (Cardiopoietic stem Cell therapy in heart failURE) multicenter randomized trial with lineage-specified biologics. J Am Coll Cardiol 61(23):2329–2338
16. Lee JW, Lee SH, Youn YJ, Ahn MS, Kim JY, Yoo BS, Yoon J, Kwon W, Hong IS, Lee K, Kwan J, Park KS, Choi D, Jang YS, Hong MK (2014) A randomized, open-label, multicenter trial for the safety and efficacy of adult mesenchymal stem cells after acute myocardial infarction. J Korean Med Sci 29(1):23–31
17. Freyman T, Polin G, Osman H, Crary J, Lu M, Cheng L, Palasis M, Wilensky RL (2006) A quantitative, randomized study evaluating three methods of mesenchymal stem cell delivery following myocardial infarction. Eur Heart J 27(9):1114–1122
18. Fischer UM, Harting MT, Jimenez F, Monzon-Posadas WO, Xue H, Savitz SI, Laine GA, Cox CS Jr (2009) Pulmonary passage is a major obstacle for intravenous stem cell delivery: the pulmonary first-pass effect. Stem Cells Dev 18(5):683–692
19. Gnecchi M, He H, Liang OD, Melo LG, Morello F, Mu H, Noiseux N, Zhang L, Pratt RE, Ingwall JS, Dzau VJ (2005) Paracrine action accounts for marked protection of ischemic heart by Akt-modified mesenchymal stem cells. Nat Med 11(4):367–368

20. Gnecchi M, He H, Noiseux N, Liang OD, Zhang L, Morello F, Mu H, Melo LG, Pratt RE, Ingwall JS, Dzau VJ (2006) Evidence supporting paracrine hypothesis for Akt-modified mesenchymal stem cell-mediated cardiac protection and functional improvement. FASEB J 20(6):661–669
21. Mirotsou M, Zhang Z, Deb A, Zhang L, Gnecchi M, Noiseux N, Mu H, Pachori A, Dzau V (2007) Secreted frizzled related protein 2 (Sfrp2) is the key Akt-mesenchymal stem cell-released paracrine factor mediating myocardial survival and repair. Proc Natl Acad Sci U S A 104(5):1643–1648
22. Lai RC, Arslan F, Lee MM, Sze NS, Choo A, Chen TS, Salto-Tellez M, Timmers L, Lee CN, El Oakley RM, Pasterkamp G, de Kleijn DP, Lim SK (2010) Exosome secreted by MSC reduces myocardial ischemia/reperfusion injury. Stem Cell Res 4(3):214–222
23. Arslan F, Lai RC, Smeets MB, Akeroyd L, Choo A, Aguor EN, Timmers L, van Rijen HV, Doevendans PA, Pasterkamp G, Lim SK, de Kleijn DP (2013) Mesenchymal stem cell-derived exosomes increase ATP levels, decrease oxidative stress and activate PI3K/Akt pathway to enhance myocardial viability and prevent adverse remodeling after myocardial ischemia/reperfusion injury. Stem Cell Res 10(3):301–312
24. Bian S, Zhang L, Duan L, Wang X, Min Y, Yu H (2014) Extracellular vesicles derived from human bone marrow mesenchymal stem cells promote angiogenesis in a rat myocardial infarction model. J Mol Med 92(4):387–397
25. Feng Y, Huang W, Wani M, Yu X, Ashraf M (2014) Ischemic preconditioning potentiates the protective effect of stem cells through secretion of exosomes by targeting Mecp2 via miR-22. PLoS One 9(2):e88685
26. Yu B, Kim HW, Gong M, Wang J, Millard RW, Wang Y, Ashraf M, Xu M (2015) Exosomes secreted from GATA-4 overexpressing mesenchymal stem cells serve as a reservoir of anti-apoptotic microRNAs for cardioprotection. Int J Cardiol 182:349–360
27. Wang K, Jiang Z, Webster KA, Chen J, Hu H, Zhou Y, Zhao J, Wang L, Wang Y, Zhong Z, Ni C, Li Q, Xiang C, Zhang L, Wu R, Zhu W, Yu H, Hu X, Wang J (2016) Enhanced cardioprotection by human endometrium mesenchymal stem cells driven by exosomal microRNA-21. Stem Cells Transl Med 6(1):209–222
28. Xin H, Li Y, Cui Y, Yang JJ, Zhang ZG, Chopp M (2013) Systemic administration of exosomes released from mesenchymal stromal cells promote functional recovery and neurovascular plasticity after stroke in rats. J Int Soc Cereb Blood Flow Metab 33(11):1711–1715
29. Xin H, Li Y, Liu Z, Wang X, Shang X, Cui Y, Zhang ZG, Chopp M (2013) MiR-133b promotes neural plasticity and functional recovery after treatment of stroke with multipotent mesenchymal stromal cells in rats via transfer of exosome-enriched extracellular particles. Stem Cells 31(12):2737–2746
30. Doeppner TR, Herz J, Gorgens A, Schlechter J, Ludwig AK, Radtke S, de Miroschedji K, Horn PA, Giebel B, Hermann DM (2015) Extracellular vesicles improve post-stroke neuroregeneration and prevent postischemic immunosuppression. Stem Cells Transl Med 4(10):1131–1143
31. Lee C, Mitsialis SA, Aslam M, Vitali SH, Vergadi E, Konstantinou G, Sdrimas K, Fernandez-Gonzalez A, Kourembanas S (2012) Exosomes mediate the cytoprotective action of mesenchymal stromal cells on hypoxia-induced pulmonary hypertension. Circulation 126(22):2601–2611
32. Wang X, Gu H, Qin D, Yang L, Huang W, Essandoh K, Wang Y, Caldwell CC, Peng T, Zingarelli B, Fan GC (2015) Exosomal miR-223 contributes to mesenchymal stem cell-elicited cardioprotection in polymicrobial sepsis. Sci Rep 5:13721

Chapter 13
Exosomes Derived from Embryonic Stem Cells as Potential Treatment for Cardiovascular Diseases

Yao-Hua Song, Lianbo Shao, Yu Zhang, Jin Zhou, Bin Liu, Xiangbin Pan, Yong-jian Geng, Xi-yong Yu, and Yangxin Li

13.1 Introduction

Cardiovascular disease (CVD) resulting from ischemic heart diseases remain to be the main causes of heart failure (HF) and death despite significant advances in medical treatment, such as percutaneous coronary intervention and bypass surgery. Patients develop heart failure due to the death of cardiomyocytes, which are replaced by fibrotic tissue. As the induction of the proliferation and differentiation of cardiac stem cells remains challenging, heart transplantation is still the best treatment option for end stage HF. However, heart transplantation is complicated by donor availability and infections resulting from immunosuppression. The development of

Y.-H. Song
Cyrus Tang Hematology Center, Collaborative Innovation Center of Hematology, Soochow University, Suzhou, Jiangsu 215123, People's Republic of China

L. Shao • Y. Zhang • Y. Li (✉)
Department of Cardiovascular Surgery, Institute for Cardiovascular Science, Soochow University, Suzhou, China
e-mail: yangxin_li@yahoo.com; yangxinli88@aliyun.com

J. Zhou
Department of General Surgery, First Affiliated Hospital of Soochow University, Suzhou, China

B. Liu
Second Affiliated Hospital of Ji Lin University, Changchun, China

X. Pan
Department of Surgery, Fuwai Hospital, Beijing, China

Y.-j. Geng
School of Medicine, University of Texas, Houston, TX, USA

X.-y. Yu
Guangdong Cardiovascular Institute, Guangzhou Medicine University, Guangzhou, China

J. Xiao, S. Cretoiu (eds.), *Exosomes in Cardiovascular Diseases*, Advances in Experimental Medicine and Biology 998, DOI 10.1007/978-981-10-4397-0_13

Table 13.1 Exosomes derived from different cell types and their roles in cardiac regeneration

Sources of exosomes	Animal model	Exosome function	Reference
Rat cardiospheres	Rat AMI	Improves cardiac function by inducing fibroblast to secrete SDF-1 and VEGF	[9]
hCSCs	Mouse dilated cardiomyopathy	Reduced apoptosis and fibrosis in the myocardium and improved heart function	[10]
mESCs	AMI	Delivery of ESC specific miRNA-294 to CPCs augment cardiac function by promoting angiogenesis and cardiomyocytes survival and, reducing fibrosis	[19]
hCPCs	AMI	Promotes angiogenesis and cell survival by releasing miRNA-210, miRNA-132, and miRNA-146a-3p	[17]
mCPCs	AMI	Contains high level expression of GATA4-responsive-miRNA-451	[20]
		Inhibited cardiomyocyte apoptosis	
hESC-derived MSCs	AMI	Reduces infarction by releasing 20S proteasome	[21]
	MI/R	Improves cardiac function by restoring bioenergetics, reducing oxidative stress and activating pro-survival signaling	[16]

CSCs cardiac stem cells, *ESCs* embroynic stem cells, *CPCs* cardiac progenitor cells, *MSCs* mesenchymal stem cells

new therapies for HF is thus required to improve the outcome in these patients, and this has led to the development of cell-based therapies.

Cell therapy is now considered as an alternative option to treat cardiovascular diseases. Animal studies showed interesting results using various cell types, which include bone marrow stem cells (BMSCs) [1], cardiac stem cells (CSCs) [2], cardiospheres [3], embryonic stem cells (ESCs) [4, 5], and induced pluripotent stem cells (iPSCs) [6]. Some stem cell based therapies have been tested in clinical trials [7, 8]. Although the results were encouraging, challenges remain. Tumorigenic potential, immune rejection, and low engraftment and survival rate of transplant cells have hindered the widespread application of stem cells in the clinic. Recent studies suggested that stem cell secrete exosomes, which can protect injured myocardium [9–11] (Table 13.1). It was shown that exosomes functions as communicators between cardiomyocytes and endothelial cells under glucose-deprived condition [12]. Exosomes are microvesicles with phospholipid bilayer, ranging between 30 and 100 nm in size [13]. Most exosomes express specific biomarkers such as CD9, CD63, CD81, and Tsg101, regardless their cellular source. Exosomes have been found in almost all body fluids, including blood, saliva, urine, and breast milk. Until recently, exosomes have been regarded as cell debris. Exosomes carries a wide variety of cargo, including mRNA, miRNA, lncRNA, proteins and cytokines (Fig. 13.1). The exosomes do not form tumor or induce immune response, therefore, have many advantages

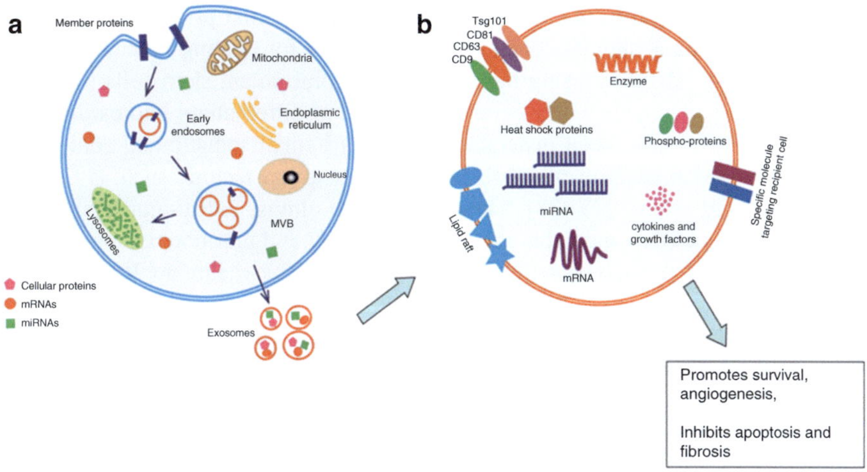

Fig. 13.1 Exosomes biogenesis. (**a**) The process of exosome biogenesis and release from parental cells. (**b**) The bioactive molecules expressed in exosomes

compared to stem cells in clinical application [14]. Exosomes protect injured heart by promoting angiogenesis [15], cell survival [16, 17] and inhibiting fibrosis [18].

In this chapter, we will discuss recent development in ESCs-derived exosomes and their potential roles to treat cardiovascular diseases. Because iPSCs have similar functions as ESC, we also address iPSCs-derived exosomes and their potential roles to treat cardiovascular diseases.

13.2 Embryonic Stem Cells in Cardiac Regeneration

Stem cells are defined as undifferentiated cells that can self-renew and differentiate to all types of somatic cells. Embryonic stem cells have the capacity to produce all types of cells present in the adult organism, therefore, hold great potential for regenerative medicine. Murine derived embryonic stem cells (ESCs) were first isolated and cultured in vitro over 30 years ago. It was shown that human embryonic stem cell-derived cardiomyocytes (hESC-CMs) can remuscularize substantial amounts of the infarcted monkey heart although non-fatal ventricular arrhythmias were observed [5]. It was shown that ESCs overexpressing Thymosin β4 can differentiate into cardiomyocytes improve cardiac function in a mouse myocardial infarction model [22]. However, ESCs can also differentiate into other cell types, which could form teratoma [23]. To solve this problem, researchers have developed a serum free system to obtain pure contracting cardiomyocytes from ESCs [4]. Other problems associated with ESCs are low differentiation efficiency and poor survival after transplantation. It was shown that tumour-necrosis factor could promote ESCs differentiation toward cardiac lineage [24]. Furthermore, the survival of ESCs derived cardiomyocyte can be enhanced by culturing the cells in a pro-survival cocktail prior transplantation [25]. Ethical issue is another concern for using ESCs for treating human diseases, but this can be solved by using human inducible pluripotent stem cells (hiPSCs) which have also been shown to have the potential to differentiate into cardiomyocytes [26]. iPSCs have been generated from both human somatic and adult mice cells by ectopic expression of a cocktail of transcription factors, such as Oct3/4, Sox2, Kfl4 and c-Myc [27, 28]. These cells were almost indistinguishable from ESCs in terms of DNA methylation, gene expression, and pluripotency, suggesting that iPSCs can be used to substitute ESCs for regenerative medicine. It was shown that iPSCs induced from H9C2 cells differentiated into cardiomyocytes and integrated into native myocardium through newly formed gap junction [29]. Induced cardiospheres (iCS) was generated from adult skin fibroblasts via somatic reprogramming by Sox2, Klf4, and Oct4 transfection and a cocktail of Gsk3β inhibitor-6-Bromoindirubin-3′-oxime and Oncostatin M. The iCS differentiated into functional cardiomyocytes and improved left ventricular ejection fraction, anterior/septal ventricular wall thickness and capillary density in the infarcted region 4 weeks after transplantation [30]. Importantly, iPSCs derived from patients' own somatic cells can overcome immunorejection issue associated with allogeneic donor cells.

In addition to direct differentiation, stem cell can repair injured tissue through paracrine mechanism [31, 32]. It was shown that ESCs derived exosomes enhances neovascularization, reduces fibrosis and promotes the survival of c-kit$^+$ cardiac progenitor cells [19].

13.3 The Exosomes Derived from ESCs

Exosomes are small secreted vesicles that contain many different types of bioactive molecules, including proteins, mRNAs, miRNAs and lncRNAs. Exosomes release bioactive molecules to recipient cells and represent a cell-free source for tissue repair. Recent studies suggest that exosomes derived from iPSCs and ESCs can regenerate injured myocardium by inducing cardiomyocyte proliferation and promoting neovascularization. These exosomes were also found to have anti-apoptotic and anti-fibrotic property (Table 13.2). Currently, the therapeutic value of these exosomes is being investigated, and if successful, will provide a novel cell free system to enhance myocardial repair.

13.3.1 The Effect of ESCs-Derived Exosomes on Cell Proliferation, Apoptosis, and Differentiation

Previous studies have demonstrated that hESC-MSCs or fetal tissues derived MSCs' conditioned medium reduced infarction size in a mouse model of myocardial ischemia reperfusion injury, and the conditioned medium contains microparticles with a

Table 13.2 The functional molecules in ESCs-derived exosomes

Molecule types	Component type	Functional molecules	Exosome source	Reference
Protein	Exosome-associated proteins	CD9, CD63, CD81, Tsg101	hESC-MSCs	[33]
	Proteasome subunit	20S proteasome	hESC-MSCs	[21]
	TGF superfamily	Lefty protein	hESCs, mESCs	[34]
mRNA	Stem cell self-renew and expansion	Wnt3	mESCs	[35]
	Transcription factors	Upregulates Oct4, Sox2, Nanog, Rex1 to maintain pluripotency	mESCs, hiPSCs	[35]
		GATA-2, GATA-4, NKX2.5, TRP63 (differentiation factors towards cardiac lineage)		[36]
miRNA	Cardioprotective miRNAs	miRNA 290–295 cluster	mESCs, hiPSCs, miPSCs	[19, 36, 37]

hESC-MSCs human embryonic stem cells derived mesenchymal stem cells, *hiPSC-MSCs* human induced pluripotent stem cells derived mesenchymal stem cells, *miPSCs* mouse induced pluripotent stem cells, *hESCs* human embryonic stem cells, *mESCs* mouse embryonic stem cells, *hiPSCs* human induced pluripotent stem cells

hydrodynamic radius of 50–65 nm [38, 39]. Subsequent studies confirmed these microparticles carry pre-miRNA [40]. Khan et al. reported that exosomes derived from mouse ESCs restored cardiac function by promoting neovascularization and cardiomyocyte survival in a mouse model of AMI. They showed that miRNA-294-3p treatment led to accumulation of CPCs in S-phase and significant reduction of the G1-phase compared to non-treated CPCs. mRNA expression of cyclins (E1, A2 and D1) was increased in CPCs treated with miRNA-294-3p mimic [19]. These results suggest that miRNA-294-3p enhanced increased S-phase transition.

MiRNA-294 is a member of the miRNA 290–295 cluster which is comprise about 70% of miRNA contents in ESCs and in ESCs-derived exosomes [41]. All miRNA 290–295 cluster members contain a common sequence (AAAGUGC). miRNA-291, miRNA-294 and miRNA-295 are involved in early development by regulating cell cycle, self-renewal [42], differentiation [43, 44], survival [45] and proliferation [42]. The expressions of miRNA-294 become undetectable as ESCs differentiate into mature cells. miRNA-294, miRNA-291-3p, and miRNA-295 also promote induced pluripotency by increasing the efficiency of reprogramming by Oct4, Sox2 and Klf4 [46].

MiRNA-21 and miRNA-210 have also been found in the exosomes derived from mouse iPSCs [37]. Previous studies have shown that miRNA-210 can be induced by hypoxia to protect cells from apoptotic stimuli by targeting E2F3 [47]. MiRNA-21 protects H9C2 cardiac cells from oxidative stress induced apoptosis by down regulating PDCD4 (programmed cell death 4) [48]. These studies suggest that iPSCs derived exosomes may protect the injured myocardium via releasing specific miroRNAs.

It was shown that mature somatic cell can be reprogrammed when cocultured with ESCs or their extracts [49]. Exosomes derived from ESCs contain mRNA and protein of several pluripotent molecules such as Oct4 and Sox2 [35]. It was confirmed that the exosomal mRNAs could be translated in their recipient cells. For example, when ESCs derived exosomes were incorporated to hematopoietic progenitor cells, Oct4 mRNA was translated into Oct4 protein [35]. Exosome mediated transfer of miRNA and mRNA may be involved in transdifferentiation of fibroblasts into induced pluripotent cells. For example, exosomes from partially reprogrammed iPSCs may reprogram nearby cells by releasing exosomes [36].

Mouse embryonic stem cell derived microvesicles (ESMV) induced dedifferentiation and pluripotency in human retinal progenitor Müller cells by transferring mRNAs for Oct4 and Sox2, and the miRNAs of the miRNA-290 cluster [50]. Exosomes isolated from murine ES cells can enhance survival and improve expansion of murine hematopoietic progenitor cells by up regulating early pluripotent (Oct-4, Nanog and Rex-1) and early hematopoietic stem cells (Scl, HoxB4 and GATA 2) markers in these cells [35]. It is possible that ESC derived exosomes might be able to select certain mRNAs that are critical for pluripotency before they are released from ESCs. Thus the ESCs derived exosomes have the potential tool to cure myocardium infarction by reprogramming somatic cells.

In a post-infarct mouse heart failure model, Kervadec et al. investigated whether post-infarction administration of exosomes released by human embryonic stem cell-derived cardiovascular progenitors (hESC-Pg) can provide equivalent benefits to administered hESC-Pg. The exosomes were delivered into the peri-infarct myocardium by percutaneous injections under echocardiographic guidance. The results showed that the exosomes were as efficient as the stem cells in improving cardiac function, which was mediated by paracrine factors that promote cell survival and proliferation [51].

13.3.2 ESCs-Derived Exosomes and Angiogenesis

Angiogenesis is a process to form new blood vessels from existing endothelial cells, and plays a central role during cardiac repair following myocardial infarction. Vascular endothelial cells are located in the inner vessel wall, and are responsible to form new blood vessels when needed. The process of angiogenesis occurs in both physiological and pathological conditions such as normal development, tumorigenesis, inflammation, cardiac repair. The process of angiogenesis starts with increased vascular permeability, degradation of basement membrane, followed by migration, proliferation of endothelial cells and formation of sprout to connect neighboring vessels, and then, vessel lumen is formed from the sprout [52]. Studies have demonstrated that exosomes secreted by human iPSCs derived MSCs promote bone regeneration and angiogenesis in critical-sized calvarial defects in ovariectomized rats in a dose dependent manner [53]. Mouse ESCs-derived exosomes stimulated neovascularization in a myocardial infarction model [19]. Exosomes released from the iPSCs derived MSCs attenuated limb ischemia by promoting angiogenesis in mice [54].

It was shown that ESC derived exosomes are enriched in mRNA of certain angiogenic cytokines such as VEGF and FGF2 compared to parental ES-D3 cells [35]. VEGF stimulate angiogenesis by banding to VEGF receptor (VEGFR) on endothekial cells. Activation of VEGFR triggers PI3K/Akt pathway to promote survival and the p38MAPK and focal adhesion kinase pathway to induce cytoskeletal reorganization, proliferation and migration of endothelial cells. FGF2 is also known as basic fibroblast growth factor, which is synergistic to VEGF in promoting angiogenesis. In addition, FGF2 is also necessary to maintain ESC cells in an undifferentiated state.

MicroRNA molecules also regulate angiogenesis. MiRNA-21 is highly expressed in mouse iPSCs derived exosomes. It was shown that miRNA-21 promote angiogenesis by targeting PTEN, leading to activation of Akt and ERK1/2 signaling pathways, which stimulate HIF-1α and VEGF expression [55] (Fig. 13.2). By contrast, another study showed that miRNA-21 overexpression inhibited tube formation of endothelial cells by targeting Rho B [56].

Using microRNA microarrays, Hu et al. discovered that miRNA-210 was highly expressed in mouse HL-1 cardiomyocytes exposed to hypoxia. They further showed that intramyocardial injections of miRNA-210 precursor improved neovascularization and cardiac function in a mouse myocardial infarction model [57]. Recent studies showed that exosomes secreted from cardiac fibroblast derived iPSC (CF-iPSC) protected against ischemia induced injury by releasing miRNA-210 [37].

Matrix metalloproteinases (MMPs) belong to a family of enzymes that are involved in the degradation of extracellular matrix proteins under physiological or pathological conditions such as cell migration, invasion, wound healing and angiogenesis. Angiogenesis requires MMPs because the vascular basement membranes need to be degraded in order to allow endothelial cells to migrate from exiting ves-

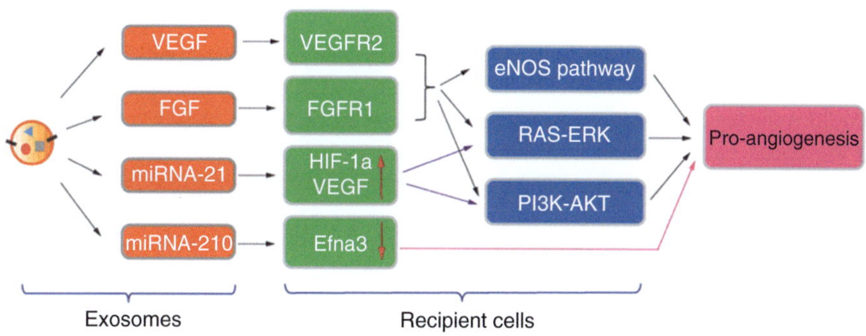

Fig. 13.2 Mechanisms involved in ESCs exosome mediated angiogenesis

sels and invade into the neighboring tissue. MMPs participate in this remodeling of basement membranes and ECM. In addition, MMPs have also been shown to induce angiogenesis by releasing ECM-bound angiogenic growth factors (Table 13.2), Johnson et al. [58] have further confirmed that MMP-9 involved in capillary branching in ischemic muscle. The pro-angiogenic factors carried by exosomes also include extracellular matrix metalloproteinase inducer (EMMPRIN). It was shown that knockdown of EMMPRIN on cardiomyocyte progenitor cells (CMPC) diminished the pro-angiogenic effect [59]. EMMPRIN stimulates angiogenesis through upregulation of MMPs and VEGF [60, 61]. Therefore, the delivery of EMMPRIN to endothelial cells by exosomes can induce angiogenesis by enhancing VEGF signaling. It was shown that exosomes from stem cells overexpressing Akt promote angiogensis through upregulation of PDGF-D [62]. Intravenous infusion of these exosomes improved cardiac function in a rat myocardial infarction model.

13.3.3 ESCs-Derived Exosomes Improve the Microenvironment After AMI

It has been shown that human heart has the capacity of continuous regeneration [63, 64]. However, the endogenous regenerative capacity of the heart is limited and the regeneration process is hampered by the hostile microenvironment resulting from inflammation after infarction. Therefore, improvement of the microenvironment of infarcted region is the key for the cardiac repair post AMI. Previous studies showed that mesenchymal stem cell derived exosomes were able to improve the microenvironment after AMI [65].

In the infarct region, cell death triggers activation of complement cascade, generation of free radicals, release of cytokines, infiltration of neutrophils which induce further damage to myocardium through the release of proteolytic enzymes. In addition, the released cytokines such as TGF-β can induce myofibroblast transdifferentiation, leading to excessive fibrosis and adverse remodeling and eventually to heart failure. Therefore, improvement of post-infarct microenvironment through the modulation of the inflammatory response can prevent or delay the development of heart failure.

The transfer of exosomal miRNAs from stem cells to recipient cells is an important process to regulate microenvironment during tissue regeneration. MiRNA-21 could improve microenvironment by regulating macrophage phenotype. MiRNA-146a could target NF-κB and reduce the expression of TNFα, IL-6 and IFN-γ. MiRNA-181 could protect host against endotoxin shock by reprogramming immune response [66–68].

In a mouse peritonitis model, an increase in miRNA-21 expression in macrophages resulted in a decrease in TNF-α and IL-6 production and an increase in IL-10, suggesting an anti-inflammatory effect [69, 70] Other have shown that the transcription of miRNA-21 requires the binding and activation by the p65 subunit of NF-κB [71].

It was shown that miRNA-210 is negative regulator of inflammation by targeting molecules down stream of IKKβ to inhibit NF-κB1 signaling. Transfection of miRNA-210 mimics to macrophages inhibited LPS-induced production of inflammatory cytokines such as IL-6 and TNF-α [72]. Both miRNA-21 and miRNA-210 have been found in mouse iPSCs derived exosomes and play an important role in regulating inflammatory response after myocardial infarction.

Although stem cell based therapy showed some promising result in animal studies, hurdles on poor cell engraftment and survival, immune response, teratoma remain to be overcome before this approach can be applied in the clinic. Recent studies suggest that exosome-based therapy can overcome these limitations of cell therapy [73]. In a murine model of hypoxic pulmonary hypertension, it was shown that exosomes suppressed the infiltration of macrophages and the release of proinflammatory and proproliferative mediators, including monocyte chemoattractant protein-1 and hypoxia-inducible mitogenic factor. The improvement of microenvironment by exosomes was mediated by miRNAs. MiRNA-204 was inhibited by STAT3 under pathological condition such as chronic hypoxia, but this inhibitory effect was reversed by exosomes. On the other hand, the pro-proliferative miRNA-17 was induced by hypoxia but inhibited by exosomes [74]. Kalani et al. [75] delivered tested the effect of exsomes on ischemic reperfusion injury (IR) in mice. The IR injury was created by inserting a silicon-rubber coated monofilament into the middle cerebral artery for 40 min. Then then delivered a nano-formulation with an anti-inflammatory molecue curcumin and embryonic stem cell exosomes (MESC-exo^{cur}) to mice by intranasal administration for 7 days. They demonstrated that MESC-exo^{cur} treatment reduced the expression levels of inflammation mediators such as ROS, TNF-α, but increased the levels of endothelial vascular junction protein (VE-cadherin).

13.4 Exosomes and Therapeutic Approaches

Though many reports have shown that ESCs-derived exosomes have the potential to enhance endogenous cardiac repair, exosomal content need to be modified to improve therapeutic efficacy.

13.4.1 Modification of ESCs-Derived Exosomes

Exosomes carry a wide variety of molecules, including mRNA, miRNA, proteins, lipid and signalling molecules. Some of the molecules act as communicators constantly shuttling between cells to control cellular signaling in recipient cells [76, 77]. Because of the specificity of exsome mediated cellular delivery and the ability to enter cells, exosomes are being exploited as a vehicle to deliver small RNAs such as siRNA and miRNA for therapeutic purpose [78]. RNAi refers to a biological process in which double stranded RNA interrupt gene expression through the degradation of complementary mRNA [78–80]. This approach has been tested in several disease models. To target brain cells, the cells were transfected with a construct to express exosomal membrane protein Lamp2b, which is fused to a neuron-specific RVG peptide. Purified exosomes were loaded with siRNA targeting BACE1, a therapeutic target in Alzheimer's disease. The intravenously injected exosomes were found in the brain, resulting in a specific gene knockdown of BACE1 [81]. This technique has been used to treat Huntington's disease. Hydrophobically modified siRNA targeting Huntingtin mRNA were loaded into exosomes purified from conditioned media of glioblastoma U87 cells. Infusion of siRNA-loaded exosomes resulted in significant bilateral silencing of up to 35% of Huntingtin mRNA [80]. Exosomes from peripheral blood have been loaded with siRNA targeting MAPK1 by electroporation to silence MAPK1 in monocytes and lymphocytes [82]. Both human embryonic kidney 293 and mesenchymal stem cell exosomes were used to knockdown PLK-1 in bladder cancer cells [83]. Exosomes carrying TGF-β1 siRNA efficiently decreased the viability and migration of tumor cells [84]. In summary, the transfer of small RNAs for gene modification represents an efficient approach to improve microenvironment after AMI.

Exosomes have been used as drug carrier. Exosome loaded with curcumin was able to inhibit inflammation in a LPS-induced septic shock mouse model [85]. Microvesicles carrying Paclitaxel demonstrated a strong anti-proliferative activity on human pancreatic cell line CFPAC-1 [86]. In order to reduce toxicity to host, exosomes have also been used to deliver doxorubicin to kill cancer cells [87]. Exosomes released from the heart under conditions of cellular stress are enriched in AT1Rs (Angiotensin II Type 1 receptors) and injection of AT1R-containing exosomes restored AngII-induced blood pressure response in AT1R knock out animals [88].

Another way to enhance exosome based delivery is to modify proteins associated with exosomal membrane. Peptides expressed on the N terminus of Lamp2b are usually cleaved from exosomes, which hampered targeted delivery of exosomes to recipient cells. To overcome this problem, Hung et al. added a glycosylation motif to the protein at various positions to protect the peptide from degradation [88]. Importantly, they showed that the glycosylated peptides enhanced the delivery of exosomes to neuroblastoma cells.

Exosomes can also be optimized through genetic modification. Trivedi M et al. [89] modified exosomal content by transfecting cells with plasmid expressing p53 (wt-p53) and miRNA-125b. They showed that the modified exosomes have a reprogramed global miRNA profile, which were mainly related to p53 signaling and genes associated with apoptosis. Such modification could create a hostile microenvironment for tumor cells. Huang and his colleagues [90] showed that a non-viral minicircle vector that express HIF-1alpha in the heart, could improve transfection efficiency, duration of transgene expression, and cardiac function [91].

Exosomes derived from stem cells over expressing GATA-4 prevent ischemia induced apoptosis [92]. The protective effect was mediated by miRNA-19a, which target PTEN, leading to activation of Akt signaling pathway. It was shown that the survival of transplanted cardiac progenitor cells (CPCs) can be improved by codelivery of CPCs with a nonviral minicircle plasmid expressing HIF1 (MC-HIF1) into the ischemic myocardium. The prosurvial benefit was provided by miRNA-126 and miRNA-210 from exsomes released by cardiac endothelial cells [93].

Therefore, the modification of exosome structure and content represents a new direction for targeted therapy. However, the yield of exosomes isolated from cells are usually very low and the procedures for isolating exosomes are tedious and time consuming, which limited the widely use of exosomes for research and clinical therapy. To overcome this problem, Jeong et al. [94], prepared nanovesicles by passing ES cells through micro-sized pores. The cells were ruptured into plasma membrane fragments, which assembled into nanovesicles due to the inherent features of the amphiphilic molecules present in the plasma membrane. The size and content of the nanovesicles were similar to exosomes. When treating primary murine skin fibroblasts with the nanovesicles, the treated cells showed enhanced cell proliferation rate, accompanied by higher expression levels of VEGF-α, TGF-β collagen I, PCNA, and Ki-67. The authors suggested that the nanovesicles could potentially contribute to tissue repair.

Yoon et al. [95] designed a cell-slicing system to generate nanovesicles (NVs) from fragments of sliced ES cells. They employed an array of 500 nm-thick low-stress silicon nitride blades along microchannels to slice living cells and the NVs contain both ICAM-1 (marker of plasma membrane protein) and Oct 3/4 (marker of ES cells). They also showed that the NVs contain miRNA, rRNA and tRNA, suggesting that the NVs can be used as vehicles for RNA delivery. Interestingly, they demonstrated that the NVs could encapsulate fluorescent beads and then delivered to recipient cells. Therefore, it is possible that the NVs can used for drug delivery.

Large quantity of cell-derived nanovesicles can be generated by using centrifugal force and a filter with micro-sized pores [96]. This technique can produce 250 times more nanovesicles than t naturally secreted exosomes. Nanovesicles generated from murine embryonic stem cells contain more intracellular contents can transfer RNAs to target cells such as NIH-3 T3 fibroblasts and mouse embryonic fibroblasts.

13.4.2 Precondition of ESCs-Derived Exosomes

It is known that ischemic pre-conditioning can produce therapeutic effect on infarcted myocardium. New evidence suggests that exosomes are involved in this process. There are no report regarding precondition of ESCs-derived exosomes, but the precondition of other stem cell-derived exosomes would provide a clue.

Exosomes purified from MSC subjected to ischemic preconditioning (IPC) are enriched with miRNA-22. Furthermore, the miRNA-22 enriched exosomes can migrate into cardiomyocytes in a co-culture system. Importantly, cardiomyocytes that have received the exosomes were more resistant to ischemic stress [97]. Yasukatsu Izumi et al. [98] first demonstrated that remote ischemic conditioning (RIC) can prevent deterioration of LV diastolic function, and attenuate LV interstitial fibrosis after myocardial infarction. In this rat model, RIC was performed by repeated bilateral hindlimb ischemia and reperfusion once a day for 4 weeks. MicroRNA-29a (miRNA-29a), a negative regulator of tissue fibrosis, was highly enriched in the exosomes found in the marginal area of the infarction from the RIC group. They further showed that exosomes derived from a myoblast line C2C12 cells subjected to ischemia also contain high levels of miRNA-29a and IGF-1R. Therefore, these results suggest that the cardiac protection by RIC was afforded by miRNA-29a and IGF-1R, which were transported by muscle derived exosomes. Ferdinandy et al. [99] also confirmed that the release of EVs from the heart after preconditioning was responsible for the transfer of remote conditioning signals for cardioprotection using hearts of male Wistar rats isolated and perfused in Langendorff mode.

Gray et al. showed that exosomes derived from cardiac progenitor cells (CPCs) subjected to hypoxic condition enhanced tube formation of endothelial cells and inhibited profibrotic gene expression in TGF-β-stimulated fibroblasts. They further demonstrated that exosomes from hypoxic CPCs decreased levels of CTGF, Vimentin, and Collagens I and III in cardiac fibroblasts. They suggested that the anti-fibrosis effects of hypoxia-derived CPC exosomes could be due to the increased levels of miRNA-17, -199a, -210, and -292, whereas the proangiogenic effects were due to increase the levels of miRNA-17 and -210 in exosomes [100].

Electrical stimulation (EleS) has been used as a pre-conditioning method to promote stem cell survival post-transplantation within the ischemic heart. Kim et al. showed that cardiac stem cells preconditioned by EleS are more resistant to apoptotic stimuli due to the release of connective tissue growth factor, which is regulated by miRNA-378 [101].

The release of exosomes are regulated by nearby cells via paracrine mechanism. Gleboy et al. showed that microglial cells express serotonin receptors and the release of exosomes from microglia are regulated by serotonin released by embryonic stem cell-derived serotonergic neurons [102].

13.5 The Benefit and Potential Risk of Exosome Based Therapies

Exosome based therapies are convenient to store, easier to use and do not trigger immune rejection. Stem cell-derived exosomes provide beneficial miRNA and growth factors to promote angiogenesis, differentiation and cell survival. In addition, exosomes can improve post infarction microenvironment by inhibiting inflammatory response.

Patients with myocardial infarction often need immediate intervention. This is the problem with cell based therapy because it takes time to isolate cells and grown to desired numbers. One alternative is to use frozen cells, but many cells do not survive the freeze/thaw procedure. Another problem is poor engraftment and survival in a proinflammatory post-infarction microenvironment. By contrast, exosomes are cell free particles which can be purified with standard techniques and stored in freezer until use. Exosomes are encircled by rigid lipid bilayer membrane which can withstand freeze-thaw cycles without the need of toxic cryopreservative agents. Although animal studies demonstrated good results using stem cell based therapy, the cells were usually injected directly into myocardium, which is not applicable for curing human infarction. An ideal intervention would be delivered intravenously. However, most cells would be trapped in the lung when injected through the vein. On the other hand, exosomes are small enough that allow them to pass the lung and reach to circulation. Importantly, exosomes have the capability to reach their target cells with precision. Another main advantage with exsomes based therapy is that the content within the exosomes can be modified by either preconditioning or genetic engineering. It was shown that intra-articular injections of 100 μg human embryonic MSC exosomes promote cartilage repair in a rat model of osteochondral defects [103]. The injections were performed weekly for a period of 12 weeks.

Despite these advantages associated with exosome based therapy, there are potential problems that need to be addressed before exosomes can be used in the clinic. The pro-angiogenic and pro-survival effects of exosomes can promote the growth of cancer cells. It was shown that MSC-exosomes enhanced VEGF expression in tumor cells by activating ERK1/2 pathway [104]. Another concern is exosomes may carry undesired molecules that may produce side effects (Fig. 13.3). It was shown that exosomes from serum of patients with cardiomyopathy induces pathological changes in gene expression in pluripotent stem cells [105]. Therefore, the function of transplanted stem cells can be influenced by host environment.

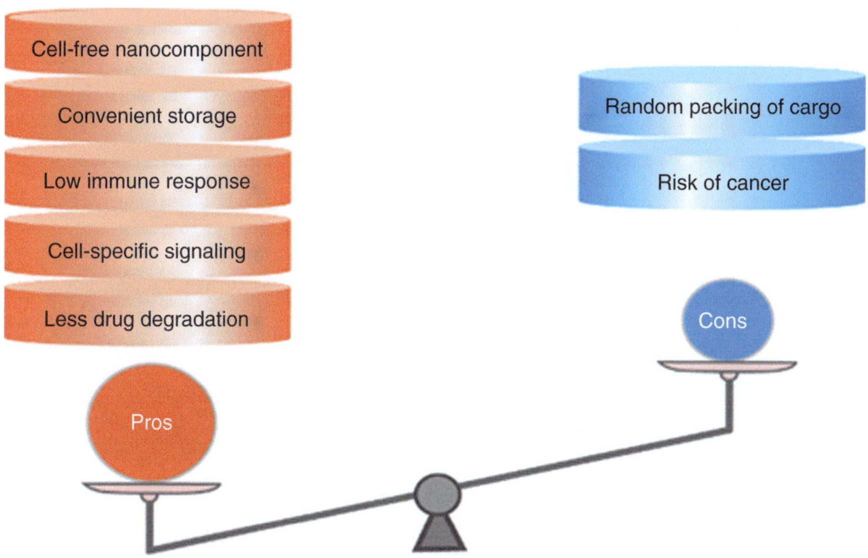

Fig. 13.3 The benefit and risk of ESCs-derived exosome in organ regeneration

13.6 Perspective

Although stem cell based therapy showed promise for myocardial regeneration, challenges remain. The main hurdles for cell therapy include poor engraftment and survival in the pro-inflammatory post-infarction microenvironment. In addition, the transplanted stem cells might generate electric signal which could become a potential source of arrhythmia. Fortunately, exosome based therapy could avoid these problems associated with cell therapy. Future research should focus on how various molecules are sorted into exosomes and this information will help to design better exosomes for treatment of cardiovascular diseases. Recent studies suggest that exosome content can vary depending on how cells are challenged. It would be important to find out exactly what types of cellular stress is needed for producing most useful exosomes. Alternatively, specific molecules can be introduced into exosomes by genetic engineering in order to treat specific conditions and to improve efficacy.

13.7 Conclusion

The main concern with ESCs based therapy is the possibility of generating teratoma. This problem can be solved by using exosomes derived from ESCs. It was confirmed that exosomes derived from ESCs contain miRNAs that can promote cell proliferation, differentiation, survival, angiogenesis and inhibit apoptosis and inflammation (Fig. 13.4). Thus ESCs derived exosomes represent a promising therapeutic modality for myocardium regeneration.

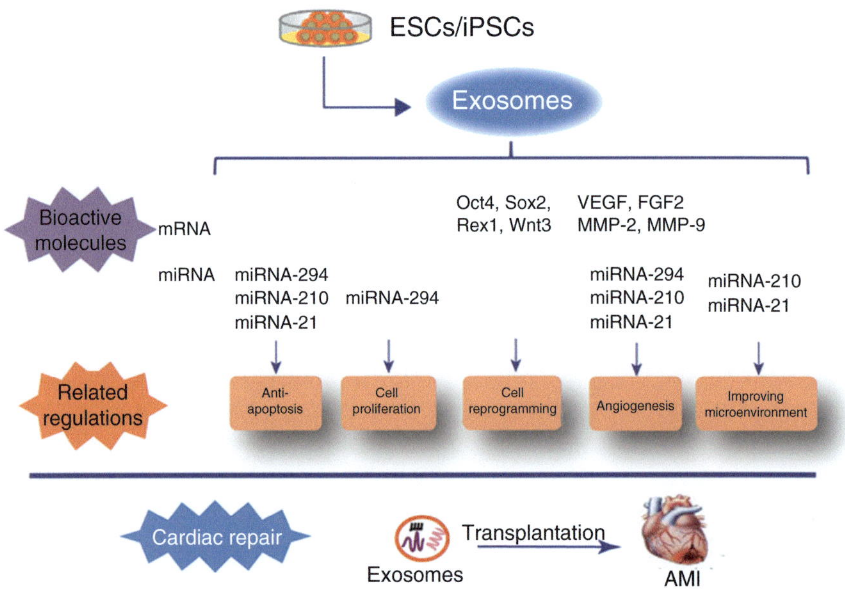

Fig. 13.4 Mechanisms exosome mediated cardiac repair. Exosomes improve myocardial function by promoting cell cycle activation, angiogenesis, improving the microenvironment, and reducing apoptosis in the infarcted region

References

1. Williams AR, Hatzistergos KE, Addicott B, McCall F, Carvalho D, Suncion V, Morales AR, Da Silva J, Sussman MA, Heldman AW, Hare JM (2013) Enhanced effect of combining human cardiac stem cells and bone marrow mesenchymal stem cells to reduce infarct size and to restore cardiac function after myocardial infarction. Circulation 127(2):213–223
2. Bolli R, Tang XL, Sanganalmath SK, Rimoldi O, Mosna F, Abdel-Latif A, Jneid H, Rota M, Leri A, Kajstura J (2013) Intracoronary delivery of autologous cardiac stem cells improves cardiac function in a porcine model of chronic ischemic cardiomyopathy. Circulation 128(2):122–131
3. Makkar RR, Smith RR, Cheng K, Malliaras K, Thomson LE, Berman D, Czer LS, Marban L, Mendizabal A, Johnston PV, Russell SD, Schuleri KH, Lardo AC, Gerstenblith G, Marban E (2012) Intracoronary cardiosphere-derived cells for heart regeneration after myocardial infarction (CADUCEUS): a prospective, randomised phase 1 trial. Lancet 379(9819):895–904
4. Moon SH, Kang SW, Park SJ, Bae D, Kim SJ, Lee HA, Kim KS, Hong KS, Kim JS, Do JT, Byun KH, Chung HM (2013) The use of aggregates of purified cardiomyocytes derived from human ESCs for functional engraftment after myocardial infarction. Biomaterials 34(16):4013–4026
5. Chong JJ, Yang X, Don CW, Minami E, Liu YW, Weyers JJ, Mahoney WM, Van Biber B, Cook SM, Palpant NJ, Gantz JA, Fugate JA, Muskheli V, Gough GM, Vogel KW, Astley CA, Hotchkiss CE, Baldessari A, Pabon L, Reinecke H, Gill EA, Nelson V, Kiem HP, Laflamme MA, Murry CE (2014) Human embryonic-stem-cell-derived cardiomyocytes regenerate non-human primate hearts. Nature 510(7504):273–277
6. Zhang L, Guo J, Zhang P, Xiong Q, SC W, Xia L, Roy SS, Tolar J, O'Connell TD, Kyba M, Liao K, Zhang J (2015) Derivation and high engraftment of patient-specific cardiomyocyte sheet using induced pluripotent stem cells generated from adult cardiac fibroblast. Circ Heart Fail 8(1):156–166
7. Sanganalmath SK, Bolli R (2013) Cell therapy for heart failure: a comprehensive overview of experimental and clinical studies, current challenges, and future directions. Circ Res 113(6):810–834
8. Prowse AB, Timmins NE, Yau TM, Li RK, Weisel RD, Keller G, Zandstra PW (2014) Transforming the promise of pluripotent stem cell-derived cardiomyocytes to a therapy: challenges and solutions for clinical trials. Can J Cardiol 30(11):1335–1349
9. Tseliou E, Fouad J, Reich H, Slipczuk L, de Couto G, Aminzadeh M, Middleton R, Valle J, Weixin L, Marban E (2015) Fibroblasts rendered Antifibrotic, Antiapoptotic, and Angiogenic by priming with cardiosphere-derived extracellular membrane vesicles. J Am Coll Cardiol 66(6):599–611
10. Vandergriff AC, de Andrade JB, Tang J, Hensley MT, Piedrahita JA, Caranasos TG, Cheng K (2015) Intravenous cardiac stem cell-derived exosomes ameliorate cardiac dysfunction in doxorubicin induced dilated cardiomyopathy. Stem Cells Int 2015:960926
11. Li Y, Shen Z, XY Y (2015) Transport of microRNAs via exosomes. Nat Rev Cardiol 12(4):198
12. Garcia NA, Moncayo-Arlandi J, Sepulveda P, Diez-Juan A (2016) Cardiomyocyte exosomes regulate glycolytic flux in endothelium by direct transfer of GLUT transporters and glycolytic enzymes. Cardiovasc Res 109(3):397–408
13. Thery C, Zitvogel L, Amigorena S (2002) Exosomes: composition, biogenesis and function. Nat Rev Immunol 2(8):569–579
14. Ibrahim AG, Cheng K, Marban E (2014) Exosomes as critical agents of cardiac regeneration triggered by cell therapy. Stem Cell Reports 2(5):606–619
15. Bian S, Zhang L, Duan L, Wang X, Min Y, Yu H (2014) Extracellular vesicles derived from human bone marrow mesenchymal stem cells promote angiogenesis in a rat myocardial infarction model. J Mol Med (Berlin, Germany) 92(4):387–397
16. Arslan F, Lai RC, Smeets MB, Akeroyd L, Choo A, Aguor EN, Timmers L, van Rijen HV, Doevendans PA, Pasterkamp G, Lim SK, de Kleijn DP (2013) Mesenchymal stem cell-derived exosomes increase ATP levels, decrease oxidative stress and activate PI3K/Akt pathway to enhance myocardial viability and prevent adverse remodeling after myocardial ischemia/reperfusion injury. Stem Cell Res 10(3):301–312

17. Barile L, Lionetti V, Cervio E, Matteucci M, Gherghiceanu M, Popescu LM, Torre T, Siclari F, Moccetti T, Vassalli G (2014) Extracellular vesicles from human cardiac progenitor cells inhibit cardiomyocyte apoptosis and improve cardiac function after myocardial infarction. Cardiovasc Res 103(4):530–541
18. Li T, Yan Y, Wang B, Qian H, Zhang X, Shen L, Wang M, Zhou Y, Zhu W, Li W, Xu W (2013) Exosomes derived from human umbilical cord mesenchymal stem cells alleviate liver fibrosis. Stem Cells Dev 22(6):845–854
19. Khan M, Nickoloff E, Abramova T, Johnson J, Verma SK, Krishnamurthy P, Mackie AR, Vaughan E, Garikipati VN, Benedict C, Ramirez V, Lambers E, Ito A, Gao E, Misener S, Luongo T, Elrod J, Qin G, Houser SR, Koch WJ, Kishore R (2015) Embryonic stem cell-derived exosomes promote endogenous repair mechanisms and enhance cardiac function following myocardial infarction. Circ Res 117(1):52–64
20. Chen L, Wang Y, Pan Y, Zhang L, Shen C, Qin G, Ashraf M, Weintraub N, Ma G, Tang Y (2013) Cardiac progenitor-derived exosomes protect ischemic myocardium from acute ischemia/reperfusion injury. Biochem Biophys Res Commun 431(3):566–571
21. Lai RC, Tan SS, Teh BJ, Sze SK, Arslan F, de Kleijn DP, Choo A, Lim SK (2012) Proteolytic potential of the MSC exosome proteome: implications for an exosome-mediated delivery of therapeutic proteasome. Int J Proteomics 2012:971907
22. Yan B, Singla RD, Abdelli LS, Singal PK, Singla DK (2013) Regulation of PTEN/Akt pathway enhances cardiomyogenesis and attenuates adverse left ventricular remodeling following thymosin beta4 overexpressing embryonic stem cell transplantation in the infarcted heart. PLoS One 8(9):e75580
23. Knoepfler PS (2009) Deconstructing stem cell tumorigenicity: a roadmap to safe regenerative medicine. Stem Cells 27(5):1050–1056
24. Hamidi S, Letourneur D, Aid-Launais R, Di Stefano A, Vainchenker W, Norol F, Le Visage C (2014) Fucoidan promotes early step of cardiac differentiation from human embryonic stem cells and long-term maintenance of beating areas. Tissue Eng Part A 20(7–8):1285–1294
25. Don CW, Murry CE (2013) Improving survival and efficacy of pluripotent stem cell-derived cardiac grafts. J Cell Mol Med 17(11):1355–1362
26. Lalit PA, Hei DJ, Raval AN, Kamp TJ (2014) Induced pluripotent stem cells for post-myocardial infarction repair: remarkable opportunities and challenges. Circ Res 114(8):1328–1345
27. Takahashi K, Yamanaka S (2006) Induction of pluripotent stem cells from mouse embryonic and adult fibroblast cultures by defined factors. Cell 126(4):663–676
28. Takahashi K, Tanabe K, Ohnuki M, Narita M, Ichisaka T, Tomoda K, Yamanaka S (2007) Induction of pluripotent stem cells from adult human fibroblasts by defined factors. Cell 131(5):861–872
29. Singla DK, Long X, Glass C, Singla RD, Yan B (2011) Induced pluripotent stem (iPS) cells repair and regenerate infarcted myocardium. Mol Pharm 8(5):1573–1581
30. Xu JY, Lee YK, Ran X, Liao SY, Yang J, KW A, Lai WH, Esteban MA, Tse HF (2016) Generation of induced cardiospheres via reprogramming of skin fibroblasts for myocardial regeneration. Stem Cells. doi:10.1002/stem.2438
31. Garbern JC, Lee RT (2013) Cardiac stem cell therapy and the promise of heart regeneration. Cell Stem Cell 12(6):689–698
32. Gnecchi M, Zhang Z, Ni A, Dzau VJ (2008) Paracrine mechanisms in adult stem cell signaling and therapy. Circ Res 103(11):1204–1219
33. Villarroya-Beltri C, Gutierrez-Vazquez C, Sanchez-Cabo F, Perez-Hernandez D, Vazquez J, Martin-Cofreces N, Martinez-Herrera DJ, Pascual-Montano A, Mittelbrunn M, Sanchez-Madrid F (2013) Sumoylated hnRNPA2B1 controls the sorting of miRNAs into exosomes through binding to specific motifs. Nat Commun 4:2980
34. Khalkhali-Ellis Z, Galat V, Galat Y, Gilgur A, Seftor EA, Hendrix MJ (2016) Lefty glycoproteins in human embryonic stem cells: extracellular delivery route and posttranslational modification in differentiation. Stem Cells Dev. doi:10.1089/scd.2016.0081

35. Ratajczak J, Miekus K, Kucia M, Zhang J, Reca R, Dvorak P, Ratajczak MZ (2006) Embryonic stem cell-derived microvesicles reprogram hematopoietic progenitors: evidence for horizontal transfer of mRNA and protein delivery. Leukemia 20(5):847–856
36. Bobis-Wozowicz S, Kmiotek K, Sekula M, Kedracka-Krok S, Kamycka E, Adamiak M, Jankowska U, Madetko-Talowska A, Sarna M, Bik-Multanowski M, Kolcz J, Boruczkowski D, Madeja Z, Dawn B, Zuba-Surma EK (2015) Human induced pluripotent stem cell-derived microvesicles transmit RNAs and proteins to recipient mature heart cells modulating cell fate and behavior. Stem Cells 33(9):2748–2761
37. Wang Y, Zhang L, Li Y, Chen L, Wang X, Guo W, Zhang X, Qin G, He SH, Zimmerman A, Liu Y, Kim IM, Weintraub NL, Tang Y (2015) Exosomes/microvesicles from induced pluripotent stem cells deliver cardioprotective miRNAs and prevent cardiomyocyte apoptosis in the ischemic myocardium. Int J Cardiol 192:61–69
38. Lai RC, Arslan F, Tan SS, Tan B, Choo A, Lee MM, Chen TS, Teh BJ, Eng JK, Sidik H, Tanavde V, Hwang WS, Lee CN, El Oakley RM, Pasterkamp G, de Kleijn DP, Tan KH, Lim SK (2010) Derivation and characterization of human fetal MSCs: an alternative cell source for large-scale production of cardioprotective microparticles. J Mol Cell Cardiol 48(6):1215–1224
39. Timmers L, Lim SK, Arslan F, Armstrong JS, Hoefer IE, Doevendans PA, Piek JJ, El Oakley RM, Choo A, Lee CN, Pasterkamp G, de Kleijn DP (2007) Reduction of myocardial infarct size by human mesenchymal stem cell conditioned medium. Stem Cell Res 1(2):129–137
40. Chen TS, Lai RC, Lee MM, Choo AB, Lee CN, Lim SK (2010) Mesenchymal stem cell secretes microparticles enriched in pre-microRNAs. Nucleic Acids Res 38(1):215–224
41. Calabrese JM, Seila AC, Yeo GW, Sharp PA (2007) RNA sequence analysis defines Dicer's role in mouse embryonic stem cells. Proc Natl Acad Sci U S A 104(46):18097–18102
42. Lichner Z, Pall E, Kerekes A, Pallinger E, Maraghechi P, Bosze Z, Gocza E (2011) The miR-290-295 cluster promotes pluripotency maintenance by regulating cell cycle phase distribution in mouse embryonic stem cells. Differentiation 81(1):11–24
43. Wang Y, Baskerville S, Shenoy A, Babiarz JE, Baehner L, Blelloch R (2008) Embryonic stem cell-specific microRNAs regulate the G1-S transition and promote rapid proliferation. Nat Genet 40(12):1478–1483
44. Zovoilis A, Smorag L, Pantazi A, Engel W (2009) Members of the miR-290 cluster modulate in vitro differentiation of mouse embryonic stem cells. Differentiation 78(2–3):69–78
45. Zheng GX, Ravi A, Calabrese JM, Medeiros LA, Kirak O, Dennis LM, Jaenisch R, Burge CB, Sharp PA (2011) A latent pro-survival function for the mir-290-295 cluster in mouse embryonic stem cells. PLoS Genet 7(5):e1002054
46. Judson RL, Babiarz JE, Venere M, Blelloch R (2009) Embryonic stem cell-specific microRNAs promote induced pluripotency. Nat Biotechnol 27(5):459–461
47. Gou D, Ramchandran R, Peng X, Yao L, Kang K, Sarkar J, Wang Z, Zhou G, Raj JU (2012) miR-210 has an antiapoptotic effect in pulmonary artery smooth muscle cells during hypoxia. Am J Physiol Lung Cell Mol Physiol 303(8):L682–L691
48. Xiao J, Pan Y, Li XH, Yang XY, Feng YL, Tan HH, Jiang L, Feng J, XY Y (2016) Cardiac progenitor cell-derived exosomes prevent cardiomyocytes apoptosis through exosomal miR-21 by targeting PDCD4. Cell Death Dis 7(6):e2277
49. Talaei-Khozani T, Kharazinejad E, Rohani L, Vojdani Z, Mostafavi Pour Z, Tabei SZ (2012) Expression of pluripotency markers in human granulosa cells after embryonic stem cell extract exposure and epigenetic modification. Iran J Reprod Med 10(3):193–200
50. Katsman D, Stackpole EJ, Domin DR, Farber DB (2012) Embryonic stem cell-derived microvesicles induce gene expression changes in Muller cells of the retina. PLoS One 7(11):e50417
51. Kervadec A, Bellamy V, El Harane N, Arakelian L, Vanneaux V, Cacciapuoti I, Nemetalla H, Perier MC, Toeg HD, Richart A, Lemitre M, Yin M, Loyer X, Larghero J, Hagege A, Ruel M, Boulanger CM, Silvestre JS, Menasche P, Renault NK (2016) Cardiovascular progenitor-

derived extracellular vesicles recapitulate the beneficial effects of their parent cells in the treatment of chronic heart failure. J Heart Lung Transplant 35(6):795–807
52. Carmeliet P (2005) Angiogenesis in life, disease and medicine. Nature 438(7070):932–936
53. Qi X, Zhang J, Yuan H, Xu Z, Li Q, Niu X, Hu B, Wang Y, Li X (2016) Exosomes secreted by human-induced pluripotent stem cell-derived Mesenchymal stem cells repair critical-sized bone defects through enhanced angiogenesis and Osteogenesis in osteoporotic rats. Int J Biol Sci 12(7):836–849
54. Hu GW, Li Q, Niu X, Hu B, Liu J, Zhou SM, Guo SC, Lang HL, Zhang CQ, Wang Y, Deng ZF (2015) Exosomes secreted by human-induced pluripotent stem cell-derived mesenchymal stem cells attenuate limb ischemia by promoting angiogenesis in mice. Stem Cell Res Ther 6:10
55. Liu LZ, Li C, Chen Q, Jing Y, Carpenter R, Jiang Y, Kung HF, Lai L, Jiang BH (2011) MiR-21 induced angiogenesis through AKT and ERK activation and HIF-1alpha expression. PLoS One 6(4):e19139
56. Sabatel C, Malvaux L, Bovy N, Deroanne C, Lambert V, Gonzalez ML, Colige A, Rakic JM, Noel A, Martial JA, Struman I (2011) MicroRNA-21 exhibits antiangiogenic function by targeting RhoB expression in endothelial cells. PLoS One 6(2):e16979
57. Hu S, Huang M, Li Z, Jia F, Ghosh Z, Lijkwan MA, Fasanaro P, Sun N, Wang X, Martelli F, Robbins RC, JC W (2010) MicroRNA-210 as a novel therapy for treatment of ischemic heart disease. Circulation 122(11 Suppl):S124–S131
58. Johnson C, Sung HJ, Lessner SM, Fini ME, Galis ZS (2004) Matrix metalloproteinase-9 is required for adequate angiogenic revascularization of ischemic tissues: potential role in capillary branching. Circ Res 94(2):262–268
59. Vrijsen KR, Maring JA, Chamuleau SA, Verhage V, Mol EA, Deddens JC, Metz CH, Lodder K, van Eeuwijk EC, van Dommelen SM, Doevendans PA, Smits AM, Goumans MJ, Sluijter JP (2016) Exosomes from Cardiomyocyte progenitor cells and Mesenchymal stem cells stimulate angiogenesis via EMMPRIN. Adv Healthc Mater 5(19):2555–2565
60. Khayati F, Perez-Cano L, Maouche K, Sadoux A, Boutalbi Z, Podgorniak MP, Maskos U, Setterblad N, Janin A, Calvo F, Lebbe C, Menashi S, Fernandez-Recio J, Mourah S (2015) EMMPRIN/CD147 is a novel coreceptor of VEGFR-2 mediating its activation by VEGF. Oncotarget 6(12):9766–9780
61. Hatanaka M, Higashi Y, Fukushige T, Baba N, Kawai K, Hashiguchi T, Su J, Zeng W, Chen X, Kanekura T (2014) Cleaved CD147 shed from the surface of malignant melanoma cells activates MMP2 produced by fibroblasts. Anticancer Res 34(12):7091–7096
62. Ma J, Zhao Y, Sun L, Sun X, Zhao X, Qian H, Xu W, Zhu W (2017) Exosomes derived from Akt-modified human umbilical cord Mesenchymal stem cells improve cardiac regeneration and promote angiogenesis via activating platelet-derived growth factor D. Stem Cells Transl Med 6(1):51–59
63. Priori SG (2001) Evidence that human cardiac myocytes divide after myocardial infarct. Ital Heart J Suppl 2(11):1248–1249
64. Bergmann O, Bhardwaj RD, Bernard S, Zdunek S, Barnabe-Heider F, Walsh S, Zupicich J, Alkass K, Buchholz BA, Druid H, Jovinge S, Frisen J (2009) Evidence for cardiomyocyte renewal in humans. Science 324(5923):98–102
65. Teng X, Chen L, Chen W, Yang J, Yang Z, Shen Z (2015) Mesenchymal stem cell-derived exosomes improve the microenvironment of infarcted myocardium contributing to angiogenesis and anti-inflammation. Cell Physiol Biochem 37(6):2415–2424
66. Ti D, Hao H, Fu X, Han W (2016) Mesenchymal stem cells-derived exosomal microRNAs contribute to wound inflammation. Sci China Life Sci 59(12):1305–1312
67. Yang Y, Cheng HW, Qiu Y, Dupee D, Noonan M, Lin YD, Fisch S, Unno K, Sereti KI, Liao R (2015) MicroRNA-34a plays a key role in cardiac repair and regeneration following myocardial infarction. Circ Res 117(5):450–459
68. Wahlgren J, Statello L, Skogberg G, Telemo E, Valadi H (2016) Delivery of small interfering RNAs to cells via exosomes. Methods Mol Biol 1364:105–125

69. Barnett RE, Conklin DJ, Ryan L, Keskey RC, Ramjee V, Sepulveda EA, Srivastava S, Bhatnagar A, Cheadle WG (2016) Anti-inflammatory effects of miR-21 in the macrophage response to peritonitis. J Leukoc Biol 99(2):361–371
70. Peacock O, Lee AC, Cameron F, Tarbox R, Vafadar-Isfahani N, Tufarelli C, Lund JN (2014) Inflammation and MiR-21 pathways functionally interact to downregulate PDCD4 in colorectal cancer. PLoS One 9(10):e110267
71. Zhou R, Hu G, Gong AY, Chen XM (2010) Binding of NF-kappaB p65 subunit to the promoter elements is involved in LPS-induced transactivation of miRNA genes in human biliary epithelial cells. Nucleic Acids Res 38(10):3222–3232
72. Qi J, Qiao Y, Wang P, Li S, Zhao W, Gao C (2012) microRNA-210 negatively regulates LPS-induced production of proinflammatory cytokines by targeting NF-kappaB1 in murine macrophages. FEBS Lett 586(8):1201–1207
73. Sahoo S, Losordo DW (2014) Exosomes and cardiac repair after myocardial infarction. Circ Res 114(2):333–344
74. Lee C, Mitsialis SA, Aslam M, Vitali SH, Vergadi E, Konstantinou G, Sdrimas K, Fernandez-Gonzalez A, Kourembanas S (2012) Exosomes mediate the cytoprotective action of mesenchymal stromal cells on hypoxia-induced pulmonary hypertension. Circulation 126(22):2601–2611
75. Kalani A, Chaturvedi P, Kamat PK, Maldonado C, Bauer P, Joshua IG, Tyagi SC, Tyagi N (2016) Curcumin-loaded embryonic stem cell exosomes restored neurovascular unit following ischemia-reperfusion injury. Int J Biochem Cell Biol 79:360–369
76. Ghayad SE, Rammal G, Ghamloush F, Basma H, Nasr R, Diab-Assaf M, Chelala C, Saab R (2016) Exosomes derived from embryonal and alveolar rhabdomyosarcoma carry differential miRNA cargo and promote invasion of recipient fibroblasts. Sci Rep 6:37088
77. Yu X, Deng L, Wang D, Li N, Chen X, Cheng X, Yuan J, Gao X, Liao M, Wang M, Liao Y (2012) Mechanism of TNF-alpha autocrine effects in hypoxic cardiomyocytes: initiated by hypoxia inducible factor 1alpha, presented by exosomes. J Mol Cell Cardiol 53(6):848–857
78. Zhou Y, Zhou G, Tian C, Jiang W, Jin L, Zhang C, Chen X (2016) Exosome-mediated small RNA delivery for gene therapy. Wiley Interdiscip Rev RNA 7(6):758–771
79. Sioud M (2015) RNA interference: mechanisms, technical challenges, and therapeutic opportunities. Methods Mol Biol 1218:1–15
80. Didiot MC, Hall LM, Coles AH, Haraszti RA, Godinho BM, Chase K, Sapp E, Ly S, Alterman JF, Hassler MR, Echeverria D, Raj L, Morrissey DV, DiFiglia M, Aronin N, Khvorova A (2016) Exosome-mediated delivery of hydrophobically modified siRNA for huntingtin mRNA silencing. Mol Ther 24(10):1836–1847
81. Alvarez-Erviti L, Seow Y, Yin H, Betts C, Lakhal S, Wood MJ (2011) Delivery of siRNA to the mouse brain by systemic injection of targeted exosomes. Nat Biotechnol 29(4):341–345
82. Wahlgren J, De LKT, Brisslert M, Vaziri Sani F, Telemo E, Sunnerhagen P, Valadi H (2012) Plasma exosomes can deliver exogenous short interfering RNA to monocytes and lymphocytes. Nucleic Acids Res 40(17):e130
83. Greco KA, Franzen CA, Foreman KE, Flanigan RC, Kuo PC, Gupta GN (2016) PLK-1 silencing in bladder cancer by siRNA delivered with exosomes. Urology 91(241):e241–e247
84. Zhang Y, Li L, Yu J, Zhu D, Li X, Gu H, Zhang CY, Zen K (2014) Microvesicle-mediated delivery of transforming growth factor beta1 siRNA for the suppression of tumor growth in mice. Biomaterials 35(14):4390–4400
85. Sun D, Zhuang X, Xiang X, Liu Y, Zhang S, Liu C, Barnes S, Grizzle W, Miller D, Zhang HG (2010) A novel nanoparticle drug delivery system: the anti-inflammatory activity of curcumin is enhanced when encapsulated in exosomes. Mol Ther 18(9):1606–1614
86. Pascucci L, Cocce V, Bonomi A, Ami D, Ceccarelli P, Ciusani E, Vigano L, Locatelli A, Sisto F, Doglia SM, Parati E, Bernardo ME, Muraca M, Alessandri G, Bondiolotti G, Pessina A (2014) Paclitaxel is incorporated by mesenchymal stromal cells and released in exosomes that inhibit in vitro tumor growth: a new approach for drug delivery. J Control Release 192:262–270

87. Tian Y, Li S, Song J, Ji T, Zhu M, Anderson GJ, Wei J, Nie G (2014) A doxorubicin delivery platform using engineered natural membrane vesicle exosomes for targeted tumor therapy. Biomaterials 35(7):2383–2390
88. Pironti G, Strachan RT, Abraham D, Mon-Wei Yu S, Chen M, Chen W, Hanada K, Mao L, Watson LJ, Rockman HA (2015) Circulating exosomes induced by cardiac pressure overload contain functional angiotensin II type 1 receptors. Circulation 131(24):2120–2130
89. Trivedi M, Talekar M, Shah P, Ouyang Q, Amiji M (2016) Modification of tumor cell exosome content by transfection with wt-p53 and microRNA-125b expressing plasmid DNA and its effect on macrophage polarization. Oncogene 5(8):e250
90. Huang M, Chen Z, Hu S, Jia F, Li Z, Hoyt G, Robbins RC, Kay MA, JC W (2009) Novel minicircle vector for gene therapy in murine myocardial infarction. Circulation 120(11 Suppl):S230–S237
91. Huang M, Nguyen P, Jia F, Hu S, Gong Y, de Almeida PE, Wang L, Nag D, Kay MA, Giaccia AJ, Robbins RC, JC W (2011) Double knockdown of prolyl hydroxylase and factor-inhibiting hypoxia-inducible factor with nonviral minicircle gene therapy enhances stem cell mobilization and angiogenesis after myocardial infarction. Circulation 124(11 Suppl):S46–S54
92. Yu B, Kim HW, Gong M, Wang J, Millard RW, Wang Y, Ashraf M, Xu M (2015) Exosomes secreted from GATA-4 overexpressing mesenchymal stem cells serve as a reservoir of anti-apoptotic microRNAs for cardioprotection. Int J Cardiol 182:349–360
93. Ong SG, Lee WH, Huang M, Dey D, Kodo K, Sanchez-Freire V, Gold JD, Wu JC (2014) Cross talk of combined gene and cell therapy in ischemic heart disease: role of exosomal microRNA transfer. Circulation 130(11 Suppl 1):S60–S69
94. Jeong D, Jo W, Yoon J, Kim J, Gianchandani S, Gho YS, Park J (2014) Nanovesicles engineered from ES cells for enhanced cell proliferation. Biomaterials 35(34):9302–9310
95. Yoon J, Jo W, Jeong D, Kim J, Jeong H, Park J (2015) Generation of nanovesicles with sliced cellular membrane fragments for exogenous material delivery. Biomaterials 59:12–20
96. Jo W, Kim J, Yoon J, Jeong D, Cho S, Jeong H, Yoon YJ, Kim SC, Gho YS, Park J (2014) Large-scale generation of cell-derived nanovesicles. Nanoscale 6(20):12056–12064
97. Feng YL, Huang W, Wani M, XY Y, Ashraf M (2014) Ischemic preconditioning potentiates the protective effect of stem cells through secretion of exosomes by targeting Mecp2 via miR-22. PLoS One 9(2):e88685
98. Yamaguchi T, Izumi Y, Nakamura Y, Yamazaki T, Shiota M, Sano S, Tanaka M, Osada-Oka M, Shimada K, Miura K, Yoshiyama M, Iwao H (2015) Repeated remote ischemic conditioning attenuates left ventricular remodeling via exosome-mediated intercellular communication on chronic heart failure after myocardial infarction. Int J Cardiol 178:239–246
99. Giricz Z, Varga ZV, Baranyai T, Sipos P, Paloczi K, Kittel A, Buzas EI, Ferdinandy P (2014) Cardioprotection by remote ischemic preconditioning of the rat heart is mediated by extracellular vesicles. J Mol Cell Cardiol 68:75–78
100. Gray WD, French KM, Ghosh-Choudhary S, Maxwell JT, Brown ME, Platt MO, Searles CD, Davis ME (2015) Identification of therapeutic covariant microRNA clusters in hypoxia-treated cardiac progenitor cell exosomes using systems biology. Circ Res 116(2):255–263
101. Kim SW, Kim HW, Huang W, Okada M, Welge JA, Wang Y, Ashraf M (2013) Cardiac stem cells with electrical stimulation improve ischaemic heart function through regulation of connective tissue growth factor and miR-378. Cardiovasc Res 100(2):241–251
102. Glebov K, Lochner M, Jabs R, Lau T, Merkel O, Schloss P, Steinhauser C, Walter J (2015) Serotonin stimulates secretion of exosomes from microglia cells. Glia 63(4):626–634
103. Zhang S, Chu WC, Lai RC, Lim SK, Hui JH, Toh WS (2016) Exosomes derived from human embryonic mesenchymal stem cells promote osteochondral regeneration. Osteoarthr Cartil 24(12):2135–2140
104. Zhu W, Huang L, Li Y, Zhang X, Gu J, Yan Y, Xu X, Wang M, Qian H, Xu W (2012) Exosomes derived from human bone marrow mesenchymal stem cells promote tumor growth in vivo. Cancer Lett 315(1):28–37
105. Jiang X, Sucharov J, Stauffer BL, Miyamoto SD, Sucharov CC (2017) Exosomes from pediatric dilated cardiomyopathy patients modulate a pathological response in cardiomyocytes. Am J Physiol Heart Circ Physiol 312(4):H818–H826. doi:10.1152/ajpheart.00673.2016

Chapter 14
Cardiac Progenitor-Cell Derived Exosomes as Cell-Free Therapeutic for Cardiac Repair

E.A. Mol, M.J. Goumans, and J.P.G. Sluijter

14.1 Introduction

Myocardial infarction (MI) is one of the leading causes of death in the western world [1]. MI is induced by occlusion of one or more coronary arteries that supply oxygen to the heart, resulting in necrosis and apoptosis of cardiomyocytes that are highly dependent of oxygen. As a result, different molecular and cellular mechanisms are activated roughly in two phases. First, as necrotic cardiomyocytes release danger signals into the myocardium, the immune system is activated via toll-like receptors and complement activation [2]. This inflammatory response causes the attraction of neutrophils and monocytes to the infarcted area and is necessary to remove cellular debris. An overactive immune system can promote further tissue damage and infarct expansion [3]. The second step is a reparative phase characterized by activated fibroblasts (myofibroblasts) that produce excessive amounts of extracellular matrix, resulting in the formation of scar tissue [4]. Initially this scar tissue replaces the lost cardiomyocytes and provides strength to the heart to maintain its integrity, however, later progressive matrix deposition by activated

E.A. Mol
Department of Cardiology, Laboratory of Experimental Cardiology, University Medical Center Utrecht, Utrecht 3584CX, The Netherlands

M.J. Goumans
Department of Molecular Cell Biology, Leiden University Medical Center, Leiden 2333ZA, The Netherlands

J.P.G. Sluijter (✉)
Department of Cardiology, Laboratory of Experimental Cardiology, University Medical Center Utrecht, Utrecht 3584CX, The Netherlands

UMC Utrecht Regenerative Medicine Center, University Medical Center, Utrecht 3584CT, The Netherlands

Netherlands Heart Institute, Utrecht 3584CX, The Netherlands
e-mail: J.Sluijter@umcutrecht.nl

J. Xiao, S. Cretoiu (eds.), *Exosomes in Cardiovascular Diseases*, Advances in Experimental Medicine and Biology 998, DOI 10.1007/978-981-10-4397-0_14

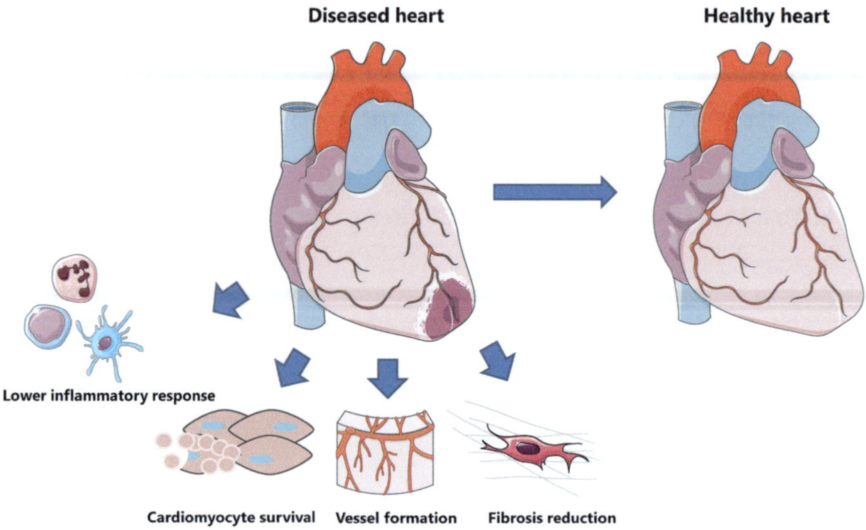

Fig. 14.1 Processes that need additional adaptations to further induce cardiac repair after myocardial infarction. Adjusted from Servier Medical Art at www.Servier.com, licensed under a Creative Commons Attribution 3.0 Unported License

myofibroblasts might lead to myocardial stiffening and impaired contraction. Since the initial myocardial damage is caused by a perfusion defect, stimulating neovessel formation or promoting arteriogenesis could contribute to cardiac regeneration [5, 6]. Cardiac repair mechanisms may be improved by interfering in these reparative mechanisms that play a role after MI by down-tuning the dentrimental processes, such as cardiomyocyte apoptosis, the inflammatory response, and fibrosis, and promoting further reparative signals like angiogenesis (Fig. 14.1).

Of all patients that suffer from MI, approximately 25% will develop heart failure within 1 year [7]. Currently, the only long-term treatment option for heart failure patients is heart transplantation, but donor availability is limited. Although patients waiting for heart transplantation can benefit from a left ventricular assist device (LVAD) taking over the pump function of the heart, this is usually a temporary solution [8]. Therefore, new treatment options are explored to replace the lost cardiomyocytes and improve contractility in cardiac diseases, especially for heart failure patients.

14.2 Cardiac Progenitor Cells as Potent Potential Cell Type for Myocardial Repair

One of the first authors describing the existence of cardiomyocyte regeneration was Oberpriller et al. [9]. Amputation of the ventricular apex of the newt heart resulted in the renewal of cardiomyocytes by re-entry into the cell cycle and proper engraftment in the myocardium. Also resection of the ventricular apex in zebrafish resulted in complete apical regeneration, mainly due to proliferation of progenitor cells in the heart and possibly also by dedifferentiation of residing cardiomyocytes [10, 11].

For decades it was believed that the mammalian heart had no regenerative capacity. Recent studies provided evidence for a limited but true regenerative potential of the heart [12–14]. Bergmann et al. demonstrated the ability of the heart to regenerate by quantifying carbon-14 incorporation into the DNA of human cardiomyocytes [13]. Approximately 1% of the cardiomyocytes is renewed at an age of 25; this capacity is fast reduced upon aging and in sharp contrast to cardiac resident non-cardiomyocytes with a renewel rate of approximately 15% [15]. Recently, a human case study of a newborn reported functional recovery of the human heart suffering from MI at this early age [16]. As a result of these observations, several new strategies have been explored to stimulate the regenerative capacity of the mammalian heart.

One of the strategies is the use of progenitor cell treatment as potential therapy to improve cardiac repair and prevent further damage in cardiac diseases. Several cell sources have been studied over the years and used to stimulate myocardial repair; these so-called first generation patient-derived cells include bone-marrow mononuclear cells (BM-MNCs) [17–19] and mesenchymal progenitor cells (MSCs) [20, 21]. The use of BM-MNCs and MSCs for cardiac repair are explored extensively due to their quick and relative easy clinical application. Furthermore, large numbers of cells could be achieved by culturing MSCs under good manufacturing practice conditions for clinical use [21, 22]. Meta-analysis of pre-clinical and clinical studies showed that injection of MSCs, in contrast to BM-MNCs, resulted in beneficial effects on cardiac function [23, 24]. MSC therapy was, however, limited to its potential to activate endogenous repair systems in the heart [25]. More recently, second-generation cells, including cardiac-derived progenitor cells (CPCs) [26–29] and induced pluripotent stem cell (iPSC)-derived cardiomyocytes [30, 31], have gained interest as a cell source for myocardial repair, mainly because of their promising regeneration capacity and their intrinsic ability to form contractile cells. iPSC-derived cardiomyocytes are cardiomyocytes generated by reprogramming fibroblasts to pluripotent stem cells using several transcription factors [30, 31]. Despite their true potential to form cardiomyocytes, the main effect of second-generation cells, observed upon cardiac transplantation, has been of paracrine origin. Excellent recent reviews describing the most relevant results and current limitations of cell-based therapies have been recently reported [32, 33].

The existence of progenitor cells in the heart was first described by Beltrami et al. [34], but since then several cardiac progenitor cell populations have been identified [26–29, 34]. CPCs are potentially the most promising adult cells for cardiac therapy as they can generate all cardiovascular lineages *in vitro* and *in vivo* [27, 35, 36]. Since they originate from the heart itself, CPCs may be destined to activate endogenous repair mechanisms. Therefore, CPCs hold greater cardiac regeneration potential compared to BM-MNCs or MSCs.

In different animal models for myocardial infarction, injection of CPCs increased cardiac performance [36–39]. However, although cardiac function was improved, cell engraftment of the injected cells in the myocardium was low, as indicated before for BM-MNC and MSCs. To stimulate cell survival upon myocardial injection, pre-treatment of CPCs with e.g. pim-1 or necrostatin-1 before CPC injection has been investigated [40, 41]. To further improve cell retention and prevent immediate flush-out [42], different approaches have been investigated, e.g. the use of cell clusters or a combination of cells with microcarriers [33, 43]. These approaches resulted in increased cell retention and survival, however, the additional beneficial effects on cardiac function was minimal.

Injection of autologous CPCs requires cell expansion *in vitro*, therefore, injection in the chronic phase is more clinically relevant. Therefore, while most studies investigate the effect of CPCs in the acute setting after MI (within a few hours), Tang et al. studied the effect of CPC treatment on an old infarct [37]. Intracoronary infusion of autologous CPCs in rats one month after MI resulted in less fibrotic tissue and improved cardiac function. The fact that CPCs still seem to have regenerative effects after a longer time period is promising for patients with chronic cardiac diseases.

14.3 Comparison of CPC Types

Although the heart has poor regenerative potential, many cardiac progenitor cell types have been identified based on marker expression/morphology, including Sca1+, c-kit+, cardiosphere-derived cells (CDCs) and cardiospheres (CSPs), and all these types can be isolated from the heart successfully [26–29, 34]. As the existence of so many CPC populations is counterintuitive, Gaetani et al. have compared the different CPC types [44]. Using their individual isolation methods, several of these progenitor cell types have been cultured and the gene expression profiles were compared to define differences between culture propagated CPCs. The gene expression profile of CSPs was most distinct from the Sca1+, c-kit+, and CDCs, most likely due to the monolayer and 3D culture conditions. Additionally, the difference between individual patients was larger than differences between different cell types from a single individual and expression partners are highly overlapping. Interestingly, when these cells were freshly isolated directly from the rodent heart some differences could be observed, indicating that c-kit positive cells were the most primitive progenitor cell [45]. However, this difference is abolished upon culture propagation. Furthermore, Zwetsloot et al. recently compared effect sizes of different types of CPCs [46], and observed that small differences in effect size can be found based on cell type; CSP treatment resulted in the largest increase in ejection fraction after injection in different animal models compared to e.g. Sca1+ and c-kit + CPCs, that showed a lower increase in cardiac function. Therefore, the mode of action of different CPC types on the myocardium is largely similar, although slight variations in effect size and transcriptome are described. Interestingly, a strong drop in functional benefit was observed upon their use in rodent and preclinical large animal models.

To date, two clinical trials have used CPCs as cell type for cardiac repair after MI. The SCIPIO (c-kit + CPC)s and CADUCEUS trial (CDCs) showed that intracoronary infusion of CPCs is safe in patients and led to enhanced cardiac function. Therefore, CPCs are a promising cell type for stem cell therapeutics [47, 48].

14.4 Paracrine Secretion

Originally, the concept for myocardial repair by progenitor cells was that they would engraft in the infarcted area and differentiate into functional cardiomyocytes upon injection. Recently, it has become more and more clear from both animal studies and clinical trials that injected progenitor cells do not engraft properly in the cardiac tissue, despite beneficial effects on cardiac function [36, 43, 47, 48]. Moreover, cardiomyocyte, endothelial, and blood vessel numbers were increased, which led to the hypothesis that the injected progenitor cells exert their effect via release of factors into their environment, called paracrine factors [36, 41, 49]. To study the effect of paracrine secretions, Timmers et al. injected MSC conditioned medium intravenously at the moment of reperfusion in pigs after MI and showed that MSC secretions could mimic the increased cardiac function [50]. This paracrine effect was observed for bone-marrow derived-, and mesenchymal progenitors, but also CPC secretions have these effects. CPC conditioned medium lowered cardiomyocyte apoptosis, stimulated endothelial cell migration, and increased tube formation of endothelial cells *in vitro* [39, 43, 51, 52].

In addition to paracrine molecules, the release of extracellular membrane vesicles such as exosomes are of increasing interest. Besides their use as biomarkers to detect early diseases [53], these nano-sized vesicles have also shown to be important mediators in repair after cardiac injury. Upon receiving stress signals, cells can influence their communication to other cells by adjusting membrane markers and vesicle content. Interestingly, Lai et al. identified the active cardioprotective component in the conditioned medium of MSCs to be exosomes [54]. They showed that upon separation of MSC conditioned medium in fractions of different sizes, the beneficial effects on ischemia/reperfusion injury observed after injection with fractionated MSC conditioned medium could only be reproduced by injecting the fraction containing complexes larger than 1000 kDa. Since progenitor-derived exosomes were found to be the paracrine factors mainly responsible for the observed beneficial effects after progenitor cell injection [54–57], the idea that CPC exosomes could be used for this purposes have emerged as potential off-the-shelf therapeutics.

CPC exosomes carry a variety of different proteins, growth factors, mRNAs, and microRNAs (miRNAs). MiRNAs are small non-coding RNAs that can inhibit or degrade mRNA, thereby preventing protein translation. Studies that investigate the effect of CPC exosomes on cardiac repair *in vitro* and *in vivo* are described below.

14.5 Functional Benefits of CPC Exosome Treatment

To study the functional benefits of CPC exosomes, CPC exosomes were intramyocardially injected in mice undergoing ischemia-reperfusion of the left coronary artery [38]. Injection of CPC exosomes reduced cardiomyocyte apoptosis by 53%. In addition, Barile et al. showed that intramyocardial injection of CPC exosomes in mice improved cardiac function after MI [39]. Morphological analysis after CPC exosome treatment in the myocardium revealed reduced scar tissue, lowered cardiomyocyte apoptosis, and increased blood vessel density.

To investigate if the release of exosomes from CPCs is critical for cardiac repair *in vivo*, Ibrahim et al. treated CPCs with GW4869, a reversible inhibitor of neutral sphingomyelinase that blocks, among others, exosome production [52]. The CPC-mediated benefits in mice after MI were completely abolished after treatment with GW4869, indicating that exosome release from CPCs is necessary to accomplish the beneficial effects on cardiac function. Altogether, these *in vivo* studies suggest that CPC exosomes induce cardiac repair, by interfering in processes such as cardiomyocyte apoptosis, fibrosis, and vessel formation. The following *in vitro* studies aim to identify the key cardioprotective processes stimulated by CPC exosomes.

14.6 Key Mechanisms Targeted by CPC Exosomes

Targeting the different processes that either prevent or reduce cardiac injury or contribute to cardiac regeneration after MI might lead to new treatment options. As described before, MI induces a cascade of molecular and cellular mechanisms in mainly two phases. The first phase is characterized by cardiomyocyte apoptosis and subsequent activation of the immune system. Cardiomyocyte apoptosis is a large contributor to impaired cardiac function after MI, as the major loss of contracting cells is responsible for the reduced contraction capacity of the heart. Preventing cardiomyocyte apoptosis could therefore be one of the mechanisms to improve cardiac injury. Interestingly, CPC exosomes have shown to have anti-apoptotic effects. Chen et al., for example, showed that CPC exosomes prevent apoptosis of H_2O_2-treated cardiomyocytes *in vitro* [38]. Caspase 3/7 activity in cardiomyocytes was lowered after treatment with CPC exosomes, which is an important mediator of H_2O_2-induced apoptosis. To further identify how CPC exosomes affect oxidative-stress related apoptosis of cardiomyocytes, Xiao et al. focused on exosomal-derived miRNAs [58]. They found that miRNA-21 is upregulated in CPC exosomes exposed to oxidative stress compared to non-exposed CPC exosomes. Interestingly, miRNA-21 targets programmed cell death 4 (PDCD4) in cardiomyocytes, thereby reducing oxidative-stress related apoptosis. Furthermore, miRNA analysis revealed that miRNA-210, miRNA-132, and miRNA-146a are highly enriched in CPC exosomes compared to fibroblast exosomes [39]. By inhibiting downstream targets such as RasGAP-p120, ephrin A3, and PTP1b, these miRNAs inhibit cardiomyocyte apoptosis and enhance endothelial migration after MI. Likewise, CDC and CSP-derived exosomes promote cardiac regeneration, as was shown after injection of these exosomes in the ischemic myocardium [52]. MiRNA analysis comparing CDC exosomes to fibroblast-derived exosomes revealed that miRNA-146a was the most highly enriched in CDC exosomes. Reduced cardiac function after MI was observed for miRNA-146a knockout mice compared to wild-type mice, indicating a role for miRNA-146a in cardiac repair. Pathway analysis revealed that miRNA-146a is involved in cell survival, cell cycle, and cellular organization, which are important processes involved in cardiac injury.

Upon MI, necrotic/apoptotic cardiomyocytes release danger signals into the environment, thereby activating the immune system via complement activation and toll-like receptors [2]. Although the immune response is required to clear tissue debris after MI, an overactive immune system might aggravate cardiac damage and infarct size [3]. Therefore, modulating the immune response might prevent/reduce cardiac injury. Progenitor exosomes might be able to modulate this balance in immune responses after MI by delivery of miRNAs, anti-inflammatory cytokines, or other molecules involved in inflammation. This anti-inflammatory response was described for MSC exosomes, as MSC exosomes were capable of switching the macrophage phenotype from the pro-inflammatory M1 to the anti-inflammatory M2 phenotype and suppress T-cell activation [59]. Until now, the immune-modulating properties of CPC exosomes have not been described in literature yet.

The second phase after MI involves myofibroblasts that are responsible for reorganizing the structure of the heart, a process called remodeling. Reducing the fibrotic tissue may be a promising way to improve cardiac repair, however, as fibrosis is initially a reparative response, a fine balance between pro- and anti-fibrotic factors is needed. Interestingly, the physiological state of CPCs can influence the secretion and cargo of CPC exosomes. Culturing CPC exosomes under hypoxic conditions resulted in higher tube formation and lowered pro-fibrotic gene expression compared to exosomes cultured under normoxic conditions [60]. Indeed, administration of hypoxic CPC exosomes in mice reduced fibrosis and increased cardiac function compared to normoxic CPC exosomes in an ischemia-reperfusion model. Microarray analysis revealed that 11 miRNAs with anti-fibrotic and pro-angiogenic properties were upregulated compared to normoxic exosomes. Whether the observed beneficial effects of hypoxic CPC exosomes on cardiac function are established through these miRNAs only or if other molecules are also involved needs to be investigated [61]. Although several *in vivo* studies indeed observed anti-fibrotic effects of CPC exosome treatment after MI [37, 60], to our knowledge there are no further studies addressing the possible anti-fibrotic mechanism of CPC exosomes so far.

Other cardioprotective mechanisms that could be important for cardiac regeneration are stimulating angiogenesis or arteriogenesis, since the initial myocardial injury is due to a perfusion defect [5, 6]. Progenitor exosomes derived from several cell sources have been described to have pro-angiogenic effects. Sahoo et al., for example, showed that exosomes from human CD34+ progenitor cells mediate their pro-angiogenic activity [55]. After adding exosomes, derived from CD34+ progenitor cells to endothelial cells *in vitro*, they observed increased viability, proliferation, and tube formation of endothelial cells. Furthermore, subcutaneous injection of a matrigel plug containing CD34+ exosomes in mice showed higher vessel formation compared to injection of a matrigel plug alone. They found that the presence of a pro-angiogenic protein in CD34+ exosomes, sonic hedgehoc, was largely responsible for the preserved cardiac function after MI [56].

This pro-angiogenic property of exosomes was also observed for CPC-derived exosomes. Vrijsen et al. reported that CPC exosomes stimulated migration of endothelial cells in a wound scratch assay [51]. Analyzing the presence of pro-angiogenic factors in CPC exosomes revealed high expression levels of extracellular matrix metalloproteinase inducer (EMMPRIN), which is present on the exosomal membrane. The migration of endothelial cells upon stimulation with CPC exosomes was not observed upon stimulation with exosomes depleted for EMMPRIN (KD EMMPRIN exosomes). Furthermore, KD EMMPRIN exosomes also inhibited angiogenesis *in vivo*, demonstrated by a reduced influx of cells into a matrigel plug compared to control exosomes after application in mice [62]. Therefore, EMMPRIN is an important mediator of the pro-angiogenic effect of CPC exosomes.

14.7 Future Perspectives

Altogether, these studies provide insights into the ability of CPC exosomes to enhance cardiac repair after injury and the involved mechanisms. The key mechanisms that are influenced by CPC exosomes described so far are neovessel formation and cardiomyocyte apoptosis (Fig. 14.2). Despite considerable efforts have been made to study the effect of CPC exosomes on cardiac repair, many challenges have to be overcome before deployment of exosomes in clinical trials. Firstly, most of the described studies investigated the effect of CPC exosomes on the acute setting after MI [38, 39]. Due to better revascularization therapy and medication planning, the survival of patients after acute MI is increased last decades. These surviving patients, however, have a higher chance to develop a more chronic disease like heart failure. From a clinical perspective it would therefore be useful to study regeneration by CPC exosomes in these more chronic phases after cardiac injury. Another important challenge is retention of exosomes after injection. Van den Akker et al. performed intramyocardial injection of stem cells and observed immediate flush-out of the cells upon injection [42]. It is thus likely that the same flush-out can be expected upon exosome injection into the myocardium, since the exosomes sizes are even smaller (30–100 nm) compared to cells (8–12 μm). Furthermore, accurate mapping of the *in vivo* biodistribution of exosomes after systemic injection is also an important objective before using exosomes in clinical trials. Lai et al. developed an excellent technique to allow multimodal imaging of exosomes *in vivo*. Membrane-bound Gaussia luciferase was combined with metabolic biotinylation to visualize exosomes after systemic injection in athymic nude mice via bioluminescent signals [63]. The highest uptake of exosomes was observed in the liver and spleen, therefore, systemic administration of exosomes might require targeted therapy towards

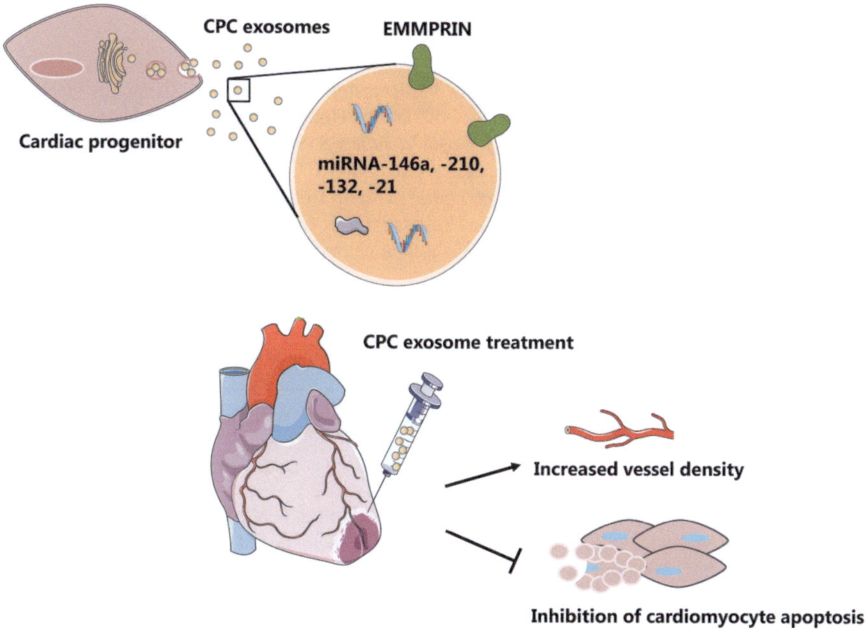

Fig. 14.2 Key mechanisms targeted by CPC exosomes. Adjusted from Servier Medical Art at www.Servier.com, licensed under a Creative Commons Attribution 3.0 Unported License

the injured heart. Aiming to target exosomes to the brain, Alvarez-Erviti et al., engineered cells to express an exosomal membrane protein (lysosome-associated membrane glycoprotein 2b) fused to a brain-specific peptide that targets the acetylcholine receptor [64]. They showed increased delivery of functional exosomes to the brain. Thus, although some achievements have been made to engineer exosomes in a way that they target tissues aimed for, by using specific ligands, non-specific accumulation of exosomes in other tissues remains an issue to be solved [63–65]. Lastly, to cover the high demand of exosomes needed for clinical application, a reproducible and standardized exosome isolation technique is required that allows for upscaling [66]. In addition, the characteristics of exosome-based therapeutics have to be defined properly, which requires more in-depth research into the mechanism of how exosomes exert their therapeutic effects. Nonetheless, CPC exosomes can be considered as potential off-the-shelf therapeutics, as they are able to stimulate the regenerative capacity of the heart mainly by increasing vessel density and lowering apoptosis of cardiomyocytes.

Acknowledgements EAM is funded by the Project SMARTCARE-II of the BioMedical Materials institute, co-funded by the ZonMw-TAS program (#116002016), the Netherlands Organization for Health Research and Development, the Dutch Ministry of Economic Affairs, Agriculture and Innovation and the Netherlands Cardio Vascular Research Initiative (CVON): the Dutch Heart Foundation, Royal Netherlands Academy of Sciences, and the Dutch Federations of University Medical Centers.JS received a Horizon 2020 ERC-2016-COG grant, called EVICARE (725229).

References

1. Thom T, Haase N, Rosamond W, Howard VJ, Rumsfeld J, Manolio T, Zheng ZJ, Flegal K, O'Donnell C, Kittner S, Lloyd-Jones D, Goff DC Jr, Hong Y, Adams R, Friday G, Furie K, Gorelick P, Kissela B, Marler J, Meigs J, Roger V, Sidney S, Sorlie P, Steinberger J, Wasserthiel-Smoller S, Wilson M, Wolf P, American Heart Association Statistics C, Stroke Statistics S (2006) Heart disease and stroke statistics—2006 update: a report from the American Heart Association Statistics Committee and Stroke Statistics Subcommittee. Circulation 113(6):e85–151
2. Frangogiannis NG (2012) Regulation of the inflammatory response in cardiac repair. Circ Res 110(1):159–173
3. Kain V, Prabhu SD, Halade GV (2014) Inflammation revisited: inflammation versus resolution of inflammation following myocardial infarction. Basic Res Cardiol 109(6):444
4. van Nieuwenhoven FA, Turner NA (2013) The role of cardiac fibroblasts in the transition from inflammation to fibrosis following myocardial infarction. Vasc Pharmacol 58(3):182–188
5. Tonnesen MG, Feng X, Clark RA (2000) Angiogenesis in wound healing. J Investig Dermatol Symp Proc 5(1):40–46
6. van der Laan AM, Piek JJ, van Royen N (2009) Targeting angiogenesis to restore the microcirculation after reperfused MI. Nat Rev Cardiol 6(8):515–523
7. Velagaleti RS, Pencina MJ, Murabito JM, Wang TJ, Parikh NI, D'Agostino RB, Levy D, Kannel WB, Vasan RS (2008) Long-term trends in the incidence of heart failure after myocardial infarction. Circulation 118(20):2057–2062
8. Birks EJ (2010) Left ventricular assist devices. Heart 96(1):63–71
9. Oberpriller JO, Oberpriller JC, Mauro A (1991) Cell division in adult newt cardiac myocytes. In: The development and regenerative potential of cardiac muscle. Harwood Academic Press, New York, London, Paris, pp 293–312
10. Poss KD, Wilson LG, Keating MT (2002) Heart regeneration in zebrafish. Science 298(5601):2188–2190
11. Lepilina A, Coon AN, Kikuchi K, Holdway JE, Roberts RW, Burns CG, Poss KD (2006) A dynamic epicardial injury response supports progenitor cell activity during zebrafish heart regeneration. Cell 127(3):607–619
12. Beltrami AP, Urbanek K, Kajstura J, Yan SM, Finato N, Bussani R, Nadal-Ginard B, Silvestri F, Leri A, Beltrami CA, Anversa P (2001) Evidence that human cardiac myocytes divide after myocardial infarction. N Engl J Med 344(23):1750–1757
13. Bergmann O, Bhardwaj RD, Bernard S, Zdunek S, Barnabe-Heider F, Walsh S, Zupicich J, Alkass K, Buchholz BA, Druid H, Jovinge S, Frisen J (2009) Evidence for cardiomyocyte renewal in humans. Science 324(5923):98–102
14. Bergmann O, Zdunek S, Alkass K, Druid H, Bernard S, Frisen J (2011) Identification of cardiomyocyte nuclei and assessment of ploidy for the analysis of cell turnover. Exp Cell Res 317(2):188–194
15. Bergmann O, Zdunek S, Felker A, Salehpour M, Alkass K, Bernard S, Sjostrom SL, Szewczykowska M, Jackowska T, Dos Remedios C, Malm T, Andra M, Jashari R, Nyengaard JR, Possnert G, Jovinge S, Druid H, Frisen J (2015) Dynamics of cell generation and turnover in the human heart. Cell 161(7):1566–1575
16. Haubner BJ, Schneider J, Schweigmann U, Schuetz T, Dichtl W, Velik-Salchner C, Stein JI, Penninger JM (2016) Functional recovery of a human neonatal heart after severe myocardial infarction. Circ Res 118(2):216–221
17. Tomita S, Li RK, Weisel RD, Mickle DA, Kim EJ, Sakai T, Jia ZQ (1999) Autologous transplantation of bone marrow cells improves damaged heart function. Circulation 100(19 Suppl):II247–II256
18. Orlic D, Kajstura J, Chimenti S, Jakoniuk I, Anderson SM, Li B, Pickel J, McKay R, Nadal-Ginard B, Bodine DM, Leri A, Anversa P (2001) Bone marrow cells regenerate infarcted myocardium. Nature 410(6829):701–705

19. Strauer BE, Brehm M, Zeus T, Kostering M, Hernandez A, Sorg RV, Kogler G, Wernet P (2002) Repair of infarcted myocardium by autologous intracoronary mononuclear bone marrow cell transplantation in humans. Circulation 106(15):1913–1918
20. Friedenstein AJ, Chailakhyan RK, Gerasimov UV (1987) Bone marrow osteogenic stem cells: in vitro cultivation and transplantation in diffusion chambers. Cell Tissue Kinet 20(3):263–272
21. Pittenger MF, Mackay AM, Beck SC, Jaiswal RK, Douglas R, Mosca JD, Moorman MA, Simonetti DW, Craig S, Marshak DR (1999) Multilineage potential of adult human mesenchymal stem cells. Science 284(5411):143–147
22. Sensebe L, Bourin P, Tarte K (2011) Good manufacturing practices production of mesenchymal stem/stromal cells. Hum Gene Ther 22(1):19–26
23. van der Spoel TI, Jansen of Lorkeers SJ, Agostoni P, van Belle E, Gyöngyösi M, Sluijter JP, Cramer MJ, Doevendans PA, Chamuleau SA (2011) Human relevance of pre-clinical studies in stem cell therapy: systematic review and meta-analysis of large animal models of ischaemic heart disease. Cardiovasc Res 91(4):649–658
24. de Jong R, Houtgraaf JH, Samiei S, Boersma E, Duckers HJ (2014) Intracoronary stem cell infusion after acute myocardial infarction: a meta-analysis and update on clinical trials. Circ Cardiovasc Interv 7(2):156–167
25. Noort WA, Feye DDAF, Stecher D, Chamuleau SA, Sluijter JP, Doevendans PA (2010) Mesenchymal stromal cells to treat cardiovascular disease: strategies to improve survival and therapeutic results. Panminerva Med 52(1):27–40
26. Oh H, Bradfute SB, Gallardo TD, Nakamura T, Gaussin V, Mishina Y, Pocius J, Michael LH, Behringer RR, Garry DJ, Entman ML, Schneider MD (2003) Cardiac progenitor cells from adult myocardium: homing, differentiation, and fusion after infarction. Proc Natl Acad Sci U S A 100(21):12313–12318
27. Goumans MJ, de Boer TP, Smits AM, van Laake LW, Van VP, Metz CH, Korfage TH, Kats KP, Hochstenbach R, Pasterkamp G (2008) TGF-beta1 induces efficient differentiation of human cardiomyocyte progenitor cells into functional cardiomyocytes in vitro. Stem Cell Res 1(2):138–149
28. Smith RR, Barile L, Cho HC, Leppo MK, Hare JM, Messina E, Giacomello A, Abraham MR, Marbán E (2007) Regenerative potential of cardiosphere-derived cells expanded from percutaneous endomyocardial biopsy specimens. Circulation 115(7):896–908
29. Messina E, De Angelis L, Frati G, Morrone S, Chimenti S, Fiordaliso F, Salio M, Battaglia M, Latronico MV, Coletta M (2004) Isolation and expansion of adult cardiac stem cells from human and murine heart. Circ Res 95:911–921
30. Zwi L, Caspi O, Arbel G, Huber I, Gepstein A, Park IH, Gepstein L (2009) Cardiomyocyte differentiation of human induced pluripotent stem cells. Circulation 120(15):1513–1523
31. Singla DK, Long X, Glass C, Singla RD, Yan B (2011) Induced pluripotent stem (iPS) cells repair and regenerate infarcted myocardium. Mol Pharm 8(5):1573–1581
32. Madonna R, Van Laake LW, Davidson SM, Engel FB, Hausenloy DJ, Lecour S, Leor J, Perrino C, Schulz R, Ytrehus K, Landmesser U, Mummery CL, Janssens S, Willerson J, Eschenhagen T, Ferdinandy P, Sluijter JP (2016) Position paper of the European Society of Cardiology Working Group Cellular Biology of the Heart: cell-based therapies for myocardial repair and regeneration in ischemic heart disease and heart failure. Eur Heart J 37(23):1789–1798
33. Feyen DA, Gaetani R, Deddens J, Van Keulen D, Van Opbergen C, Poldervaart M, Alblas J, Chamuleau S, van Laake LW, Doevendans PA (2016) Gelatin microspheres as vehicle for cardiac progenitor cells delivery to the myocardium. Adv Healthc Mater 5(9):1071–1079
34. Beltrami AP, Barlucchi L, Torella D, Baker M, Limana F, Chimenti S, Kasahara H, Rota M, Musso E, Urbanek K (2003) Adult cardiac stem cells are multipotent and support myocardial regeneration. Cell 114(6):763–776
35. van Vliet P, Roccio M, Smits AM, van Oorschot AA, Metz CH, van Veen TA, Sluijter JP, Doevendans PA, Goumans MJ (2008) Progenitor cells isolated from the human heart: a potential cell source for regenerative therapy. Neth Hear J 16(5):163–169
36. Smits AM, van Laake LW, den Ouden K, Schreurs C, Szuhai K, van Echteld CJ, Mummery CL, Doevendans PA, Goumans MJ (2009) Human cardiomyocyte progenitor cell transplan-

tation preserves long-term function of the infarcted mouse myocardium. Cardiovasc Res 83(3):527–535
37. Tang XL, Rokosh G, Sanganalmath SK, Yuan F, Sato H, Mu J, Dai S, Li C, Chen N, Peng Y, Dawn B, Hunt G, Leri A, Kajstura J, Tiwari S, Shirk G, Anversa P, Bolli R (2010) Intracoronary administration of cardiac progenitor cells alleviates left ventricular dysfunction in rats with a 30-day-old infarction. Circulation 121(2):293–305
38. Chen L, Wang Y, Pan Y, Zhang L, Shen C, Qin G, Ashraf M, Weintraub N, Ma G, Tang Y (2013) Cardiac progenitor-derived exosomes protect ischemic myocardium from acute ischemia/reperfusion injury. Biochem Biophys Res Commun 431(3):566–571
39. Barile L, Lionetti V, Cervio E, Matteucci M, Gherghiceanu M, Popescu LM, Torre T, Siclari F, Moccetti T, Vassalli G (2014) Extracellular vesicles from human cardiac progenitor cells inhibit cardiomyocyte apoptosis and improve cardiac function after myocardial infarction. Cardiovasc Res 103(4):530–541
40. Fischer KM, Cottage CT, Wu W, Din S, Gude NA, Avitable D, Quijada P, Collins BL, Fransioli J, Sussman MA (2011) Enhancement of myocardial regeneration through genetic engineering of cardiac progenitor cells expressing Pim-1 kinase. Circulation 120(120):2077–2087
41. Feyen D, Gaetani R, Liu J, Noort W, Martens A, Den OK, Doevendans PA, Sluijter JP (2013) Increasing short-term cardiomyocyte progenitor cell (CMPC) survival by necrostatin-1 did not further preserve cardiac function. Cardiovasc Res 99(1):83–91
42. van den Akker F, Feyen DA, van den Hoogen P, van Laake LW, van Eeuwijk EC, Hoefer I, Pasterkamp G, Chamuleau SA, Grundeman PF, Doevendans PA, Sluijter JP (2017) Intramyocardial stem cell injection: go(ne) with the flow. Eur Heart J 38(3):184–186
43. Chimenti I, Smith RR, Li TS, Gerstenblith G, Messina E, Giacomello A, Marban E (2010) Relative roles of direct regeneration versus paracrine effects of human cardiosphere-derived cells transplanted into infarcted mice. Circ Res 106(5):971–U304
44. Gaetani R, Feyen DAM, Doevendans PA, Gremmels H, Forte E, Fledderus JO, Ramjankhan FZ, Messina E, Sussman MA, Giacomello A, Sluijter JPG (2014) Different types of cultured human adult cardiac progenitor cells have a high degree of transcriptome similarity. J Cell Mol Med 18(11):2147–2151
45. Dey D, Han L, Bauer M, Sanada F, Oikonomopoulos A, Hosoda T, Unno K, Almeida PD, Leri A, Wu JC (2013) Dissecting the molecular relationship among various cardiogenic progenitor cells novelty and significance. Circ Res 112(9):1253
46. Zwetsloot PP, Vegh AM, Jansen of Lorkeers SJ, van Hout GP, Currie GL, Sena ES, Gremmels H, Buikema JW, Goumans MJ, Macleod MR, Doevendans PA, Chamuleau SA, Sluijter JP (2016) Cardiac stem cell treatment in myocardial infarction: a systematic review and meta-analysis of preclinical studies. Circ Res 118(8):1223–1232
47. Bolli R, Chugh AR, D'Amario D, Loughran JH, Stoddard MF, Ikram S, Beache GM, Wagner SG, Leri A, Hosoda T, Sanada F, Elmore JB, Goichberg P, Cappetta D, Solankhi NK, Fahsah I, Rokosh DG, Slaughter MS, Kajstura J, Anversa P (2011) Cardiac stem cells in patients with ischaemic cardiomyopathy (SCIPIO): initial results of a randomised phase 1 trial. Lancet 378(9806):1847–1857
48. Malliaras K, Makkar RR, Smith RR, Cheng K, Wu E, Bonow RO, Marban L, Mendizabal A, Cingolani E, Johnston PV, Gerstenblith G, Schuleri KH, Lardo AC, Marban E (2014) Intracoronary cardiosphere-derived cells after myocardial infarction: evidence of therapeutic regeneration in the final 1-year results of the CADUCEUS trial (CArdiosphere-Derived aUtologous stem CEIls to reverse ventricUlar dySfunction). J Am Coll Cardiol 63(2):110–122
49. den Haan MC, Grauss RW, Smits AM, Winter EM, van Tuyn J, Pijnappels DA, Steendijk P, Gittenberger-De Groot AC, van der Laarse A, Fibbe WE, de Vries AA, Schalij MJ, Doevendans PA, Goumans MJ, Atsma DE (2012) Cardiomyogenic differentiation-independent improvement of cardiac function by human cardiomyocyte progenitor cell injection in ischaemic mouse hearts. J Cell Mol Med 16(7):1508–1521
50. Timmers L, Lim SK, Hoefer IE, Arslan F, Lai RC, van Oorschot AA, Goumans MJ, Strijder C, Sze SK, Choo A, Piek JJ, Doevendans PA, Pasterkamp G, de Kleijn DP (2011) Human

mesenchymal stem cell-conditioned medium improves cardiac function following myocardial infarction. Stem Cell Res 6(3):206–214
51. Vrijsen KR, Sluijter JP, Schuchardt MW, van Balkom BW, Noort WA, Chamuleau SA, Doevendans PA (2010) Cardiomyocyte progenitor cell-derived exosomes stimulate migration of endothelial cells. J Cell Mol Med 14(5):1064–1070
52. Ibrahim AGE, Cheng K, Marban E (2014) Exosomes as critical agents of cardiac regeneration triggered by cell therapy. Stem Cell Reports 2(5):606–619
53. Deddens JC, Colijn JM, Oerlemans MIFJ, Pasterkamp G, Chamuleau SA, Doevendans PA, Sluijter JPG (2013) Circulating microRNAs as novel biomarkers for the early diagnosis of acute coronary syndrome. J Cardiovasc Transl Res 6(6):884–898
54. Lai RC, Arslan F, Lee MM, Sze NS, Choo A, Chen TS, Salto-Tellez M, Timmers L, Lee CN, El Oakley RM, Pasterkamp G, de Kleijn DP, Lim SK (2010) Exosome secreted by MSC reduces myocardial ischemia/reperfusion injury. Stem Cell Res 4(3):214–222
55. Sahoo S, Klychko E, Thorne T, Misener S, Schultz KM, Millay M, Ito A, Liu T, Kamide C, Agrawal H, Perlman H, Qin GJ, Kishore R, Losordo DW (2011) Exosomes from human CD34(+) stem cells mediate their proangiogenic paracrine activity. Circ Res 109(7):724–U735
56. Mackie AR, Klyachko E, Thorne T, Schultz KM, Millay M, Ito A, Kamide CE, Liu T, Gupta R, Sahoo S (2012) Sonic hedgehog modified human CD34+ cells preserve cardiac function following acute myocardial infarction. Circ Res 111(3):312–321
57. Arslan F, Lai RC, Smeets MB, Akeroyd L, Choo A, Aguor EN, Timmers L, van Rijen HV, Doevendans PA, Pasterkamp G, Lim SK, de Kleijn DP (2013) Mesenchymal stem cell-derived exosomes increase ATP levels, decrease oxidative stress and activate PI3K/Akt pathway to enhance myocardial viability and prevent adverse remodeling after myocardial ischemia/reperfusion injury. Stem Cell Res 10(3):301–312
58. Xiao J, Pan Y, Li XH, Yang XY, Feng YL, Tan HH, Jiang L, Feng J, Yu XY (2015) Cardiac progenitor cell-derived exosomes prevent cardiomyocytes apoptosis through exosomal miR-21 by targeting PDCD4. Cell Death Dis 7(6):e2277
59. François M, Romieu-Mourez R, Li M, Galipeau J (2011) Human MSC suppression correlates with cytokine induction of indoleamine 2,3-dioxygenase and bystander M2 macrophage differentiation. Mol Ther 20(20):187–195
60. Gray WD, French KM, Ghosh-Choudhary S, Maxwell JT, Brown ME, Platt MO, Searles CD, Davis ME (2015) Identification of therapeutic covariant microRNA clusters in hypoxia-treated cardiac progenitor cell exosomes using systems biology. Circ Res 116(2):255–U119
61. Sluijter JP, van Rooij E (2015) Exosomal microRNA clusters are important for the therapeutic effect of cardiac progenitor cells. Circ Res 116(2):219–221
62. Vrijsen KR, Maring JA, Chamuleau SA, Verhage V, Mol EA, Deddens JC, Metz CH, Lodder K, van Eeuwijk EC, van Dommelen SM, Doevendans PA, Smits AM, Goumans MJ, Sluijter JP (2016) Exosomes from cardiomyocyte progenitor cells and mesenchymal stem cells stimulate angiogenesis via EMMPRIN. Adv Healthc Mater 5(19):2555–2565
63. Lai CP, Mardini O, Ericsson M, Prabhakar S, Maguire CA, Chen JW, Tannous BA, Breakefield XO (2014) Dynamic biodistribution of extracellular vesicles in vivo using a multimodal imaging reporter. ACS Nano 8(1):483–494
64. Alvarez-Erviti L, Seow Y, Yin H, Betts C, Lakhal S, Wood MJ (2011) Delivery of siRNA to the mouse brain by systemic injection of targeted exosomes. Nat Biotechnol 29(4):341–345
65. Zhuang X, Xiang X, Grizzle W, Sun D, Zhang S, Axtell RC, Ju S, Mu J, Zhang L, Steinman L, Miller D, Zhang HG (2011) Treatment of brain inflammatory diseases by delivering exosome encapsulated anti-inflammatory drugs from the nasal region to the brain. Mol Ther 19(10):1769–1779
66. Mol EA, Goumans MJ, Doevendans PA, Sluijter JPG, Vader P (2017) Higher functionality of extracellular vesicles isloated using size-exclusion chromatography compared to ultracetrifugation. Nanomedicine. doi:10.1016/j.nano.2017.03.011

Chapter 15
Therapeutic Potential of Hematopoietic Stem Cell-Derived Exosomes in Cardiovascular Disease

Jana Radosinska and Monika Bartekova

15.1 Introduction

Blood and the system that forms it, known as the hematopoietic system, consist of many cell types with specialized functions. Many blood cells are short-living and need to be replenished continuously; the average human requires approximately one hundred billion new hematopoietic cells each day. The continued production of these cells depends directly on the presence of hematopoietic stem cells (HSCs) as the ultimate, and only, source of all blood cells. HSCs are present primarily in the bone marrow, but also in the peripheral blood and umbilical cord blood [1]. HSCs exert a number of unique properties, the combination of which defines them as such. Among the core properties, the ability to choose between self-renewal (remain a stem cell after cell division) or differentiation (start the path towards becoming a mature hematopoietic cell) is one of the most prominent. In addition, HSCs migrate in regulated fashion, and are subjected to regulation by apoptosis (programmed cell death). The balance between these processes determines the number of stem cells that are present in the body at the moment. As many other cell types, HSCs are able to produce extracellular vesicles (EVs) including exosomes and microvesicles and release them primarily into the bone marrow, but they may penetrate into the peripheral blood circulation consequently [2]. EVs released from the HSCs may play pleiotropic effects in the body: they contribute to the cell-to-cell communication, are proposed

J. Radosinska (✉)
Faculty of Medicine, Institute of Physiology, Comenius University in Bratislava, Sasinkova 2, 811 08 Bratislava, Slovak Republic

Institute for Heart Research, Slovak Academy of Sciences, Dúbravská cesta 9, 840 05 Bratislava, Slovak Republic
e-mail: jana.radosinska@fmed.uniba.sk

M. Bartekova
Institute for Heart Research, Slovak Academy of Sciences, Dúbravská cesta 9, 840 05 Bratislava, Slovak Republic

J. Xiao, S. Cretoiu (eds.), *Exosomes in Cardiovascular Diseases*, Advances in Experimental Medicine and Biology 998, DOI 10.1007/978-981-10-4397-0_15

to be biomarkers of the condition of the bone marrow, and are suggested also to be potentially used for the prevention and/or treatment of different diseases including cardiovascular as an alternative type of stem cells-based therapies.

This chapter will summarize the knowledge about the production of EVs by the HSCs in the organism, their role in the intercellular communication, and will discuss the cargo of these EVs as well as the protective effects of HSCs-derived exosomes and microvesicles in cardiovascular diseases (CVD) including cardiac ischemia-reperfusion injury and acute myocardial infarction.

15.2 Production of Extracellular Vesicles in HSCs

As known, EVs such as exosomes and microvesicles are produced in many types of cells and released into the extracellular environment. EVs can consequently release carried substances and thus function as paracrine mediators, and also may be potentially taken up by other cells and act in the remote parts of the body. In fact, most of cells including endothelial cells, immune cells, cancer cells, hematopoietic cells, platelets, and erythrocytes are able to produce EVs [3] which are consequently released into the corresponding extracellular compartment.

The direct evidence of HSCs-derived exosomes and microvesicles documented by Aoki et al. [2] suggested that hematopoietic precursor cells (HPCs)-derived mRNAs in plasma exosomes and microvesicles may represent new biomarkers for the assessment of bone marrow condition. Bone marrow-derived EVs has been reported previously also by other authors; the bone marrow cells in these reports were stromal cells [4], stem cells [5], dendritic cells [6], and mesenchymal stem cells [7]. Several types of hematopoietic cells have been shown to secrete exosomes in culture [8]. Among them, exosome secretion has first been reported for reticulocytes during their differentiation [9, 10]. Exosomes produced during reticulocyte maturation into erythrocytes are formed in the endosomal compartment and released in the extracellular medium. The evidence of their origin is referred by HBB gene expression, the gene coding the subunit of hemoglobin. Hemoglobin protein is abundantly expressed in red blood cells (RBC), but owing to the lack of a nucleus in RBCs, HBB gene mRNA cannot be produced in RBC. Thus the origin of HBB mRNA is the erythroblasts in bone marrow and reticulocytes [11]. This is in concordance with the fact that HSCs first grow in bone marrow to produce various HPCs, then after a series of maturation processes, white blood cells, red blood cells and platelets are released into peripheral blood circulation. More recently, other hematopoietic cells including B lymphocytes [12], dendritic cells (DCs) [13–15], T lymphocytes [16–18] and mast cells [19, 20] have been shown to secrete exosomes. EVs can be produced also by erythroid and myeloid precursor cells and megakaryocytes. It has been shown in an ex vivo co-culture system consisting of human primary hematopoietic stem and progenitor cells growing on multipotent mesenchymal stromal cells that EVs which are released into the peripheral blood are produced rather by myeloid and erythroid precursor cells and megakaryocytes than by mature white and red blood cells or platelets. These EVs influence the mesenchymal stem cells in the bone marrow and the cargo of these vesicles contains the stem cell marker prominin-1, also known as CD133 antigen [21]. In addition, the internalization of prominin-1-EVs by feeder cells has been demonstrated suggesting an additional function of these vesicles in intercellular communication either by stimulating target cells as signaling devices via cell surface expressed ligands or by transferring surface receptors/adhesion molecules or small RNAs between cells [22–25]. Such a phenomenon is in line with exosome-like vesicles acting as vesicular carriers for intercellular communication [26, 27]. However, additional investigation is urged to reveal whether prominin-1-EVs modify the biochemistry of the recipient mesenchymal stem cells [28] as it was proposed upon the direct contact of Hematopoietic Stem-Progenitor Cells (HSPCs) with osteoblasts [29]. Thus, the intercellular communication of HSPCs with their bone marrow microenvironment is a novel research field that requires further investigation [30] and the development of animal models to demonstrate its in vivo impact.

15.3 Therapeutic Potential of HSCs-Derived EVs

It is generally accepted that the therapeutic potential of stem cells may be mediated largely by the paracrine factors. Thus EVs produced by stem cells are the hot candidates to be carriers of stem cell-related therapies [31].

Most of the reported active ingredients in EVs are largely in two classes: nucleic acids including mRNAs and miRNAs and proteins, especially surface receptors and intravesicular enzymes or transcription factors [32]. In addition to tissue factor (CD142) [33] and various mRNAs [2] exosomes secreted by HSCs may contain the stem cell marker prominin-1 (CD133) which plays important roles in maintaining the properties of stem cells as well as in the endocytic-exocytic pathway [21]. Thus $CD133^+$ cells purified from hematopoietic tissues represent a potential source of stem cells. EVs derived from these cells were shown to express mRNAs of several pro-angiopoietic and anti-apoptotic factors which promote angiogenesis, providing a theoretical basis for application of $CD133^+$ cells in regenerative medicine [34]. Exosomes from granulocyte colony-stimulating factor (GCSF)-mobilized bone marrows were shown to contain abundant microRNA-126 (miR-126) and GCSF, promoting the accumulation of exosomes in the bone marrow [35]. Exosomes-delivered miR-126 reduces the expression of vascular cell adhesion molecule-1 (VCAM-1) which is crucial to the retention of HSPCs in the bone marrow. The reduced level of VCAM-1 leads to the mobilization of HSCs from the bone marrow to the peripheral blood. In addition, chronic myeloid leukemia-derived exosomes promote the proliferation and survival of tumor cells via the anti-apoptotic effects mediated by selectively expressed miRNAs [36, 37].

When referring to therapeutic effect, it has been found that HSCs-secreted exosomes express mRNAs of several anti-apoptotic and pro-angiopoietic factors like the vascular endothelial growth factor (VEGF), insulin growth factor-1, basic fibroblast growth factor, and interleukin-8 [34]. These mRNAs exert anti-apoptotic effects, increase the proliferation and survival of endothelial cells, and thus stimulate tube formation. Since exosomes are produced in chronic myeloid leukemia, vaccines based on leukemia cells-derived exosomes might be a promising strategy for enhancing survival in patients suffering from HSCs transplantation and chemotherapy [36]. Accordingly, improvements are suggested to be seen in terms of therapeutic effects of HSCs-derived EVs toward the blood diseases such as the chronic myeloid leukemia in the near future.

15.3.1 HSCs-Derived EVs in the Regenerative Therapy of the Heart

Stem cell-based therapies aimed to regenerate the heart after pathological insults such as myocardial infarction have been studied extensively [38]. It should be mentioned that there are only limited data documenting cardioprotective therapeutic potential of HSCs and HSCs-derived EVs exclusively, thus the potential therapeutic effects of bone marrow-derived stem cells and vesicles in general will be partially discussed as well.

The early promise of stem cells in cardiac regeneration generated much excitement for their potential to improve function by differentiating into new cardiomyocytes. However, huge amount of data indicate that most cells transplanted into the heart do not survive long time, thus the concept of paracrine effects by substances released from injected cells has become popular, despite only indirect evidence for this theory [39, 40]. An intriguing possibility is that some of these paracrine effects may be mediated by EVs [41]. In concordance with this hypothesis are findings showing that in addition to the in vitro modulation of the extracellular milieu [42] the effect of transplanted bone-marrow-derived cells on improving cardiac function may be primarily due to the paracrine effects [40, 43–45]. It has been shown in genetically engineered mice that intramyocardial delivery of bone marrow-derived cells after myocardial infarction improves ventricular function of the heart. This protective effect could not be explained by direct transdifferentiation of injected cells into cardiomyocytes, but suggest that activation of endogenous progenitors may depend on paracrine communication between donor and recipient cells [40]. Also in the case of human cardiosphere-derived cells (derived from human myocardium) the benefits of cell-based therapy may due to paracrine effects. The factors secreted or released from injected cells that benefit cardiac function remain to be identified; however, due to the ability of EVs to be released from the cell and be re-uptaken by other cells, make the stem cells-derived exosomes and microvesicles the prime candidates for carriers of cardioprotective substances which are mediators of beneficial effects of injected stem cells into the heart. The mechanism by which exosomes exert cardioprotection is almost entirely unknown. It seems to involve a direct interaction with cells in the heart, rather than blood components, because cardioprotection has been observed both in vitro and in vivo [41]. At least in specific cases, exosomes have been demonstrated to be capable of direct transfer of RNA [46–48] or protein [49]. If there is a specific combination of multiple factors from a defined population of cells, then unraveling the paracrine cocktail may be very challenging. Furthermore, as improved methods to enhance cell survival and engraftment are developed, distinguishing between independent cell effects and paracrine effects will become even more difficult.

As mentioned, exosomes-delivered miR-126 leads to the mobilization of HSCs from the bone marrow to the peripheral blood via reduced expression of VCAM-1. Interestingly, miR-126 seems to act not only as intrinsic regulator of cell function but is also released and acts as a paracrine factor enhancing neovascularization [50]. The release of exosomes and microvesicles, which among other substances contain miRNAs, is known to mediate cell-to-cell communication between different cell types [47, 51, 52].

Different markers, such as CD34, CD71, or CD235a were used to characterize EVs derived from human HSCs [53]. Among them, CD34 is the main hematopoietic stem cell marker. It has been shown that exosomes derived from $CD34^+$ cells improve neovascularization after ischemia [54]. A recent study demonstrated that miR-126 is preferentially enriched in microvesicles derived from $CD34^+$ cells compared with other cell types, and contributes to the proangiogenic activity of the cell supernatants [55]. In addition, release of miR-126 and miR-296–containing microvesicles by proangiogenic cells can induce angiogenesis in endothelial cells [56]. It has been also reported that miR-126-knockouts had an impaired angiogenic response in the heart in the setting of myocardial ischemia and promote endothelial cell tube formation [57]. In patients with diabetes mellitus, an impaired expression of miR-126 was demonstrated in microvesicles isolated from supernatants of peripheral blood $CD34^+$ cells, and overexpression of miR-126 restored the proangiogenic activity of diabetic $CD34^+$ cells [55]. These findings are consistent with the lower circulating levels of miR-126 detected in patients with diabetes mellitus or coronary artery disease, and may contribute to the known impaired neovascularization capacity observed in patients with diabetes mellitus [58, 59].

In the BOOST randomised controlled clinical trial [60], intracoronary autologous bone-marrow cell transfer after myocardial infarction were performed in 30 patients, and the same number of patients served as controls non-treated by bone marrow cells. All patients received optimum post-infarction medical treatment independently on the presence or absence of stem cell transfer therapy. The results of the study have shown that intracoronary transfer of autologous bone-marrow-cells promotes improvement of left-ventricular systolic function in patients after acute myocardial infarction. Even though the study was not designed to assess underlying mechanisms, it is apparent that transdifferentiation of bone-marrow-derived HSCs to cardiomyocytes cannot account for the beneficial effects [61, 62]. Recent studies have highlighted the potential of bone-marrow cells to promote paracrine effects in ischemic tissues which may include also release of EVs, and suggest that paracrine signaling promotes functional recovery rather than cell incorporation [63–66].

Myocardial infarction is followed by mobilization of the bone marrow stem cells, and subsequently increases levels of circulating endothelial progenitor cells (EPCs), HSCs, CD34+ positive stem cells, as well as mononuclear cells. These cells express early cardiac and endothelial markers as the injury signal produced by the damaged heart attracts reparative cells and promotes their migration into the target tissue. The most likely candidates for the cardiac repair are bone marrow-derived stem cells, a heterogenous population of cells that can be divided into HSCs, EPCs and mesenchymal stem cells [67–70]. Ischemic preconditioning as an endogenous adaptive mechanism protecting the heart against ischemia/reperfusion injury induced a rapid and permanent decrease in the number of circulating mesenchymal stem cells. High number of recruited cells was found mainly in the border zone of infarction but also in the infarcted area suggesting a rapid homing of the mesenchymal stem cells in the ischemic heart, imbalanced by their slower mobilization from the bone marrow. In contrast, the early phase of ischemic preconditioning led to amoderately higher level of HSCs mobilization into the blood stream as compared with the infarction without preventive intervention [71]. Additionally, the number of recruited but not the mobilized stem cells (both hematopoietic and mesenchymal) correlated significantly with the infarct size in the infarcted as well as borderline areas of the infarcted hearts. In a human study performed on patients with primary percutaneous coronary angioplasty and patients post myocardial infarction (MI) with ST segment elevation (post-STEMI) an increase in HSCs and decrease in circulating mesenchymal stem cells days after acute myocardial infarction was demonstrated, parallel with increase in plasma levels of VEGF and SDF-1 [72]. The discrepancy between the results of experimental study [71] and clinical data regarding the changes in circulating mobilizing factors [72] might be explained by the relatively short (120 min) follow-up post-infarction reperfusion in the experimental study.

Pre-clinical and clinical studies have documented the beneficial effects of human bone marrow-derived CD34+ stem cells in treatment of cardiovascular diseases such as myocardial ischemia including refractory angina [73, 74] and acute myocardial infarction [75]. A therapeutic paracrine mechanism of CD34+ cells has been identified recently as mediated at least in part by the secretion of extracellular vesicles particularly exosomes [54] known to carry RNAs, miRNAs and different proteins [76]. Previous studies documented that the use of CD34+ cells as a strategy to enhance perfusion preserved and/or improved cardiac function [77–79] thus improving the quality of life for the patients. However, patient diversity in age, disease burden and environmental factors may alter the therapeutic efficacy of isolated CD34+ cells [80–82]. A study revealing the therapeutic efficacy of CD34+ cells modified to express an established angiogenic protein, sonic hedgehog (Shh), was aimed to circumvent age and health related declines in CD34+ cell function [49]. The modified cells were observed to improve functional preservation of cardiac tissue as compared to control cells. Results of this study provide a novel insights including findings that modification with Shh improves the short-term retention of CD34+ cells, that $CD34^{Shh}$ deposit a proportionately greater amount of Shh in exosomes as compared to other Shh modified cell types, that Shh-containing exosomes derived from $CD34^{Shh}$ are capable of transferring Shh to other cell types and that exosomes containing Shh activate Shh signaling pathways in other cell types. Similarly, T-lymphocyte-derived microvesicles enriched with the morphogen Shh increased neoangiogenesis and restored endothelial function after injection in mice by stimulating the NO synthesis pathway [83, 84]. From a clinical perspective, the findings presented above point toward the advantages of applying the concept of EVs as therapeutic devices for the treatment of myocardial injury: the ability to carry and deliver specific content to the target, the possibility for in vitro expansion of the EVs, and the avoidance of damaging clearance mechanisms after transplantation. Thus, the potential of EVs in enhancing stem cell activity after genetic engineering may provide a key tool for developing novel therapeutic strategies to improve cardiac remodeling, function of the heart, and prognosis of patients after myocardial infarction. However, only gaining further insights into the complexity of the molecular interactions may allow the identification of responsible mechanisms, their connections, and how these mechanisms can be modulated for development of reliable therapies [85].

15.3.2 Therapeutic Potential of EVs Released from HSCs-Derived Cells in CVD

Not only HSCs themselves but also cells derived from HSCs such as dendritic cells (DCs) or endothelial progenitor cells (EPCs) are able to produce and release EVs including exosomes and microvesicles with promising therapeutic potential in treatment of CVD including ischemic heart disease and myocardial infarction. As therapeutic potential of these HSCs-derived cells and EVs produced by these cells is not discussed in other chapters of this book, but should be mentioned, we summarize these effect here.

Dendritic cells (DCs), as one type of cells derived from HSCs, are antigen-presenting cells (also known as accessory cells) of the mammalian immune system [86]. Their main function is to process antigen material and present it on the cell surface to the T cells of the immune system. Recent studies have shown that exosomes secreted from DCs which express immune-stimulatory molecules, deliver antigen-specific signals and are regarded as inert vehicles that target and activate T cells [87–89]. In addition to the ability of DCs to modulate immune responses, the injection of exosomes derived from donor bone marrow DCs was found to modulate response to heart transplantation [90]. Moreover, exosomes derived from DCs have been shown to improve cardiac function after myocardial infarction via activation of $CD4^+$ T lymphocytes [91] which is in concordance with finding that $CD4^+$ T cell activation plays a key role in improving myocardial wound healing post myocardial infarction [92].

It is believed that the majority of EPCs originate from the bone marrow. The hematopoietic and vascular systems develop in parallel but in an interdependent manner during embryogenesis. Multi-lineage hematopoietic progenitors are derived from the endothelium within the yolk sac [93] and embryo proper [94]. Whether the same is true in adults has not been ruled out; however, there is a close physical association between endothelial and hematopoietic stem cells (EPCs and HSCs) in postnatal bone marrow [95]. As vascular endothelial and hematopoietic cell types share many cell surface markers [96] it is not surprising that several identified EPCs populations exhibit hematopoietic characteristics, although non-hematopoietic sources of EPCs have also been identified [97].

Endothelial progenitor cells (EPCs) reside in the bone marrow and are mobilized into the circulation by specific stimuli such as certain drugs, ischemia, and exercise training. Although the understanding of the molecular pathways leading to a mobilization of EPCs from the bone marrow is not fully clear, several studies in the current literature demonstrate that VEGF is one of the most potent molecules triggering EPCs release [98]. VEGF expression is dramatically upregulated by hypoxia, which represents a critical force that drives adult vasculogenesis. Erythropoietin is another factor influencing EPCs mobilization [99]. It was demonstrated that the amount and function of EPCs is significantly impaired in CVD and that the level of circulating EPCs predicts the occurrence of cardiovascular events and death from cardiovascular causes. Recently, the beneficial effect of applied EPCs in treatment of CVD was demonstrated in several animal experiments, and later these cells were also used to treat humans with different types of CVD [100].

Regarding therapeutic potential of EPCs-derived EVs in the treatment of CVD, the study of Ratajczak et al. [25] as well as the study of Deregibus et al. [101] opened research perspectives on the use of EVs to transfer RNA-based information from stem cells/precursors to target differentiated cells. In particular, the later study indicated that microvesicles derived from EPCs may activate an angiogenic program in quiescent endothelial cells. Moreover, it has been shown that EVs, particularly exosomes, are an active component of the paracrine secretion of human EPCs and can promote vascular repair in rat models of balloon injury by up-regulating endothelial cells function [102]. EPCs-derived exosomes delivered into ischemic myocardium via an injectable hydrogel enhanced peri-infarct angiogenesis and myocardial hemodynamics in a rat model of myocardial infarction. The shear-thinning gel greatly increased therapeutic efficiency and efficacy of exosome-mediated myocardial preservation [103].

EPCs-based approaches to repair the heart after ischemic events by neovascularization can result in improvement of vascular perfusion and might offer clinical benefit. However, although functional improvement is observed, the lack of long-term engraftment of EPCs into neovessels has raised controversy regarding their mechanism of action. It has been hypothesized that after ischemic injury, EPCs induce neovascularization through the secretion of cytokines and growth factors, which act in a paracrine fashion and induce sprouting angiogenesis by the surrounding endothelium. Thus the demonstration that human EPCs secrete paracrine signals that induce neovascularization offers great therapeutic potential [104]. As EVs have been shown to be an active component of the paracrine secretion of human EPCs, there is a possibility that at least some of paracrine effects of EPCs may be mediated by EVs, and that exosomes and/or microvesicles act as carriers of pro-angiogenic and potentially cardioprotective substances released from EPCs.

15.4 Conclusions

Hematopoietic stem cells (HSCs) and HSCs-derived progenitor cells have been reported to exert beneficial effects in prevention/treatment of several cardiovascular diseases including ischemic heart disease and acute myocardial infarction. As the cardioprotective action of injected cells could not be explained by direct transdifferentiation of injected cells into cardiomyocytes, this effect is suggested to be mediated via paracrine communication between donor and recipient cells. Since it was evidenced that HSCs release EVs including exosomes and microvesicles into the surrounding environment that could be transferred and re-uptake by recipient cells, it might be assumed that at least some of these paracrine effects may be mediated by EVs. Among the cargo molecules of HSCs-derived vesicles several miRNAs, particularly miR-126, and pro-angiogenic and anti-apoptotic proteins are proposed to me the mediators of heart regeneration, mostly via neovascularization. However, the direct evidence of cardioprotective effects of HSCs-derived exosomes and microvesicles is still lacking in the literature, and the theory is based mostly on the indirect evidences of paracrine action of HSCs. On the other hand, EVs produced in HSCs-derived cells, specifically DCs and EPCs, have been shown to provide direct cardioprotective effects in CVD. Anyway, further studies are needed to be performed to assess the therapeutic potential of HSCs-derived EVs-based cardiac regenerative therapies.

References

1. Verfaillie CM (2002) Hematopoietic stem cells for transplantation. Nat Immunol 3(4):314–317
2. Aoki J, Ohashi K, Mitsuhashi M, Murakami T, Oakes M, Kobayashi T, Doki N, Kakihana K, Sakamaki H (2014) Posttransplantation bone marrow assessment by quantifying hematopoietic cell-derived mRNAs in plasma exosomes/microvesicles. Clin Chem 60(4):675–682
3. Fais S, Logozzi M, Lugini L, Federici C, Azzarito T, Zarovni N, Chiesi A (2013) Exosomes: the ideal nanovectors for biodelivery. Biol Chem 394(1):1–15
4. Katakowski M, Buller B, Zheng X, Lu Y, Rogers T, Osobamiro O, Shu W, Jiang F, Chopp M (2013) Exosomes from marrow stromal cells expressing miR-146b inhibit glioma growth. Cancer Lett 335(1):201–204
5. He J, Wang Y, Sun S, Yu M, Wang C, Pei X, Zhu B, Wu J, Zhao W (2012) Bone marrow stem cells-derived microvesicles protect against renal injury in the mouse remnant kidney model. Nephrology (Carlton) 17(5):493–500
6. Yang X, Meng S, Jiang H, Zhu C, Wu W (2011) Exosomes derived from immature bone marrow dendritic cells induce tolerogenicity of intestinal transplantation in rats. J Surg Res 171(2):826–832
7. Collino F, Deregibus MC, Bruno S, Sterpone L, Aghemo G, Viltono L, Tetta C, Camussi G (2010) Microvesicles derived from adult human bone marrow and tissue specific mesenchymal stem cells shuttle selected pattern of miRNAs. PLoS One 5(7):e11803
8. Caby MP, Lankar D, Vincendeau-Scherrer C, Raposo G, Bonnerot C (2005) Exosomal-like vesicles are present in human blood plasma. Int Immunol 17(7):879–887
9. Pan BT, Johnstone RM (1983) Fate of the transferrin receptor during maturation of sheep reticulocytes in vitro: selective externalization of the receptor. Cell 33(3):967–978

10. Harding C, Heuser J, Stahl P (1984) Endocytosis and intracellular processing of transferrin and colloidal gold-transferrin in rat reticulocytes: demonstration of a pathway for receptor shedding. Eur J Cell Biol 35(2):256–263
11. Rieu S, Geminard C, Rabesandratana H, Sainte-Marie J, Vidal M (2000) Exosomes released during reticulocyte maturation bind to fibronectin via integrin alpha4beta1. Eur J Biochem 267(2):583–590
12. Raposo G, Nijman HW, Stoorvogel W, Liejendekker R, Harding CV, Melief CJ, Geuze HJ (1996) B lymphocytes secrete antigen-presenting vesicles. J Exp Med 183(3):1161–1172
13. Théry C, Amigorena S (1999) Molecular characterization of dendritic cell-derived exosomes: selective accumulation of the heat shock protein hsc73. J Cell Biol 147(3):599–610
14. Thery C, Boussac M, Veron P, Ricciardi-Castagnoli P, Raposo G, Garin J, Amigorena S (2001) Proteomic analysis of dendritic cell-derived exosomes: a secreted subcellular compartment distinct from apoptotic vesicles. J Immunol (Baltimore, Md: 1950) 166(12):7309–7318
15. Zitvogel L, Regnault A, Lozier A, Wolfers J, Flament C, Tenza D, Ricciardi-Castagnoli P, Raposo G, Amigorena S (1998) Eradication of established murine tumors using a novel cell-free vaccine: dendritic cell-derived exosomes. Nat Med 4(5):594–600
16. Peters PJ, Geuze HJ, van der Donk HA, Borst J (1990) A new model for lethal hit delivery by cytotoxic T lymphocytes. Immunol Today 11(1):28–32
17. Peters PJ, Borst J, Oorschot V, Fukuda M, Krahenbuhl O, Tschopp J, Slot JW, Geuze HJ (1991) Cytotoxic T lymphocyte granules are secretory lysosomes, containing both perforin and granzymes. J Exp Med 173(5):1099–1109
18. Blanchard N, Lankar D, Faure F, Regnault A, Dumont C, Raposo G, Hivroz C (2002) TCR activation of human T cells induces the production of exosomes bearing the TCR/CD3/zeta complex. J Immunol (Baltimore, Md: 1950) 168(7):3235–3241
19. Raposo G, Tenza D, Mecheri S, Peronet R, Bonnerot C, Desaymard C (1997) Accumulation of major histocompatibility complex class II molecules in mast cell secretory granules and their release upon degranulation. Mol Biol Cell 8(12):2631–2645
20. Skokos D, Le Panse S, Villa I, Rousselle JC, Peronet R, David B, Namane A, Mecheri S (2001) Mast cell-dependent B and T lymphocyte activation is mediated by the secretion of immunologically active exosomes. J Immunol 166(2):868–876
21. Bauer N, Wilsch-Brauninger M, Karbanova J, Fonseca AV, Strauss D, Freund D, Thiele C, Huttner WB, Bornhauser M, Corbeil D (2011) Haematopoietic stem cell differentiation promotes the release of prominin-1/CD133-containing membrane vesicles—a role of the endocytic-exocytic pathway. EMBO Mol Med 3(7):398–409
22. Bakhti M, Winter C, Simons M (2011) Inhibition of myelin membrane sheath formation by oligodendrocyte-derived exosome-like vesicles. J Biol Chem 286(1):787–796
23. Eldh M, Ekstrom K, Valadi H, Sjostrand M, Olsson B, Jernas M, Lotvall J (2010) Exosomes communicate protective messages during oxidative stress; possible role of exosomal shuttle RNA. PLoS One 5(12):e15353
24. Janowska-Wieczorek A, Majka M, Kijowski J, Baj-Krzyworzeka M, Reca R, Turner AR, Ratajczak J, Emerson SG, Kowalska MA, Ratajczak MZ (2001) Platelet-derived microparticles bind to hematopoietic stem/progenitor cells and enhance their engraftment. Blood 98(10):3143–3149
25. Ratajczak J, Wysoczynski M, Hayek F, Janowska-Wieczorek A, Ratajczak MZ (2006) Membrane-derived microvesicles: important and underappreciated mediators of cell-to-cell communication. Leukemia 20(9):1487–1495
26. Simons M, Raposo G (2009) Exosomes—vesicular carriers for intercellular communication. Curr Opin Cell Biol 21(4):575–581
27. Thery C, Ostrowski M, Segura E (2009) Membrane vesicles as conveyors of immune responses. Nat Rev Immunol 9(8):581–593
28. Freund D, Fonseca AV, Janich P, Bornhauser M, Corbeil D (2010) Differential expression of biofunctional GM1 and GM3 gangliosides within the plastic-adherent multipotent mesenchymal stromal cell population. Cytotherapy 12(2):131–142

29. Gillette JM, Larochelle A, Dunbar CE, Lippincott-Schwartz J (2009) Intercellular transfer to signalling endosomes regulates an ex vivo bone marrow niche. Nat Cell Biol 11(3):303–311
30. Mendez-Ferrer S, Michurina TV, Ferraro F, Mazloom AR, Macarthur BD, Lira SA, Scadden DT, Ma'ayan A, Enikolopov GN, Frenette PS (2010) Mesenchymal and haematopoietic stem cells form a unique bone marrow niche. Nature 466(7308):829–834
31. Han C, Sun X, Liu L, Jiang H, Shen Y, Xu X, Li J, Zhang G, Huang J, Lin Z, Xiong N, Wang T (2016) Exosomes and their therapeutic potentials of stem cells. Stem Cells Int 2016:7653489
32. Riazifar M, Pone EJ, Lotvall J, Zhao W (2017) Stem cell extracellular vesicles: extended messages of regeneration. Annu Rev Pharmacol Toxicol 57:125–154
33. Chou J, Mackman N, Merrill-Skoloff G, Pedersen B, Furie BC, Furie B (2004) Hematopoietic cell-derived microparticle tissue factor contributes to fibrin formation during thrombus propagation. Blood 104(10):3190–3197
34. Ratajczak J, Kucia M, Mierzejewska K, Marlicz W, Pietrzkowski Z, Wojakowski W, Greco NJ, Tendera M, Ratajczak MZ (2013) Paracrine proangiopoietic effects of human umbilical cord blood-derived purified CD133+ cells—implications for stem cell therapies in regenerative medicine. Stem Cells Dev 22(3):422–430
35. Salvucci O, Jiang K, Gasperini P, Maric D, Zhu J, Sakakibara S, Espigol-Frigole G, Wang S, Tosato G (2012) MicroRNA126 contributes to granulocyte colony-stimulating factor-induced hematopoietic progenitor cell mobilization by reducing the expression of vascular cell adhesion molecule 1. Haematologica 97(6):818–826
36. Yao Y, Wang C, Wei W, Shen C, Deng X, Chen L, Ma L, Hao S (2014) Dendritic cells pulsed with leukemia cell-derived exosomes more efficiently induce antileukemic immunities. PLoS One 9(3):e91463
37. Raimondo S, Saieva L, Corrado C, Fontana S, Flugy A, Rizzo A, De Leo G, Alessandro R (2015) Chronic myeloid leukemia-derived exosomes promote tumor growth through an autocrine mechanism. Cell Commun Signal 13:8
38. Garbern JC, Lee RT (2013) Cardiac stem cell therapy and the promise of heart regeneration. Cell Stem Cell 12(6):689–698
39. Govaert JA, Swijnenburg RJ, Schrepfer S, Xie X, van der Bogt KE, Hoyt G, Stein W, Ransohoff KJ, Robbins RC, Wu JC (2009) Poor functional recovery after transplantation of diabetic bone marrow stem cells in ischemic myocardium. J Heart Lung Transplant 28(11):1158–1165. e1151
40. Loffredo FS, Steinhauser ML, Gannon J, Lee RT (2011) Bone marrow-derived cell therapy stimulates endogenous cardiomyocyte progenitors and promotes cardiac repair. Cell Stem Cell 8(4):389–398
41. Yellon DM, Davidson SM (2014) Exosomes: nanoparticles involved in cardioprotection? Circ Res 114(2):325–332
42. Baffour R, Pakala R, Hellinga D, Joner M, Okubagzi P, Epstein SE, Waksman R (2006) Bone marrow-derived stem cell interactions with adult cardiomyocytes and skeletal myoblasts in vitro. Cardiovasc Revasc Med 7(4):222–230
43. Gnecchi M, He H, Liang OD, Melo LG, Morello F, Mu H, Noiseux N, Zhang L, Pratt RE, Ingwall JS, Dzau VJ (2005) Paracrine action accounts for marked protection of ischemic heart by Akt-modified mesenchymal stem cells. Nat Med 11(4):367–368
44. Iso Y, Spees JL, Serrano C, Bakondi B, Pochampally R, Song YH, Sobel BE, Delafontaine P, Prockop DJ (2007) Multipotent human stromal cells improve cardiac function after myocardial infarction in mice without long-term engraftment. Biochem Biophys Res Commun 354(3):700–706
45. Williams AR, Hare JM (2011) Mesenchymal stem cells: biology, pathophysiology, translational findings, and therapeutic implications for cardiac disease. Circ Res 109(8):923–940
46. Valadi H, Ekstrom K, Bossios A, Sjostrand M, Lee JJ, Lotvall JO (2007) Exosome-mediated transfer of mRNAs and microRNAs is a novel mechanism of genetic exchange between cells. Nat Cell Biol 9(6):654–659

47. Hergenreider E, Heydt S, Treguer K, Boettger T, Horrevoets AJ, Zeiher AM, Scheffer MP, Frangakis AS, Yin X, Mayr M, Braun T, Urbich C, Boon RA, Dimmeler S (2012) Atheroprotective communication between endothelial cells and smooth muscle cells through miRNAs. Nat Cell Biol 14(3):249–256
48. Montecalvo A, Larregina AT, Shufesky WJ, Stolz DB, Sullivan ML, Karlsson JM, Baty CJ, Gibson GA, Erdos G, Wang Z, Milosevic J, Tkacheva OA, Divito SJ, Jordan R, Lyons-Weiler J, Watkins SC, Morelli AE (2012) Mechanism of transfer of functional microRNAs between mouse dendritic cells via exosomes. Blood 119(3):756–766
49. Mackie AR, Klyachko E, Thorne T, Schultz KM, Millay M, Ito A, Kamide CE, Liu T, Gupta R, Sahoo S, Misener S, Kishore R, Losordo DW (2012) Sonic hedgehog-modified human CD34+ cells preserve cardiac function after acute myocardial infarction. Circ Res 111(3):312–321
50. Seeger FH, Zeiher AM, Dimmeler S (2013) MicroRNAs in stem cell function and regenerative therapy of the heart. Arterioscler Thromb Vasc Biol 33(8):1739–1746
51. Ranghino A, Cantaluppi V, Grange C, Vitillo L, Fop F, Biancone L, Deregibus MC, Tetta C, Segoloni GP, Camussi G (2012) Endothelial progenitor cell-derived microvesicles improve neovascularization in a murine model of hindlimb ischemia. Int J Immunopathol Pharmacol 25(1):75–85
52. Camussi G, Deregibus MC, Bruno S, Grange C, Fonsato V, Tetta C (2011) Exosome/microvesicle-mediated epigenetic reprogramming of cells. Am J Cancer Res 1(1):98–110
53. Macri S, Pavesi E, Crescitelli R, Aspesi A, Vizziello C, Botto C, Corti P, Quarello P, Notari P, Ramenghi U, Ellis SR, Dianzani I (2015) Immunophenotypic profiling of erythroid progenitor-derived extracellular vesicles in diamond-blackfan anaemia: a new diagnostic strategy. PLoS One 10(9):e0138200
54. Sahoo S, Klychko E, Thorne T, Misener S, Schultz KM, Millay M, Ito A, Liu T, Kamide C, Agrawal H, Perlman H, Qin G, Kishore R, Losordo DW (2011) Exosomes from human CD34(+) stem cells mediate their proangiogenic paracrine activity. Circ Res 109(7):724–728
55. Mocharla P, Briand S, Giannotti G, Dorries C, Jakob P, Paneni F, Luscher T, Landmesser U (2013) AngiomiR-126 expression and secretion from circulating CD34(+) and CD14(+) PBMCs: role for proangiogenic effects and alterations in type 2 diabetics. Blood 121(1):226–236
56. Cantaluppi V, Biancone L, Figliolini F, Beltramo S, Medica D, Deregibus MC, Galimi F, Romagnoli R, Salizzoni M, Tetta C, Segoloni GP, Camussi G (2012) Microvesicles derived from endothelial progenitor cells enhance neoangiogenesis of human pancreatic islets. Cell Transplant 21(6):1305–1320
57. Fish JE, Santoro MM, Morton SU, Yu S, Yeh RF, Wythe JD, Ivey KN, Bruneau BG, Stainier DY, Srivastava D (2008) miR-126 regulates angiogenic signaling and vascular integrity. Dev Cell 15(2):272–284
58. Zampetaki A, Kiechl S, Drozdov I, Willeit P, Mayr U, Prokopi M, Mayr A, Weger S, Oberhollenzer F, Bonora E, Shah A, Willeit J, Mayr M (2010) Plasma microRNA profiling reveals loss of endothelial miR-126 and other microRNAs in type 2 diabetes. Circ Res 107(6):810–817
59. Fichtlscherer S, De Rosa S, Fox H, Schwietz T, Fischer A, Liebetrau C, Weber M, Hamm CW, Roxe T, Muller-Ardogan M, Bonauer A, Zeiher AM, Dimmeler S (2010) Circulating microRNAs in patients with coronary artery disease. Circ Res 107(5):677–684
60. Wollert KC, Meyer GP, Lotz J, Ringes-Lichtenberg S, Lippolt P, Breidenbach C, Fichtner S, Korte T, Hornig B, Messinger D, Arseniev L, Hertenstein B, Ganser A, Drexler H (2004) Intracoronary autologous bone-marrow cell transfer after myocardial infarction: the BOOST randomised controlled clinical trial. Lancet 364(9429):141–148
61. Murry CE, Soonpaa MH, Reinecke H, Nakajima H, Nakajima HO, Rubart M, Pasumarthi KB, Virag JI, Bartelmez SH, Poppa V, Bradford G, Dowell JD, Williams DA, Field LJ (2004) Haematopoietic stem cells do not transdifferentiate into cardiac myocytes in myocardial infarcts. Nature 428(6983):664–668

62. Balsam LB, Wagers AJ, Christensen JL, Kofidis T, Weissman IL, Robbins RC (2004) Haematopoietic stem cells adopt mature haematopoietic fates in ischaemic myocardium. Nature 428(6983):668–673
63. Kamihata H, Matsubara H, Nishiue T, Fujiyama S, Tsutsumi Y, Ozono R, Masaki H, Mori Y, Iba O, Tateishi E, Kosaki A, Shintani S, Murohara T, Imaizumi T, Iwasaka T (2001) Implantation of bone marrow mononuclear cells into ischemic myocardium enhances collateral perfusion and regional function via side supply of angioblasts, angiogenic ligands, and cytokines. Circulation 104(9):1046–1052
64. Ziegelhoeffer T, Fernandez B, Kostin S, Heil M, Voswinckel R, Helisch A, Schaper W (2004) Bone marrow-derived cells do not incorporate into the adult growing vasculature. Circ Res 94(2):230–238
65. Kinnaird T, Stabile E, Burnett MS, Lee CW, Barr S, Fuchs S, Epstein SE (2004) Marrow-derived stromal cells express genes encoding a broad spectrum of arteriogenic cytokines and promote in vitro and in vivo arteriogenesis through paracrine mechanisms. Circ Res 94(5):678–685
66. Heil M, Ziegelhoeffer T, Mees B, Schaper W (2004) A different outlook on the role of bone marrow stem cells in vascular growth: bone marrow delivers software not hardware. Circ Res 94(5):573–574
67. Massa M, Rosti V, Ferrario M, Campanelli R, Ramajoli I, Rosso R, De Ferrari GM, Ferlini M, Goffredo L, Bertoletti A, Klersy C, Pecci A, Moratti R, Tavazzi L (2005) Increased circulating hematopoietic and endothelial progenitor cells in the early phase of acute myocardial infarction. Blood 105(1):199–206
68. Wojakowski W, Tendera M, Michalowska A, Majka M, Kucia M, Maslankiewicz K, Wyderka R, Ochala A, Ratajczak MZ (2004) Mobilization of CD34/CXCR4+, CD34/CD117+, c-met+ stem cells, and mononuclear cells expressing early cardiac, muscle, and endothelial markers into peripheral blood in patients with acute myocardial infarction. Circulation 110(20):3213–3220
69. Leone AM, Rutella S, Bonanno G, Abbate A, Rebuzzi AG, Giovannini S, Lombardi M, Galiuto L, Liuzzo G, Andreotti F, Lanza GA, Contemi AM, Leone G, Crea F (2005) Mobilization of bone marrow-derived stem cells after myocardial infarction and left ventricular function. Eur Heart J 26(12):1196–1204
70. Aicher A, Zeiher AM, Dimmeler S (2005) Mobilizing endothelial progenitor cells. Hypertension 45(3):321–325
71. Gyongyosi M, Posa A, Pavo N, Hemetsberger R, Kvakan H, Steiner-Boker S, Petrasi Z, Manczur F, Pavo IJ, Edes IF, Wojta J, Glogar D, Huber K (2010) Differential effect of ischaemic preconditioning on mobilisation and recruitment of haematopoietic and mesenchymal stem cells in porcine myocardial ischaemia-reperfusion. Thromb Haemost 104(2):376–384
72. Wang Y, Johnsen HE, Mortensen S, Bindslev L, Ripa RS, Haack-Sorensen M, Jorgensen E, Fang W, Kastrup J (2006) Changes in circulating mesenchymal stem cells, stem cell homing factor, and vascular growth factors in patients with acute ST elevation myocardial infarction treated with primary percutaneous coronary intervention. Heart 92(6):768–774
73. Losordo DW, Schatz RA, White CJ, Udelson JE, Veereshwarayya V, Durgin M, Poh KK, Weinstein R, Kearney M, Chaudhry M, Burg A, Eaton L, Heyd L, Thorne T, Shturman L, Hoffmeister P, Story K, Zak V, Dowling D, Traverse JH, Olson RE, Flanagan J, Sodano D, Murayama T, Kawamoto A, Kusano KF, Wollins J, Welt F, Shah P, Soukas P, Asahara T, Henry TD (2007) Intramyocardial transplantation of autologous CD34+ stem cells for intractable angina: a phase I/IIa double-blind, randomized controlled trial. Circulation 115(25):3165–3172
74. Losordo DW, Henry TD, Davidson C, Sup Lee J, Costa MA, Bass T, Mendelsohn F, Fortuin FD, Pepine CJ, Traverse JH, Amrani D, Ewenstein BM, Riedel N, Story K, Barker K, Povsic TJ, Harrington RA, Schatz RA, Investigators AC (2011) Intramyocardial, autologous CD34+ cell therapy for refractory angina. Circ Res 109(4):428–436
75. Quyyumi AA, Waller EK, Murrow J, Esteves F, Galt J, Oshinski J, Lerakis S, Sher S, Vaughan D, Perin E, Willerson J, Kereiakes D, Gersh BJ, Gregory D, Werner A, Moss T, Chan WS,

Preti R, Pecora AL (2011) CD34(+) cell infusion after ST elevation myocardial infarction is associated with improved perfusion and is dose dependent. Am Heart J 161(1):98–105
76. Chaput N, Thery C (2011) Exosomes: immune properties and potential clinical implementations. Semin Immunopathol 33(5):419–440
77. Kawamoto A, Tkebuchava T, Yamaguchi J, Nishimura H, Yoon YS, Milliken C, Uchida S, Masuo O, Iwaguro H, Ma H, Hanley A, Silver M, Kearney M, Losordo DW, Isner JM, Asahara T (2003) Intramyocardial transplantation of autologous endothelial progenitor cells for therapeutic neovascularization of myocardial ischemia. Circulation 107(3):461–468
78. Kawamoto A, Iwasaki H, Kusano K, Murayama T, Oyamada A, Silver M, Hulbert C, Gavin M, Hanley A, Ma H, Kearney M, Zak V, Asahara T, Losordo DW (2006) CD34-positive cells exhibit increased potency and safety for therapeutic neovascularization after myocardial infarction compared with total mononuclear cells. Circulation 114(20):2163–2169
79. Shintani S, Kusano K, Ii M, Iwakura A, Heyd L, Curry C, Wecker A, Gavin M, Ma H, Kearney M, Silver M, Thorne T, Murohara T, Losordo DW (2006) Synergistic effect of combined intramyocardial CD34+ cells and VEGF2 gene therapy after MI. Nat Clin Pract Cardiovasc Med 3(Suppl 1):S123–S128
80. Vasa M, Fichtlscherer S, Aicher A, Adler K, Urbich C, Martin H, Zeiher AM, Dimmeler S (2001) Number and migratory activity of circulating endothelial progenitor cells inversely correlate with risk factors for coronary artery disease. Circ Res 89(1):E1–E7
81. Vasa M, Fichtlscherer S, Adler K, Aicher A, Martin H, Zeiher AM, Dimmeler S (2001) Increase in circulating endothelial progenitor cells by statin therapy in patients with stable coronary artery disease. Circulation 103(24):2885–2890
82. Valgimigli M, Rigolin GM, Fucili A, Porta MD, Soukhomovskaia O, Malagutti P, Bugli AM, Bragotti LZ, Francolini G, Mauro E, Castoldi G, Ferrari R (2004) CD34+ and endothelial progenitor cells in patients with various degrees of congestive heart failure. Circulation 110(10):1209–1212
83. Agouni A, Mostefai HA, Porro C, Carusio N, Favre J, Richard V, Henrion D, Martinez MC, Andriantsitohaina R (2007) Sonic hedgehog carried by microparticles corrects endothelial injury through nitric oxide release. FASEB J 21(11):2735–2741
84. Benameur T, Soleti R, Porro C, Andriantsitohaina R, Martinez MC (2010) Microparticles carrying Sonic hedgehog favor neovascularization through the activation of nitric oxide pathway in mice. PLoS One 5(9):e12688
85. Liehn EA, Bucur O, Weber C (2012) Role of microparticles as messengers enhancing stem cell activity after genetic engineering. Circ Res 111(3):265–267
86. Liu K, Nussenzweig MC (2010) Origin and development of dendritic cells. Immunol Rev 234(1):45–54
87. Gutierrez-Vazquez C, Villarroya-Beltri C, Mittelbrunn M, Sanchez-Madrid F (2013) Transfer of extracellular vesicles during immune cell-cell interactions. Immunol Rev 251(1):125–142
88. Pitt JM, Charrier M, Viaud S, Andre F, Besse B, Chaput N, Zitvogel L (2014) Dendritic cell-derived exosomes as immunotherapies in the fight against cancer. J Immunol 193(3):1006–1011
89. Naslund TI, Gehrmann U, Qazi KR, Karlsson MC, Gabrielsson S (2013) Dendritic cell-derived exosomes need to activate both T and B cells to induce antitumor immunity. J Immunol 190(6):2712–2719
90. Peche H, Heslan M, Usal C, Amigorena S, Cuturi MC (2003) Presentation of donor major histocompatibility complex antigens by bone marrow dendritic cell-derived exosomes modulates allograft rejection. Transplantation 76(10):1503–1510
91. Liu H, Gao W, Yuan J, Wu C, Yao K, Zhang L, Ma L, Zhu J, Zou Y, Ge J (2016) Exosomes derived from dendritic cells improve cardiac function via activation of CD4(+) T lymphocytes after myocardial infarction. J Mol Cell Cardiol 91:123–133
92. Hofmann U, Beyersdorf N, Weirather J, Podolskaya A, Bauersachs J, Ertl G, Kerkau T, Frantz S (2012) Activation of CD4+ T lymphocytes improves wound healing and survival after experimental myocardial infarction in mice. Circulation 125(13):1652–1663

93. Goldie LC, Nix MK, Hirschi KK (2008) Embryonic vasculogenesis and hematopoietic specification. Organogenesis 4(4):257–263
94. Zovein AC, Hofmann JJ, Lynch M, French WJ, Turlo KA, Yang Y, Becker MS, Zanetta L, Dejana E, Gasson JC, Tallquist MD, Iruela-Arispe ML (2008) Fate tracing reveals the endothelial origin of hematopoietic stem cells. Cell Stem Cell 3(6):625–636
95. Morrison SJ, Spradling AC (2008) Stem cells and niches: mechanisms that promote stem cell maintenance throughout life. Cell 132(4):598–611
96. Jackson KA, Majka SM, Wang H, Pocius J, Hartley CJ, Majesky MW, Entman ML, Michael LH, Hirschi KK, Goodell MA (2001) Regeneration of ischemic cardiac muscle and vascular endothelium by adult stem cells. J Clin Invest 107(11):1395–1402
97. Chao H, Hirschi KK (2010) Hemato-vascular origins of endothelial progenitor cells? Microvasc Res 79(3):169–173
98. Khakoo AY, Finkel T (2005) Endothelial progenitor cells. Annu Rev Med 56:79–101
99. Heeschen C, Aicher A, Lehmann R, Fichtlscherer S, Vasa M, Urbich C, Mildner-Rihm C, Martin H, Zeiher AM, Dimmeler S (2003) Erythropoietin is a potent physiologic stimulus for endothelial progenitor cell mobilization. Blood 102(4):1340–1346
100. Mobius-Winkler S, Hollriegel R, Schuler G, Adams V (2009) Endothelial progenitor cells: implications for cardiovascular disease. Cytometry A 75(1):25–37
101. Deregibus MC, Cantaluppi V, Calogero R, Lo Iacono M, Tetta C, Biancone L, Bruno S, Bussolati B, Camussi G (2007) Endothelial progenitor cell derived microvesicles activate an angiogenic program in endothelial cells by a horizontal transfer of mRNA. Blood 110(7):2440–2448
102. Li X, Chen C, Wei L, Li Q, Niu X, Xu Y, Wang Y, Zhao J (2016) Exosomes derived from endothelial progenitor cells attenuate vascular repair and accelerate reendothelialization by enhancing endothelial function. Cytotherapy 18(2):253–262
103. Chen CW, Venkataraman CM, Wang LL, Hung G, Gaffey AC, Chung JJ, Liccardi A, Trubelja A, Seeholzer SH, Burdick JA (2016) Sustained release of endothelial progenitor cell derived exosomes from shear-thinning hydrogel improves angiogenesis and promotes function after myocardial infarction. American Heart Association A19338
104. Krenning G, van Luyn MJ, Harmsen MC (2009) Endothelial progenitor cell-based neovascularization: implications for therapy. Trends Mol Med 15(4):180–189

Chapter 16
Cardiac Telocyte-Derived Exosomes and Their Possible Implications in Cardiovascular Pathophysiology

Mirca Marini, Lidia Ibba-Manneschi, and Mirko Manetti

16.1 Introduction

Intercellular crosstalk is essential to survival and maintenance of tissue and organ homeostasis within a multicellular system [1]. The communication between living cells may occur by different modalities, which include either intercellular contacts, such as adhesion molecules, gap junctions and nanotubes, or the exchange of a variety of cell-released factors including cytokines, growth factors and hormones acting in an autocrine, paracrine, or endocrine manner [2].

Of note, an additional intercellular signaling mechanism that can act over both short and long distances has recently emerged, based on the release and uptake of membrane-bound vesicles which are referred to as extracellular vesicles (EVs) [1–4]. These circular membrane fragments enriched for mRNAs, small, single-stranded RNAs called microRNAs (miRNAs), long non-coding RNAs, proteins, and bioactive lipids may be released by exocytosis from the intracellular endosomal compartment or are formed by budding from the cell surface membrane [1–5]. Increasing evidence indicates that EVs may play important roles in a variety of physiological processes, including stem cell self-renewal and differentiation, tissue repair, immune surveillance and vascular homeostasis [2, 6, 7]. Furthermore, EVs appear to be implicated in several pathologies, such as cancer, neurodegenerative, cardiovascular, and metabolic disorders [1–3, 6, 7]. Nowadays, the importance of EVs is further highlighted by the evidence that they can also be considered as disease biomarkers, as well as possible drug, vaccine, or gene vector delivery tools with potential therapeutic applications [2, 3, 6–10]. Among the different types of EVs, the term 'exosomes' specifically refers to nano-sized EVs deriving from the endosomal compartment [3, 5, 6]. Exosomes are released and taken up by most cell

M. Marini • L. Ibba-Manneschi • M. Manetti (✉)
Section of Anatomy and Histology, Department of Experimental and Clinical Medicine, University of Florence, Largo Brambilla 3, Florence 50134, Italy
e-mail: mirkomanetti@yahoo.it; mirko.manetti@unifi.it

J. Xiao, S. Cretoiu (eds.), *Exosomes in Cardiovascular Diseases*, Advances in Experimental Medicine and Biology 998, DOI 10.1007/978-981-10-4397-0_16

types, thereby playing a pivotal role in the maintenance of tissue and organ homeostasis via horizontal transfer of cargos between cells [3, 5, 6]. Thus, the properties and roles of exosomes are now being increasingly investigated in a variety of physiological and pathological settings, with a main focus on their possible diagnostic and therapeutic utility in different conditions [1, 6, 9–11].

In this context, EVs and, in particular, exosomes are being increasingly implicated in multiple biological effects possibly exerted by a recently identified interstitial (stromal) cell type known as telocytes (TCs) [2, 12]. TCs, firstly identified by Popescu's group in 2005 as interstitial Cajal-like cells and officially renamed in 2010, have been described in the stromal compartment of many organs in humans and other vertebrates [13, 14]. As distinctive morphological features, TCs are characterized by a small cell body from which extremely long and slender processes, named telopodes (Tps), originate [13, 14]. The latter typically display a moniliform silhouette conferred by the alternation of thin segments (podomers) and small dilations (podoms) which accommodate caveolae, mitochondria and endoplasmic reticulum cisternae [13, 14]. Within the stromal compartment, Tps make a three-dimensional labyrinthine system establishing multiple intercellular communications by direct homocellular and heterocellular junctions [14, 15]. Moreover, Tps may release different types of EVs either in vivo or in vitro suggesting that TCs may profoundly influence the activity of neighboring cells by vesicular paracrine signals [2, 12].

In the heart, TCs have been reported to be ubiquitously distributed in the epicardium, myocardial interstitium, endocardium and in cardiac valves, where they are supposed to participate in the regulation of cardiac homeostasis and regeneration [16–22]. TCs appear to be in close contact with virtually all cell types in the human heart, such as cardiomyocytes, cardiac stem cells, blood capillaries, nerve endings and other cells found in the stromal compartment [16, 23, 24]. Noteworthy, it has been demonstrated that cardiac TCs are able to release at least three different types of EVs, namely exosomes, ectosomes and the so-called multivesicular cargos [2, 12, 25]. Indeed, the heart seems characterized by a complex intercellular shuttle mechanism which involves EV-mediated bidirectional paracrine signals either between TCs and tissue-resident stem cells or between TCs and cardiomyocytes [26]. In particular, TC-released exosomes, containing a cell-specific cargo of proteins, lipids and nucleic acids, seem to play a pivotal role in the crosstalk between TCs and other cardiac cells, thus making substantial contribution to cardiac physiology and response to injury [26]. In addition, TCs have been proposed to guide or 'nurse' putative stem cells and cardiomyocyte progenitors within cardiac stem cell niches [27, 28]. Of note, a number of studies have indicated that the TC interstitial network is reduced and impaired during myocardial infarction either in humans or in animal models [29]. Interestingly, there is also experimental evidence that transplantation of cardiac TCs in the infarcted and border zones of the heart may be effective in decreasing the infarction size and improving myocardial function [30].

This review summarizes the recent research findings on cardiac TCs and their EVs. We first provide an overview of the general features of TCs, including their morphological traits and immunophenotypes, intercellular signaling mechanisms and possible functional roles. Thereafter, we describe the distribution of TCs in the cardiac stromal compartment and the emerging role of cardiac TCs as intercellular communicators via the release of different EVs with particular focus on exosomes. Finally, the involvement of TCs in cardiovascular diseases and the potential utility of TC-derived exosomes in cardiac regeneration and repair are discussed.

16.2 General Characteristics of Telocytes

16.2.1 Morphological Features and Immunophenotypes of Telocytes

TCs are a novel type of stromal cells widely distributed in the interstitium of many tissues and organs [14, 31]. The shortest possible definition of TCs is 'cells with Tps' [14]. In fact, TCs display unique ultrastructural features characterized by a small piriform-, spindle- or triangular-shaped cell body (9–15 μm) giving rise to a variable number of extremely long (10–1000 μm) and thin prolongations which have been named Tps and distinguish them from 'classical' stromal cells, such as fibroblasts [13, 14, 32]. The cellular nucleus occupies about 25% of the cell body and contains clusters of heterochromatin attached to the nuclear envelope, while the surrounding scarce cytoplasm accommodates mitochondria, endoplasmic reticulum and Golgi apparatus. The Tps display an uneven caliber (mostly below 0.2 μm under light microscopy, and about 0.1–0.5 μm under transmission electron microscopy) with a distinctive moniliform appearance due to the alternation of thin segments (podomers) (~80 nm) and small dilated portions (podoms) (250–300 nm) containing mitochondria, endoplasmic reticulum cisternae and caveolae [14, 32]. In the interstitial space, Tps are typically organized to form a three-dimension labyrinthine network and establish multiple intercellular communications either between TCs through homocelullar junctions or between TCs and other cell types through heterocellular junctions [14, 15, 32]. Moreover, TCs and their Tps can release different types of EVs, which act as important transporters involved in intercellular signaling, including the transfer of genetic material consisting mainly of miRNAs [2, 12, 33].

Electron microscopy is commonly considered the gold standard method to identify TCs [14]. However, double immunolabeling for CD34 and c-kit/CD117, vimentin, platelet-derived growth factor receptor (PDGFR)-α or PDGFR-β may help in distinguishing TCs from other stromal cells under light microscopy [14, 26, 32]. Even if TCs do not display a unique antigenic profile, CD34 and PDGFR-α are currently regarded as the most suitable markers for their in situ identification by immunohistochemistry [14]. In fact, coexpression of CD34 and PDGFR-α has been extensively found in TCs from different tissues and organs [14, 34, 35]. However, there is also increasing evidence that the immunophenotypical features of TCs may vary among different organs/systems and that TC subtypes characterized by the expression of different markers may even coexist within the same organ [36, 37]. For instance, TCs may exhibit either CD34, PDGFR-α or c-kit/CD117 in some organs, such as the heart, while they are CD34/PDGFR-α double-positive and c-kit-negative in others, such as the gastrointestinal tract [34, 35, 38]. A growing number of studies also indicate that TCs display gene expression and proteomic profiles and miRNA signatures that are rather different from those of 'classical' fibroblasts [14, 39–43].

16.2.2 Telocytes as Intercellular Communicators: Telocyte's Contacts and Telocyte-Derived Extracellular Vesicles

In a variety of either cavitary or non-cavitary organs, TCs make a three-dimensional interstitial network which consists of their long Tps establishing either homocellular contacts between Tps or heterocellular contacts with other neighboring cell types, such as tissue-specific parenchymal cells, vessels, nerve endings, stem/progenitor cells, and other stromal cells including fibroblasts and immunoreactive cells like macrophages and mast cells [14, 15].

In particular, the homocellular contacts may be of several types and are formed by either simple appositions of the plasma membranes of contiguous TCs or by complex junctional areas accomplishing mechanical functions or allowing functional intercellular exchanges [15]. Junctional complexes with a mechanical function can be found in all the TCs and, since they resemble various types of the adherens junctions, have been named 'puncta adhaerentia minima' and 'processus adhaerens', which usually connect the overlapping Tps, and 'recessus adhaerens' or 'manubria adhaerentia' having a cuff-like appearance [15]. Instead, junctional complexes that functionally allow intercellular exchanges and signaling are mostly represented by gap junctions. Heterocellular contacts between TCs and other cell types consist mainly of minute junctions (e.g. point contacts, nanocontacts and planar contacts) typically with an inter-membrane distance of 10–30 nm, but more often by variably extended simple apposition of the contiguous TC plasma membranes [15].

Furthermore, increasing evidence indicates that TCs may participate in intercellular signaling through the release of a variety of EVs which regulate multiple neighboring cell functions [2, 12]. In fact, EVs are currently regarded as a new important way of communication for either short- or long-distance intercellular signaling events. EVs, characterized by a lipid bilayer membrane, carry a rich cargo of various bioactive materials including DNAs, RNAs, proteins, and lipids that are released into the extracellular space during both physiological and pathological processes [3, 5, 7, 44]. These EVs can interact with different cell types by ligand-receptor interactions, membrane fusion, and subsequent internalization via receptor-mediated endocytosis or macropinocytosis [44]. According to the classification criteria based on biogenesis mechanisms, different types of EVs can be distinguished, namely exosomes, ectosomes (also known as shedding microvesicles, microparticles or plasma membrane-derived vesicles), apoptotic bodies as well as a recently described novel EV type termed multivesicular cargos [1–3, 5, 7, 12, 25, 44].

Among the aforementioned EV subtypes, exosomes have been most extensively studied and characterized in recent years. These nano-sized vesicles are originated from the fusion of the plasma membrane with the multivesicular bodies, which are large cytoplasmic endosomal structures characterized by multiple intraluminal vesicles [1, 3, 5, 6, 44]. In fact, multivesicular bodies can either traffic to lysosomes for degradation or, alternatively, to the plasma membrane where, upon fusion, they may release their contents into the surrounding extracellular space. Once released into the extracellular space by exocytosis, the multivesicular body-derived vesicles are referred to as 'exosomes' [1, 3, 5, 6]. These exosomes, released into tissue interstitial spaces and bodily fluids, appear as multiple homogenous vesicles of around 30–150 nm in diameter containing numerous macromolecules including mRNAs, miRNAs, long non-coding RNAs, cytokines, chemokines, growth factors and various endosomal proteins such as tetraspanins (e.g. CD9, CD63, CD81), ALG-2 interacting protein X (Alix), tumour susceptibility gene 101 (TSG101), and annexin A5 (ANXA5), which are commonly used as markers for exosomal identification [1, 3, 5, 6, 12].

Unlike exosomes, ectosomes are small EVs with a diameter of about 50–1000 nm which originate directly from the plasma membrane by outward budding [1, 2, 5, 7, 44]. The molecular composition of ectosomes is still poorly characterized, although they seem to contain matrix metalloproteinases, glycoproteins (e.g. GPIb, GPIIb–IIIa and P-selectin), and integrins (e.g. Mac-1), depending on the ectosome-releasing cell type. Apoptotic bodies are instead heterogeneous vesicles (50 nm–5 μm) released upon programmed cell death via outward blebbing of the cell membrane [1, 2]. Finally, multivesicular cargos (0.4–1 μm) are large EVs which contain tightly packed endomembrane-bound smaller vesicles and have been recently reported to be secreted by cardiac TCs [2, 12, 25, 33]. Under transmission electron microscopy, multivesicular cargos appear frequently clustered in the subplasmallemal space of TCs, bulging from the plasma membrane of either the cell body or Tps, and released in an envelope formed by the plasma membrane [25]. The subsequent disruption of this envelope results in the release of individual or grouped small vesicles into the extracellular space [25]. Consistent with these electron microscopy observations, a peculiar structure with a cup-shaped or ellipsoid morphology, usually containing between 60 and 500 tightly packed endomembrane vesicles of varying shapes and dimensions, has been highlighted for multivesicular cargos by electron tomography [25].

16.2.3 Potential Roles of Telocytes

According to the distinctive morphological features, distribution and intercellular communications of the three-dimensional network-building Tps either in normal or in diseased tissues, multiple potential biological functions have been suggested for the TCs [14, 26, 29]. It is commonly believed that TCs may be functionally committed to the maintenance of local tissue homeostasis, as well as the regulation of tissue differentiation and renewal by short- and long-distance intercellular crosstalk mechanisms [14, 26, 29]. In particular, it has been proposed that during organ morphogenesis TCs might act as inductors and regulators of cell differentiation due to their capability to release paracrine molecular signals and to structurally build the three-dimensional scaffold driving parenchymal organization, while in the adulthood, these cells might behave as mesenchymal stromal cells with stemness properties and the potential to differentiate toward different mature cell types [14, 16, 21, 26, 45].

TCs have also been proposed to participate in immunomodulation and immunosurveillance, and possibly in the regulation of the activity of neighboring stromal cells, such as fibroblasts [29]. Moreover, TCs might be essential for the maintenance, proliferation, differentiation, maturation and guidance of the local stem/progenitor cells found within the niches of various organs, eventually stimulating and sustaining tissue regenerative and reparative processes [14, 26, 28]. Interestingly, increasing evidence also suggests that TCs may be involved in different pathologies including cancers, liver fibrosis, systemic sclerosis, inflammatory bowel diseases, and cardiovascular diseases [29, 45–52]. Therefore, currently there is growing research interest on the possible applications of TCs in regenerative medicine [26, 53].

16.3 Cardiac Telocytes and Their Extracellular Vesicles

The cardiac stroma plays a fundamental role in the building and maintenance of the normal heart architecture, as well as in any changes occurring in a variety of cardiac diseases [16]. Numerous electron microscopy studies have demonstrated that the heart contains typical TCs (Fig. 16.1) which are found in the epicardium, myocardial interstitium, endocardium, and cardiac valves [16–24]. Noteworthy, TCs were also identified in epicardial stem cell niches, where they appear located in close relationship with tissue-resident stem cells and putative cardiomyocyte progenitors, possibly contributing to form an interstitial scaffold which supports cardiomyocyte precursors during their self-renewal process and differentiation into new mature cardiomyocytes [27, 28]. During heart morphogenesis, TCs may even guide the process of cardiac tissue compaction from the embryonic myocardial trabeculae [16, 21, 54]. In addition, considering that cardiac TCs and epicardial progenitor cells share the expression of some stemness markers (e.g. c-kit/CD117), it has also

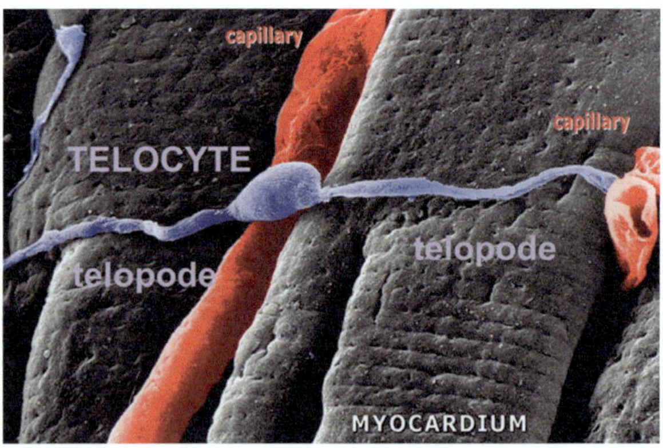

Fig. 16.1 Representative scanning electron micrograph of monkey left ventricular myocardium. The image shows a typical telocyte located across the cardiomyocytes. The three-dimensional view reveals close interconnections of the telocyte body and telopodes with cardiomyocytes and blood capillaries (Reproduced with permission from Kostin and Popescu [19])

been suggested that TCs might represent a subpopulation of progenitor cells which could therefore be directly implicated in cardiac development and regenerative processes [21, 53]. Of note, it also appears that both TCs and tissue-resident stem cells are decreased in the adult heart compared with newborns, which might contribute to the reduced cardiac regenerative capacity during aging [26, 55]. A recent experimental study in mice also reported that the number of cardiac TCs was significantly increased following physical exercise training, which is consistent with the evidence that exercise-induced cardiac growth is an important way to promote cardiac regeneration and repair [56].

At present, cardiac TCs are clearly the best ex vivo, in vitro and in vivo characterized TCs. In the adult heart, TCs display immunopositivity for different markers, such as CD34, c-kit/CD117 (Fig. 16.2), vimentin, PDGFR-α or PDGFR-β [26, 30, 35, 38]. Moreover, cardiac TCs in primary culture have been reported to express the embryonic stem cell marker Nanog and the myocardial stem cell marker Sca-1, suggesting that these cells may possess pluripotent properties [26, 38]. In addition, cardiac TCs exhibit a distinctive miRNA signature that further differentiates them from other interstitial cells. In particular, at variance with cardiac fibroblasts, cardiac TCs do not express miR-193, which has been shown to repress the expression of c-kit/CD117 [42]. Interestingly, this seems consistent with the evidence that cardiac TCs display c-kit/CD117 immunopositivity either ex vivo or in vitro [30, 38, 42]. Furthermore, miRNAs which are usually expressed by cardiomyocytes and other muscle cells (e.g. miR-133a, miR-208a) are undetectable in cardiac TCs [42]. Cultured cardiac TCs also behave differently from fibroblasts in terms of adherence, spreading, and extension of their cell prolongations when seeded on various matrix proteins [57]. Overall, these data clearly support the notion that TCs are a unique type of cardiac interstitial cells definitely distinct from 'classical' fibroblasts [58].

Although it appears that TCs represent a small fraction of interstitial cells in the human heart, their very long and convoluted Tps form a dynamic and extensive three-dimensional network within the cardiac stroma [24, 53, 55]. In addition to transmission electron microscopy studies, three-dimensional reconstruction of cardiac TCs has been recently performed by focused ion beam scanning electron microscopy (FIB-SEM) tomography, which confirmed that these cells have very long, slender and flattened (ribbon-like) Tps, with humps along their length due to the presence of podoms [59]. FIB-SEM tomography also highlighted that TCs make a network in the cardiac interstitium through wide adherens junctions connecting Tps [59]. Moreover, TCs build a supportive network in the myocardial interstitium and may communicate with the surrounding cells, namely cardiomyocytes, stem/progenitor cells, blood vessels, nerve endings, fibroblasts and immune cells (Fig. 16.3) [26, 55]. In particular, heterocellular connections between Tps and cardiomyocytes consist mainly of small point junctions with electron-dense nanocontacts, presumably forming a 'functional unit' which might help in mediating the electrical coupling of cardiomyocytes [26, 60–62]. Consistent with the well-documented spatial relationship between TCs and stem cell niches in cardiac tissue, different types of junctions have also been observed between cardiac TCs and cardiac stem cells in vitro [27, 28, 63].

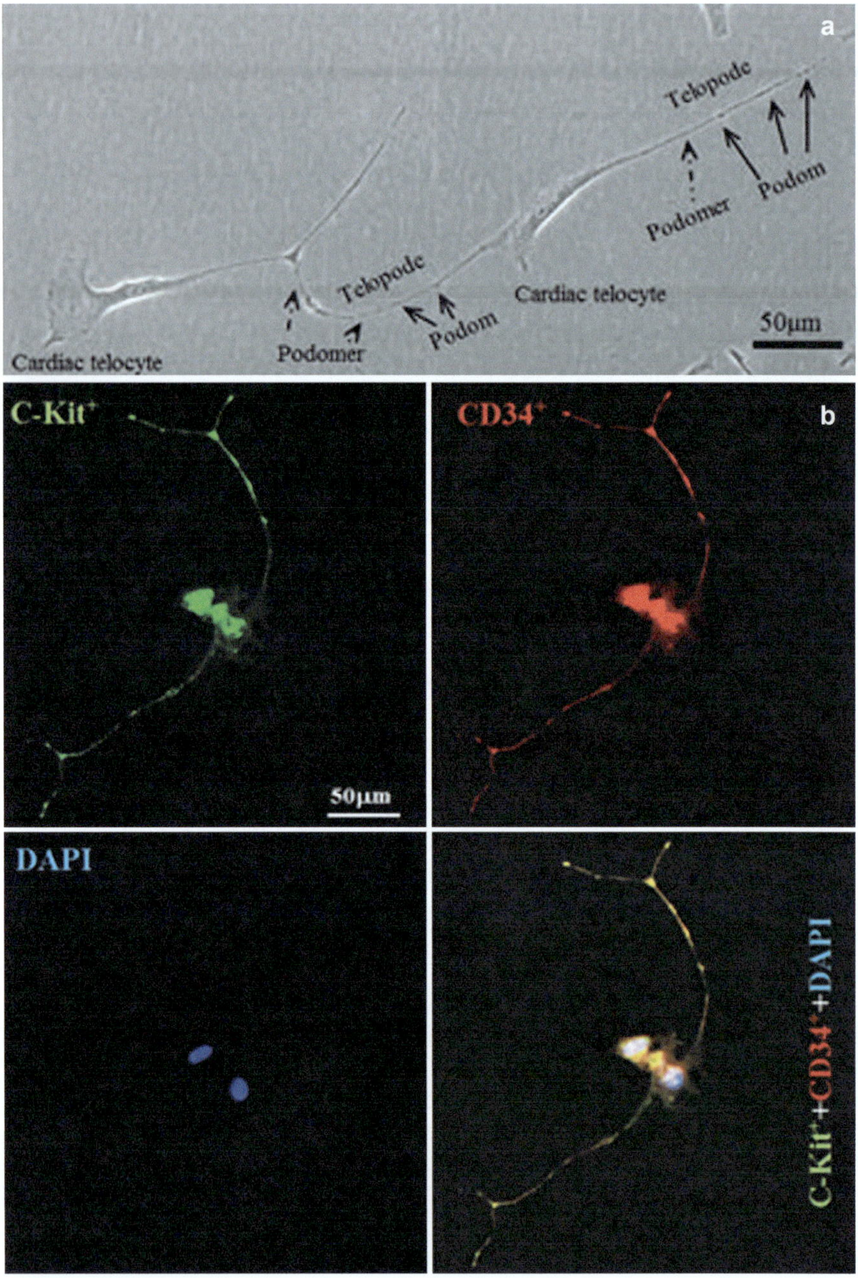

Fig. 16.2 Identification of rat cardiac telocytes in culture. (**a**) Primary culture of isolated cardiac telocytes reveals that under phase-contrast microscopy cardiac telocytes display piriform/spindle/triangular cell bodies and very long and slender telopodes formed by the alternation of small dilated segments (podoms, *arrows*) and thin segments (podomers, *dotted line arrows*). (**b**) Cardiac telocytes with unique morphology are c-kit+ and CD34+ (Adapted with permission from Zhao et al. [30])

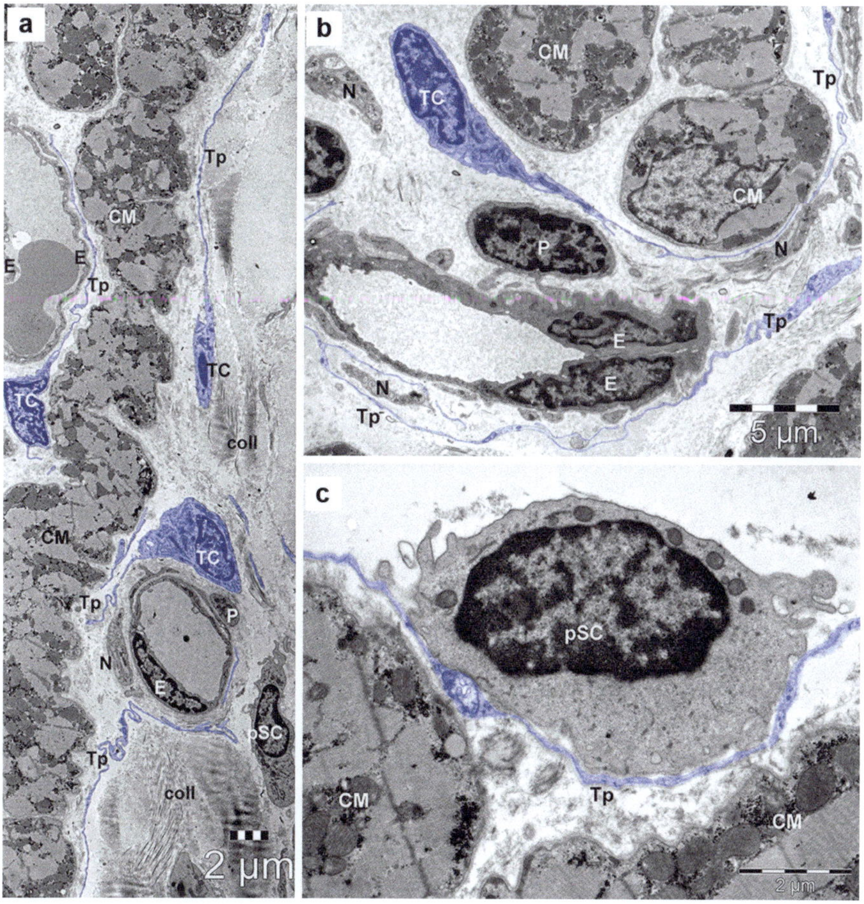

Fig. 16.3 (**a–c**) Representative transmission electron micrographs of human atrial interstitium. (**a**, **b**) General views of human atrial interstitium showing the distribution of telocytes and their telopodes. (**c**) A telopode is enfolding a putative stem cell with very few mitochondria and numerous ribosomes in the cytoplasm. *TC* telocyte, *Tp* telopode, *CM* cardiomyocyte, *E* endothelial cell, *P* pericyte, *N* nerve ending, *pSC* putative stem cell, *coll* collagen (Adapted with permission from Popescu et al. [55])

Besides intercellular contacts, paracrine signaling also plays an important role in the crosstalk between different cardiac cells, contributing substantially to cardiac physiology, responses to injury, regeneration and repair [26]. One pivotal component of this paracrine signaling machinery is represented by different specialized subtypes of EVs, such as exosomes and shedding microvesicles [2, 12, 26]. Indeed, growing evidence indicates that both types of EVs may function as shuttles to translocate genetic material (e.g. mRNAs and miRNAs) between cells over a long distance thereby modulating the gene expression and phenotype of the recipient cells [2, 12, 26]. In particular, recent studies demonstrated the importance of exosome-encased miRNAs in multiple intercellular communications within the cardiovascular system [6]. This specific exchange mechanism may be of crucial importance in cardiac tissue regeneration and remodeling [6, 11, 26]. For instance, it has been shown that post-mytotic cardiomyocytes are capable to deliver miRNAs to cardiac stem cells promoting their differentiation [64]. Moreover, stem cell-derived exosomes contain cardioprotective enzymes, which may exert beneficial effects on cardiomyocytes as demonstrated in a rat model of myocardial infarction and reperfusion [6, 65]. Noteworthy, it also appears that different types of TC-released EVs may act as important transporters for paracrine molecular signal exchange between cardiac TCs and cardiomyocytes or tissue-resident progenitor cells [25, 26, 33]. Using transmission electron microscopy and electron tomography, it could be demonstrated that cardiac TCs in culture release at least three different types of EVs, namely exosomes released from intracellular endosomes, ectosomes budding directly from the plasma membrane, and multivesicular cargos, these latter containing tightly packaged endomembrane-bound vesicles (Fig. 16.4) [25]. Electron tomography further highlighted that such endomembrane vesicles are released into the extracellular space as a cargo enclosed by plasma membranes [25]. These differ-

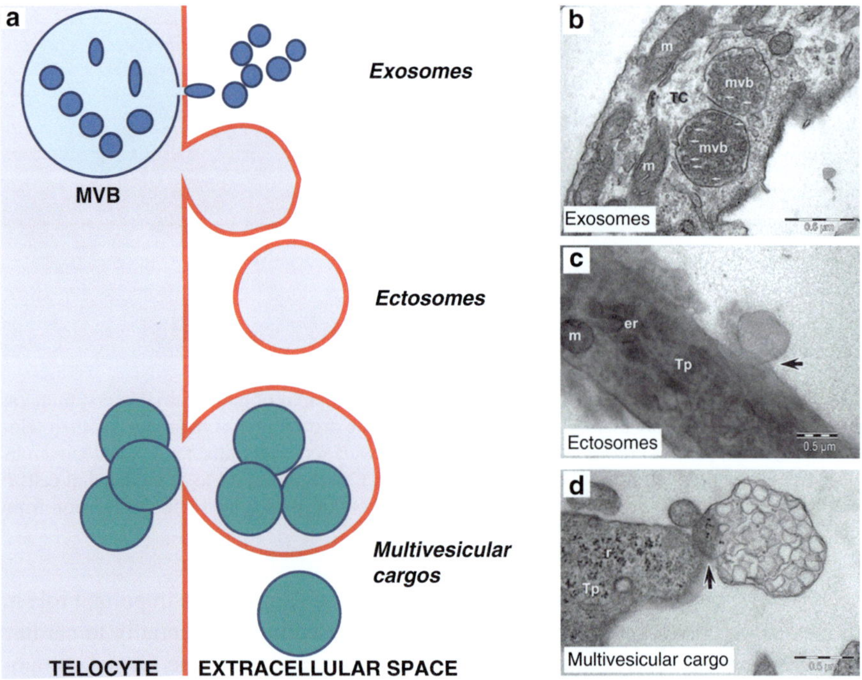

Fig. 16.4 (**a**) Schematic representation of the three types of extracellular vesicles released by cardiac telocytes, namely exosomes, ectosomes and multivesicular cargos. (**b–d**) Electron microscopy of cardiac telocytes in culture demonstrates: (**b**) the presence of numerous intraluminal vesicles (*small arrows*) in two multivesicular bodies, precursors of exosomes; (**c**) the ectosome budding (*arrow*) from the plasma membrane of a telopode; (**d**) a multivesicular cargo emerging (*arrow*) from a telopode. *TC* telocyte, *Tp* telopode, *mvb* multivesicular bodies, *m* mitochondria, *er* endoplasmic reticulum, *r* ribosome (Adapted with permission from Fertig et al. [25])

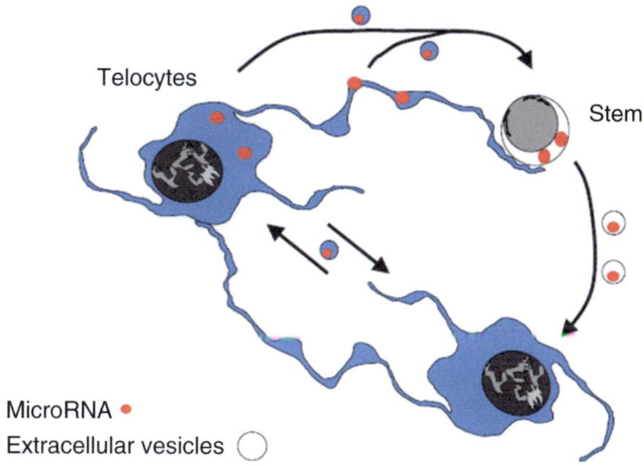

Fig. 16.5 Cardiac telocytes and stem cells exchange microRNAs by extracellular vesicles (Reproduced with permission from Cismaşiu and Popescu [33])

ent types of EVs, which are also released in situ by TCs within the cardiac tissue, likely represent an essential component of the intercellular signaling machinery of cardiac TCs and may be directly involved in the complex physiological and regenerative mechanisms of the heart [26]. Of note, in vitro studies have shown that the secretome of myocardial TCs may modulate the activity and increase the self-renewal capacity of cardiac stem cells [41]. Using fluorescent labeling of cells and EVs with calcein and Cy5-miR-21 oligos, it could be demonstrated that cardiac TCs deliver EVs loaded with miRNAs to cardiac stem cells [33]. Similarly, cardiac stem cells were found to deliver miRNA-loaded EVs to TCs, suggesting the existence of a reciprocal (bidirectional) post-transcriptional signaling between cardiac TCs and stem cells (Fig. 16.5) [33]. Collectively, the aforementioned observations support the notion that cardiac TCs may exert an epigenetic control over stem and progenitor cells, thus contributing to the regulation of post-natal cardiac tissue homeostasis and renewal. Depending of the specific types of miRNAs delivered by EVs, cardiac TCs might contribute substantially to the local balance between quiescent and proliferative states of stem cells, as well as between self-renewal and differentiation of putative cardiomyocyte progenitors [26, 33].

16.4 Telocytes in Cardiovascular Diseases

The infarcted myocardium experiences a loss of cardiomyocytes via ischemia-induced necrosis and apoptosis followed by neoangiogenesis and fibrotic changes, resulting into pathological tissue remodeling and, eventually, end-stage organ failure [26]. Recently, it has been demonstrated that the number of myocardial TCs is also dramatically decreased during heart failure due to dilated, ischemic or inflammatory cardiomyopathy [29, 66]. In particular, in the failing human heart TCs exhibit several ultrastructural degenerative changes culminating into apoptotic cell death [66]. It was also shown that the composition of the extracellular matrix may substantially influence the distribution of TCs within the cardiac interstitium [66].

Indeed, in fibrotic areas of the failing myocardium which were characterized by the deposition of tightly packed collagen fibers, the TCs and Tps were severely reduced or even almost completely undetectable. Moreover, the few remaining TCs exhibited a variety of ultrastructural alterations, such as cytoplasmic vacuolization and shrinkage/shortening of the Tps along with the loss of the typical Tp labyrinthine arrangement [66]. On the contrary, in interstitial areas rich in amorphous material, TCs were more numerous and displayed typical morphological features and organization of Tps. As further evidenced by semiquantitative analysis, the number of cardiac TCs and Tps was negatively correlated to the amount of mature fibrillar collagens [66]. Therefore, the interstitial distribution of TCs and Tps appear to closely reflect any quantitative and qualitative changes in the extracellular matrix composition of the failing human myocardium. Of note, TCs might also be involved in the formation of cardiac amyloid deposits in patients with long-standing atrial fibrillation [67]. In particular, Tps were found to intimately surround the amyloid deposits, likely in the attempt to prevent their expansion in the adjacent areas of the cardiac interstitium [67].

The pathophysiological consequences of the TC reduction and loss in the failing human heart are not completely understood, but it has been proposed that such a severe impairment of the TC interstitial network could largely contribute to the disruption of the normal three-dimensional myocardial organization and complex intercellular signaling mechanisms [29, 66]. Besides building a supportive structural network within the myocardial stroma, in the adult heart TCs have also been detected in the cardiogenic niches, where they establish close communications and may exchange paracrine signals through EVs with the tissue-resident stem cells possibly acting as nursing and guiding cells [27, 28, 33]. Therefore, the extensive damage and reduction of TCs occurring in the failing heart may profoundly hamper the TC ability to maintain stem cell niches with consequent impairment and loss of the pool of cardiac stem cells and putative cardiomyocyte progenitors [29]. Of note, experimental studies in a rat model of myocardial infarction showed that TCs were strongly reduced in fibrotic zones of the myocardium, while exogenous transplantation of cardiac TCs in the infarcted and border zones effectively decreased the infarction size with significant improvement of post-infarcted cardiac function [30, 68]. Histological analyses further revealed that a reconstruction of the stromal network of TCs paralleled by an impressive reduction in tissue fibrosis occurred in the exogenous TC-injected myocardium [30, 68]. The aforementioned beneficial effects could also depend on the ability of transplanted cardiac TCs to promote the expansion, recruitment and differentiation of local cardiomyocyte progenitors [26, 30, 68]. In another study, transplantation of human induced pluripotent stem cell-derived mesenchymal stem cells was able to reduce myocardial infarction improving cardiac function in mice, and this positive effect was accompanied by the rebuilding of the interstitial network of TCs within the infarcted myocardium [69]. Interestingly, TCs may also behave as key players in neoangiogenesis after experimental acute myocardial infarction [70]. In fact, it was demonstrated that TCs are markedly increased in the border zone of the infarcted myocardium during the post-infarction neoangiogenesis phase, with multiple Tps exhibiting numerous close intercellular connections either with pre-existing or neoformed microvessels [70]. Besides such physical contacts, TCs may presumably contribute to neoangiognesis through paracrine secretion of proangiogenic factors, including VEGF, NOS2 and several proangiogenic miRNAs (e.g. let-7e, 10a, 21, 27b, 100, 126-3p, 130a, 143, 155, and 503) [70].

Overall, the currently available experimental data support the possible therapeutic application of exogenous TC transplantation in the treatment of cardiac diseases. Nevertheless, further preclinical in vivo studies and the use of in vitro cardiac tissue engineering will help to better decipher the specific roles exerted by TCs during cardiac repair and regenerative processes [53, 62]. In this context, it is worth mentioning that a recent study highlighted the importance of TCs in the architectural organization of three-dimensional engineered heart tissue [71]. Indeed, electron microscopy revealed that typical TCs surrounded and contacted the cardiomyocytes with their long Tps exhibiting cardiomyocyte nursing properties during the construction of engineered heart tissue. Thus, engineered heart tissues may represent a very useful model system to clarify the specific functions of TCs during cardiac morphogenesis and post-injury regeneration [71].

16.5 The Potential Utility of Telocyte-Derived Exosomes in Cardiac Homeostasis, Regeneration and Repair

Exosomes seem to play a preferential role in the paracrine crosstalk between different cardiac cells, making substantial contribution to cardiac physiology, response to injury, regeneration and repair [6, 11, 26]. Therefore, exosomes are now being increasingly investigated for their possible diagnostic and therapeutic use in cardiovascular diseases [6, 11, 26, 72]. Interestingly, intramyocardial delivery of stem cell-derived exosomes resulted in the reduction of cardiomyocyte apoptosis and fibrosis, stimulated neoangiogenesis, and ameliorated cardiac function after experimental myocardial infarction [73–76]. In this experimental setting, the cardioprotective effects of exosomes were mostly ascribed to their enriched content in a variety of angiogenic and anti-apoptotic miRNAs [73, 74].

Besides tissue-resident stem cell-derived exosomes, either electron microscopy or electron tomography studies provided direct evidence that cardiac TCs are an additional important source of exosomes within the heart microenvironment [25, 26, 33]. As already mentioned, increased numbers of cardiac TCs were observed in the proangiogenic phase of the post-infarcted heart, with numerous secreted exosomes being detectable around the TC cell bodies and Tps [70]. Of note, TC-derived exosomes bear a cocktail of molecular signals (e.g. proangiogenic miRNAs) which may regulate the activity of neighboring vascular endothelial cells with consequent promotion of angiogenesis via an epigenetic paracrine mechanism [70]. In vitro studies further suggest that cardiac TCs could influence mesenchymal stem cell functions via exosomes [33, 41]. Considering that intramyocardial injection of cardiac TCs showed therapeutic utility in reducing the infarction size and myocardial fibrosis in rodent models, further research is required to elucidate the possible implication of TC-secreted exosomes in cardiac response to injury [30, 68]. An in-depth molecular characterization of cardiac TC-derived exosomes will also be of fundamental importance to decipher the TC paracrine machinery and its possible targeting in cardiovascular diseases. Anticipating a new avenue for potential therapeutic applications, TC-derived exosomes could be employed in the not-too-distant future as novel therapeutic nanovectors to deliver specific biological signals that may foster cardiomyocyte survival, cardiac neovascularization, and tissue-resident progenitor cell activation/differentiation to promote myocardial regeneration and repair [26, 72, 77].

16.6 Conclusions and Future Perspectives

Growing evidence supports a pivotal role of TCs in cardiac pathophysiology [16, 26, 29, 47]. Among cardiac interstitial cells, TCs appear to possess the unique ability to organize a proper three-dimensional scaffold consisting of their cell bodies and very long and convoluted Tps, and stimulate the growth and differentiation of putative cardiomyocyte progenitors to build the complex multicellular architecture of the heart [16, 21]. In fact, due to their close spatial relationship and intimate connections with other cell types, TCs are seen as 'connecting cells' mostly specialized to orchestrate the intercellular signaling mechanisms that constitute the basis for either a proper heart development or the maintenance of cardiac homeostasis in post-natal life [16, 21, 26]. In this context, recent studies have highlighted that multiple paracrine signaling effects possibly exerted by TCs in the adult heart largely depend on the secretion of EVs and, in particular, exosomes [2, 25, 26, 33, 41]. On the basis of the current knowledge, it is believed that the exogenous transplantation of TCs or the delivery of TC-derived exosomes might have great potential as future therapeutic strategies to foster cardiac regeneration and repair [26, 30, 33, 41, 68]. Given their biophysical properties, among the various types of EVs exosomes are particularly easy to isolate and their mRNA, miRNA and protein contents can be easily manipulated for therapeutic purposes [6]. Therefore, the possible use of exosomes, either natural exosomes or exosome-mimetic nanovesicles, as carriers/vectors of biological or synthetic therapeutics might be a promising strategy to allow efficient delivery of drugs across different physiological barriers to specific target cells [6]. Interestingly, beneficial effects of stem cell-derived exosomes for ischemic myocardial tissue regeneration and repair have already been reported in different preclinical studies [6, 78, 79]. As far as cardiac TCs are concerned, the specific molecular cargo of their exosomes and the mechanisms that promote their secretion still require a thorough characterization. The exact biodistribution of cardiac TC-derived exosomes also remains to be established. Finally, whether the manipulation of cardiac TC-released exosomes might represent a novel therapeutic strategy to counteract heart failure and other cardiovascular diseases will need to be comprehensively addressed in future translational studies.

References

1. Vader P, Breakefield XO, Wood MJ (2014) Extracellular vesicles: emerging targets for cancer therapy. Trends Mol Med 20(7):385–393
2. Cretoiu D, Xu J, Xiao J, Cretoiu SM (2016) Telocytes and their extracellular vesicles—evidence and hypotheses. Int J Mol Sci 17(8):1322
3. Corrado C, Raimondo S, Chiesi A, Ciccia F, De Leo G, Alessandro R (2013) Exosomes as intercellular signaling organelles involved in health and disease: basic science and clinical applications. Int J Mol Sci 14(3):5338–5366
4. Ratajczak J, Wysoczynski M, Hayek F, Janowska-Wieczorek A, Ratajczak MZ (2006) Membrane-derived microvesicles: important and underappreciated mediators of cell-to-cell communication. Leukemia 20(9):1487–1495
5. Camussi G, Deregibus MC, Bruno S, Cantaluppi V, Biancone L (2010) Exosomes/microvesicles as a mechanism of cell-to-cell communication. Kidney Int 78(9):838–848
6. Emanueli C, Shearn AI, Angelini GD, Sahoo S (2015) Exosomes and exosomal miRNAs in cardiovascular protection and repair. Vasc Pharmacol 71:24–30
7. EL Andaloussi S, Mäger I, Breakefield XO, Wood MJ (2013) Extracellular vesicles: biology and emerging therapeutic opportunities. Nat Rev Drug Discov 12(5):347–357
8. Zhang B, Yeo RW, Tan KH, Lim SK (2016) Focus on extracellular vesicles: therapeutic potential of stem cell-derived extracellular vesicles. Int J Mol Sci 17(2):174
9. Batrakova EV, Kim MS (2015) Using exosomes, naturally-equipped nanocarriers, for drug delivery. J Control Release 219:396–405
10. Ren J, He W, Zheng L, Duan H (2016) From structures to functions: insights into exosomes as promising drug delivery vehicles. Biomater Sci 4(6):910–921
11. Zhang Y, Hu YW, Zheng L, Wang Q (2017) Characteristics and roles of exosomes in cardiovascular disease. DNA Cell Biol 36(3):202–211
12. Ratajczak MZ, Ratajczak D, Pedziwiatr D (2016) Extracellular microvesicles (ExMVs) in cell to cell communication: a role of telocytes. Adv Exp Med Biol 913:41–49
13. Popescu LM, Faussone-Pellegrini MS (2010) Telocytes—a case of serendipity: the winding way from interstitial cells of Cajal (ICC), via interstitial Cajal-like cells (ICLC) to telocytes. J Cell Mol Med 14(4):729–740
14. Cretoiu SM, Popescu LM (2014) Telocytes revisited. Biomol Concepts 5(5):353–369
15. Faussone-Pellegrini MS, Gherghiceanu M (2016) Telocyte's contacts. Semin Cell Dev Biol 55:3–8
16. Bani D (2016) Telocytes in cardiac tissue architecture and development. Adv Exp Med Biol 913:127–137
17. Popescu LM, Manole CG, Gherghiceanu M, Ardelean A, Nicolescu MI, Hinescu ME, Kostin S (2010) Telocytes in human epicardium. J Cell Mol Med 14(8):2085–2093
18. Gherghiceanu M, Manole CG, Popescu LM (2010) Telocytes in endocardium: electron microscope evidence. J Cell Mol Med 14(9):2330–2334
19. Kostin S, Popescu LM (2009) A distinct type of cell in myocardium: interstitial Cajal-like cells (ICLCs). J Cell Mol Med 13(2):295–308
20. Kostin S (2010) Myocardial telocytes: a specific new cellular entity. J Cell Mol Med 14(7):1917–1921
21. Bani D, Formigli L, Gherghiceanu M, Faussone-Pellegrini MS (2010) Telocytes as supporting cells for myocardial tissue organization in developing and adult heart. J Cell Mol Med 14(10):2531–2538
22. Yang Y, Sun W, Wu SM, Xiao J, Kong X (2014) Telocytes in human heart valves. J Cell Mol Med 18(5):759–765
23. Gherghiceanu M, Popescu LM (2012) Cardiac telocytes—their junctions and functional implications. Cell Tissue Res 348(2):265–279
24. Rusu MC, Pop F, Hostiuc S, Curca GC, Jianu AM, Paduraru D (2012) Telocytes form networks in normal cardiac tissues. Histol Histopathol 27(6):807–816

25. Fertig ET, Gherghiceanu M, Popescu LM (2015) Extracellular vesicles release by cardiac telocytes: electron microscopy and electron tomography. J Cell Mol Med 18(10):1938–1943
26. Bei Y, Zhou Q, Sun Q, Xiao J (2016) Telocytes in cardiac regeneration and repair. Semin Cell Dev Biol 55:14–21
27. Gherghiceanu M, Popescu LM (2010) Cardiomyocyte precursors and telocytes in epicardial stem cell niche: electron microscope images. J Cell Mol Med 14(4):871–877
28. Popescu LM, Gherghiceanu M, Manole CG, Faussone-Pellegrini MS (2009) Cardiac renewing: interstitial Cajal-like cells nurse cardiomyocyte progenitors in epicardial stem cell niches. J Cell Mol Med 13(5):866–886
29. Ibba-Manneschi L, Rosa I, Manetti M (2016) Telocyte implications in human pathology: an overview. Semin Cell Dev Biol 55:62–69
30. Zhao B, Chen S, Liu J, Yuan Z, Qi X, Qin J, Zheng X, Shen X, Yu Y, Qnin TJ, Chan JY, Cai D (2013) Cardiac telocytes were decreased during myocardial infarction and their therapeutic effects for ischaemic heart in rat. J Cell Mol Med 17(1):123–133
31. Xiao J, Bei Y (2016) Decoding telocytes. Adv Exp Med Biol 913:23–39
32. Faussone-Pellegrini MS, Popescu LM (2011) Telocytes. Biomol Concepts 2(6):481–489
33. Cismaşiu VB, Popescu LM (2015) Telocytes transfer extracellular vesicles loaded with microRNAs to stem cells. J Cell Mol Med 19(2):351–358
34. Vannucchi MG, Traini C, Manetti M, Ibba-Manneschi L, Faussone-Pellegrini MS (2013) Telocytes express PDGFRα in the human gastrointestinal tract. J Cell Mol Med 17(9):1099–1108
35. Zhou Q, Wei L, Zhong C, Fu S, Bei Y, Huică RI, Wang F, Xiao J (2015) Cardiac telocytes are double positive for CD34/PDGFR-α. J Cell Mol Med 19(8):2036–2042
36. Vannucchi MG, Faussone-Pellegrini MS (2016) The telocyte subtypes. Adv Exp Med Biol 913:115–126
37. Cretoiu D, Radu BM, Banciu A, Banciu DD, Cretoiu SM (2016) Telocytes heterogeneity: from cellular morphology to functional evidence. Semin Cell Dev Biol 64:26–39
38. Chang Y, Li C, Lu Z, Li H, Guo Z (2015) Multiple immunophenotypes of cardiac telocytes. Exp Cell Res 338(2):239–244
39. Sun X, Zheng M, Zhang M, Qian M, Zheng Y, Li M, Cretoiu D, Chen C, Chen L, Popescu LM, Wang X (2014) Differences in the expression of chromosome 1 genes between lung telocytes and other cells: mesenchymal stem cells, fibroblasts, alveolar type II cells, airway epithelial cells and lymphocytes. J Cell Mol Med 18(5):801–810
40. Zheng Y, Cretoiu D, Yan G, Cretoiu SM, Popescu LM, Wang X (2014) Comparative proteomic analysis of human lung telocytes with fibroblasts. J Cell Mol Med 18(4):568–589
41. Albulescu R, Tanase C, Codrici E, Popescu DI, Cretoiu SM, Popescu LM (2015) The secretome of myocardial telocytes modulates the activity of cardiac stem cells. J Cell Mol Med 19(8):1783–1794
42. Cismaşiu VB, Radu E, Popescu LM (2011) miR-193 expression differentiates telocytes from other stromal cells. J Cell Mol Med 15(5):1071–1074
43. Zheng Y, Zhang M, Qian M, Wang L, Cismasiu VB, Bai C, Popescu LM, Wang X (2013) Genetic comparison of mouse lung telocytes with mesenchymal stem cells and fibroblasts. J Cell Mol Med 17(4):567–577
44. Yoon YJ, Kim OY, Gho YS (2014) Extracellular vesicles as emerging intercellular communicasomes. BMB Rep 47(10):531–539
45. Diaz-Flores L, Gutierrez R, Garcia MP, Sáez FJ, Díaz-Flores L Jr, Valladares F, Madrid JF (2014) CD34+ stromal cells/fibroblasts/fibrocytes/telocytes as a tissue reserve and a principal source of mesenchymal cells. Location, morphology, function and role in pathology. Histol Histopathol 29(7):831–870
46. Ibba-Manneschi L, Rosa I, Manetti M (2016) Telocytes in chronic inflammatory and fibrotic diseases. Adv Exp Med Biol 913:51–76
47. Kostin S (2016) Cardiac telocytes in normal and diseased hearts. Semin Cell Dev Biol 55:22–30
48. Zheng Y, Bai C, Wang X (2012) Telocyte morphologies and potential roles in diseases. J Cell Physiol 227(6):2311–2317
49. Manetti M, Rosa I, Messerini L, Ibba-Manneschi L (2015) Telocytes are reduced during fibrotic remodelling of the colonic wall in ulcerative colitis. J Cell Mol Med 19(1):62–73

50. Milia AF, Ruffo M, Manetti M, Rosa I, Conte D, Fazi M, Messerini L, Ibba-Manneschi L (2013) Telocytes in Crohn's disease. J Cell Mol Med 17(12):1525–1536
51. Manetti M, Guiducci S, Ruffo M, Rosa I, Faussone-Pellegrini MS, Matucci-Cerinic M, Ibba-Manneschi L (2013) Evidence for progressive reduction and loss of telocytes in the dermal cellular network of systemic sclerosis. J Cell Mol Med 17(4):482–496
52. Fu S, Wang F, Cao Y, Huang Q, Xiao J, Yang C, Popescu LM (2015) Telocytes in human liver fibrosis. J Cell Mol Med 19(3):676–683
53. Bei Y, Wang F, Yang C, Xiao J (2015) Telocytes in regenerative medicine. J Cell Mol Med 19(7):1441–1454
54. Faussone-Pellegrini MS, Bani D (2010) Relationships between telocytes and cardiomyocytes during pre- and post-natal life. J Cell Mol Med 14(5):1061–1063
55. Popescu LM, Curici A, Wang E, Zhang H, Hu S, Gherghiceanu M (2015) Telocytes and putative stem cells in ageing human heart. J Cell Mol Med 19(1):31–45
56. Xiao J, Chen P, Qu Y, Yu P, Yao J, Wang H, Fu S, Bei Y, Chen Y, Che L, Xu J (2016) Telocytes in exercise-induced cardiac growth. J Cell Mol Med 20(5):973–979
57. Niculite CM, Regalia TM, Gherghiceanu M, Huica R, Surcel M, Ursaciuc C, Leabu M, Popescu LM (2015) Dynamics of telopodes (telocyte prolongations) in cell culture depends on extracellular matrix protein. Mol Cell Biochem 398(1–2):157–164
58. Bei Y, Zhou Q, Fu S, Lv D, Chen P, Chen Y, Wang F, Xiao J (2015) Cardiac telocytes and fibroblasts in primary culture: different morphologies and immunophenotypes. PLoS One 10(2):e0115991
59. Cretoiu D, Hummel E, Zimmermann H, Gherghiceanu M, Popescu LM (2014) Human cardiac telocytes: 3D imaging by FIB-SEM tomography. J Cell Mol Med 18(11):2157–2164
60. Gherghiceanu M, Popescu LM (2011) Heterocellular communication in the heart: electron tomography of telocyte-myocyte junctions. J Cell Mol Med 15(4):1005–1011
61. Shim W (2016) Myocardial telocytes: a new player in electric circuitry of the heart. Adv Exp Med Biol 913:241–251
62. Zhou J, Zhang Y, Wen X, Cao J, Li D, Lin Q, Wang H, Liu Z, Duan C, Wu K, Wang C (2010) Telocytes accompanying cardiomyocyte in primary culture: two- and three-dimensional culture environment. J Cell Mol Med 14(11):2641–2645
63. Popescu LM, Fertig ET, Gherghiceanu M (2016) Reaching out: junctions between cardiac telocytes and cardiac stem cells in culture. J Cell Mol Med 20(2):370–380
64. Hosoda T, Zheng H, Cabral-da-Silva M, Sanada F, Ide-Iwata N, Ogórek B, Ferreira-Martins J, Arranto C, D'Amario D, del Monte F, Urbanek K, D'Alessandro DA, Michler RE, Anversa P, Rota M, Kajstura J, Leri A (2011) Human cardiac stem cell differentiation is regulated by a mircrine mechanism. Circulation 123(12):1287–1296
65. Lai RC, Arslan F, Lee MM, Sze NS, Choo A, Chen TS, Salto-Tellez M, Timmers L, Lee CN, El Oakley RM, Pasterkamp G, de Kleijn DP, Lim SK (2010) Exosome secreted by MSC reduces myocardial ischemia/reperfusion injury. Stem Cell Res 4(3):214–222
66. Richter M, Kostin S (2015) The failing human heart is characterized by decreased numbers of telocytes as result of apoptosis and altered extracellular matrix composition. J Cell Mol Med 19(11):2597–2606
67. Mandache E, Gherghiceanu M, Macarie C, Kostin S, Popescu LM (2010) Telocytes in human isolated atrial amyloidosis: ultrastructural remodelling. J Cell Mol Med 14(12):2739–2747
68. Zhao B, Liao Z, Chen S, Yuan Z, Yilin C, Lee KK, Qi X, Shen X, Zheng X, Quinn T, Cai D (2014) Intramyocardial transplantation of cardiac telocytes decreases myocardial infarction and improves post-infarcted cardiac function in rats. J Cell Mol Med 18(5):780–789
69. Miao Q, Shim W, Tee N, Lim SY, Chung YY, Ja KP, Ooi TH, Tan G, Kong G, Wei H, Lim CH, Sin YK, Wong P (2014) iPSC-derived human mesenchymal stem cells improve myocardial strain of infarcted myocardium. J Cell Mol Med 18(8):1644–1654
70. Manole CG, Cismaşiu V, Gherghiceanu M, Popescu LM (2011) Experimental acute myocardial infarction: telocytes involvement in neo-angiogenesis. J Cell Mol Med 15(11):2284–2296
71. Zhou J, Wang Y, Zhu P, Sun H, Mou Y, Duan C, Yao A, Lv S, Wang C (2014) Distribution and characteristics of telocytes as nurse cells in the architectural organization of engineered heart tissues. Sci China Life Sci 57(2):241–247

72. Ong SG, Wu JC (2015) Exosomes as potential alternatives to stem cell therapy in mediating cardiac regeneration. Circ Res 117(1):7–9
73. Barile L, Lionetti V, Cervio E, Matteucci M, Gherghiceanu M, Popescu LM, Torre T, Siclari F, Moccetti T, Vassalli G (2014) Extracellular vesicles from human cardiac progenitor cells inhibit cardiomyocyte apoptosis and improve cardiac function after myocardial infarction. Cardiovasc Res 103(4):530–541
74. Wang Y, Zhang L, Li Y, Chen L, Wang X, Guo W, Zhang X, Qin G, He SH, Zimmerman A, Liu Y, Kim IM, Weintraub NL, Tang Y (2015) Exosomes/microvesicles from induced pluripotent stem cells deliver cardioprotective miRNAs and prevent cardiomyocyte apoptosis in the ischemic myocardium. Int J Cardiol 192:61–69
75. Khan M, Nickoloff E, Abramova T, Johnson J, Verma SK, Krishnamurthy P, Mackie AR, Vaughan E, Garikipati VN, Benedict C, Ramirez V, Lambers E, Ito A, Gao E, Misener S, Luongo T, Elrod J, Qin G, Houser SR, Koch WJ, Kishore R (2015) Embryonic stem cell-derived exosomes promote endogenous repair mechanisms and enhance cardiac function following myocardial infarction. Circ Res 117(1):52–64
76. Teng X, Chen L, Chen W, Yang J, Yang Z, Shen Z (2015) Mesenchymal stem cell-derived exosomes improve the microenvironment of infarcted myocardium contributing to angiogenesis and anti-inflammation. Cell Physiol Biochem 37(6):2415–2424
77. Chistiakov DA, Orekhov AN, Bobryshev YV (2016) Cardiac extracellular vesicles in normal and infarcted heart. Int J Mol Sci 17(1):63
78. Ibrahim AG, Cheng K, Marban E (2014) Exosomes as critical agents of cardiac regeneration triggered by cell therapy. Stem Cell Reports 2(5):606–619
79. Ong SG, Lee WH, Huang M, Dey D, Kodo K, Sanchez-Freire V, Gold JD, Wu JC (2014) Cross talk of combined gene and cell therapy in ischemic heart disease: role of exosomal microRNA transfer. Circulation 130(11 Suppl 1):S60–S69

Chapter 17
Circulating Exosomes in Cardiovascular Diseases

Yihua Bei, Ting Chen, Daniel Dumitru Banciu, Dragos Cretoiu, and Junjie Xiao

17.1 Circulating Exosomes and Exosomal Cargos

Numerous studies demonstrated that exosomes in the early phase are formed into a structure which is regarded as a multivesicular body (MVB) through endocytic invagination [1, 2]. Subsequently, the MVB fuses with the cytoplasmic membrane and is secreted with its cargos of lipids, proteins, functional mRNAs, and microRNAs (miRNAs, miRs) into the extracellular environment. The Rab-family GTPases, Annexins, SNAREs, and Endosomal Sorting Complexes Required for Transport (ESCRT) associated proteins are essentially involved in the formation and secretion of exosomes [2, 3]. Some of the exosomes are eventually released into the circulation, known as circulating exosomes [4]. Circulating exosomes could arrive in

Y. Bei • J. Xiao (✉)
Cardiac Regeneration and Ageing Lab, School of Life Science, Shanghai University, 333 Nan Chen Road, Shanghai 200444, China
e-mail: junjiexiao@shu.edu.cn

T. Chen
Laboratory for Advanced Interdisciplinary Research, Center for Personalized Medicine/ Institutes of Translational Medicine, First Affiliated Hospital, Wenzhou Medical University, Wenzhou 325000, China

D.D. Banciu
Research Beyond Limits SRL, 15 Dimitrie Cantemir Blvd., Bl.9, Sc.C, Ap.107, Sector 4, 040234 Bucharest, Romania

University of Bucharest, 36-46 Mihail. Kogălniceanu Bld., Sector 5, 050107 Bucharest, Romania

D. Cretoiu
Victor Babes National Institute of Pathology, Bucharest 050096, Romania

Division of Cellular and Molecular Biology and Histology, Carol Davila University of Medicine and Pharmacy, Bucharest 050474, Romania

J. Xiao, S. Cretoiu (eds.), *Exosomes in Cardiovascular Diseases*, Advances in Experimental Medicine and Biology 998, DOI 10.1007/978-981-10-4397-0_17

distant tissues via blood circulation, thus directly communicating with target cells and rapidly regulating intracellular signalings.

In various physiological and pathological conditions, different patterns of proteins, lipids, and non-coding RNAs such as miRNAs can be detected in the circulation [5, 6]. The cell-free non-coding RNAs could be stably present in blood circulation via being packaged into exosomes [7]. The circulating exosomes can be uptaken by recipient cells, whereby transferring the composite cargos or activating the signaling pathways [8–11]. Particularly, the various types of cargos loaded in exosomes and the signaling diversity are closely related to the different tissue and cell types from which exosomes are originated [12–15]. Among the diverse exosomal cargos, miRNAs can effectively regulate the target genes and influence the biological functions of target cells. miRNAs are a large group of small (18–25 nucleotides in length) noncoding RNAs that regulate target gene expressions at post transcriptional level [16, 17]. It has been increasingly reported that exosomal components, especially miRNAs, play important roles in regulating cardiac function and protecting the heart against acute myocardial infarction (AMI) and ischemia reperfusion injury (IRI) [18, 19]. For example, exosomes derived from chemokine receptor CXCR4-overexpressing mesenchymal stem cells (MSCs) were reported to activate the IGF-1/PI3K/Akt signaling pathway in cardiomyocytes, thereby reducing myocardial apoptosis, promoting angiogenesis, decreasing ventricular remodeling, and protecting cardiac function after MI [20]. Since it is difficult to obtain cardiac tissue samples from patients, detecting changes of circulating exosomes from peripheral blood might be useful strategy to attain information about the pathophysiological processes of cardiovascular diseases [21–23] as well as to guide the treatment for patients [24–26].

17.2 Circulating Exosomes in Cardiovascular Pathophysiology

Intercellular communication is one of the essential mechanisms for cells exerting their biological functions in all multicellular organisms. Almost all cells exchange messages by direct interaction or the secretion of signaling molecules. Studies have revealed that circulating exosomes can mediate comprehensive interactions among various cell types and exert biological functions by transmitting exosomal cargos to recipient cells [2, 27]. Exosomes were proved to be secreted from the injured heart and participate in cardiovascular pathophysiology [28–30]. Although real success has been achieved in experimental studies of exosomes in cardiovascular physiological and pathological progresses, the molecular mechanisms remain incompletely understood [2, 31, 32].

Exosomes derived from cardiomyocytes are initially found under the hypoxia and reoxygenation condition, which may contain biological molecules such as HSP70 [33–35]. Likewise, exosomes function as messenger of intercellular

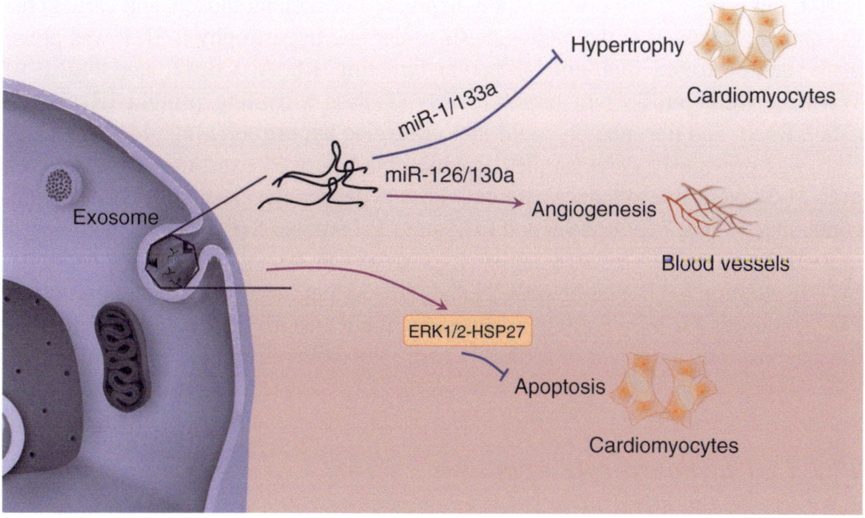

Fig. 17.1 Circulating exosomes regulate cardiomyocyte hypertrophy, apoptosis, and angiogenesis

communication among cardiomyocytes, fibroblasts, smooth muscle cells, and endothelial cells, and participate in the regulation of cardiac regeneration, ventricular remodeling, and angiogenesis in cardiovascular diseases [31]. Due to the perfect peculiarity as carriers of signal molecules, circulating exosomes deliver both protective and detrimental information [36–39]. Circulating exosomes generally regulate cardiovascular pathophysiology, such as cardiomyocyte hypertrophy, apoptosis, and angiogenesis (Fig. 17.1).

17.2.1 Cardiomyocyte Hypertrophy

Various forms of stress in the heart can contribute to activate cardiac myocyte hypertrophy [40, 41]. The general cardiac hypertrophy is characterized by myocyte enlargement and the re-expression of embryonic genes. Cardiomyocyte hypertrophy is a common response upon the increased heart hemodynamic state (such as high blood pressure or valvular stenosis), myocardial injury, and neurohormonal stress in the compensatory period. Early compensatory cardiac hypertrophy can be adapted to the enhanced post-ventricular load and maintain normal cardiac output. However, sustained cardiac hypertrophy will eventually lead to cardiac ventricular dilatation, reverse remodeling, and heart failure [40].

Circulating exosomes were reported to be involved in the regulation of pathological cardiac hypertrophy. Circulating exosomes loaded with miR-1 and miR-133a were found to be significantly increased in the serum of patients with AMI [42]. miR-1 and miR-133 are preferentially expressed in skeletal muscle and cardiac tissue and are involved in the pathogenesis of cardiac hypertrophy [43]. It was previously demonstrated that miR-133a via targeting RhoA, Cdc42, and NELF-A/WHSC2, while miR-1 via targeting Ras GTPase-activating protein (RasGAP), Cdk9, Rheb, and fibronectin, could inhibit cardiac hypertrophy [42, 44–46].

It was previously demonstrated that fibroblast-derived exosomes enriched with miR-21-3p were able to induce cardiomyocyte hypertrophy via targeting SH3 domain-containing protein 2 (SORBS2) and PDZ and LIM domain 5 (PDLIM5). Inhibition of miR-21-3p resulted in reduced cardiac hypertrophy in Angiotensin II-treated animals [47]. In addition to circulating miR-29 and miR-30 that have been identified as possible biomarkers for left ventricle hypertrophy, the relevance of circulating miR-21 in the diagnosis and prognosis of cardiac hypertrophy deserves further investigation [48].

17.2.2 Cardiomyocyte Apoptosis

Cardiomyocyte apoptosis is a significant issue underlying ischemic cardiac diseases [49], and occurs with dilated cardiomyopathy [50] and aging-related cardiac dysfunction [51]. Myocardial ischemic injury is associated with a shared characteristic patterns of cell death and metabolic changes which could result in irreversible myocardial injury [52, 53]. Apoptosis is involved in the whole process of myocardial ischemic injury, which could range from the initial phase after myocardial infarction to reperfusion stage [54, 55]. However, the specific molecular mechanisms underlying cardiomyocyte apoptosis are not fully understood.

Inhibition of miR-155 was previously demonstrated to inhibit cardiomyocyte apoptosis and cardiac dysfunction in lipopolysaccharide (LPS)-treated mice, via targeting Pea15a. Furthermore, increased circulating miR-155 was found to be associated with cardiac dysfunction in sepsis patients [56]. In this regard, the increased circulating miRNA-155, whether packaged in circulating exosomes or not, deserves further investigation in sepsis-induced cardiac dysfunction [56]. Notably, plasma exosomes isolated from healthy human and rats were recently demonstrated to be able to protect against cardiomyocyte apoptosis and ischemia reperfusion injury, indicating that endogenous circulating exosomes at baseline have protective effect for the heart [57].

17.2.3 Angiogenesis

Angiogenesis is a biological process of growing new vessels from the existing vascular structure and promoting endothelial cell proliferation to form vascular network. Many factors, such as fibroblast growth factor (FGF) and vascular endothelial growth factor (VEGF) can stimulate the formation of new vessels. Exosomes were reported to participate in the regulation of angiogenesis which is an essential process contributing to cardiac repair after injury. The CD34-positive stem cell-derived exosomes enriched with angiogenesis-related miR-126 and miR-130a were found to be significantly reduced in the peripheral blood of patients with chronic heart failure [58]. miR-126 and miR-130a were previously reported to stimulate angiogenesis by down-regulating the angiogenic negative regulator SPRED1 and HOXA5, respectively [59–61]. SPRED1, the member of Sprouty protein family, blocks angiogenesis through negatively regulating the VEGF-C/VEGFR-3 signaling [62]. HOXA5 also suppresses angiogenesis by upregulating the anti-angiogenic gene Thrombospondin-2. Besides that, HOXA5 also downregulates many pro-angiogenic genes including VEGFR2, Ephrin A1, HIF1alpha, and COX-2 [63].

17.3 Circulating Exosomes in Myocardial Ischemia Reperfusion Injury

The early reperfusion of the myocardium is considered as an important intervention in the treatment of myocardial ischemia which can efficiently attenuate further damage to the myocardium [64]. However, some infarct areas could be expanded when the blood flow regains after ischemia, which is known as myocardial ischemia reperfusion injury (MIRI) [65]. Ultimately, MIRI can lead to ventricular remodeling and even progressive heart failure [66, 67]. MIRI is associated with a complexity of multiple pathophysiological features [68], such as calcium overload, accumulation of oxygen free radicals, endothelial dysfunction, immune activation, mitochondrial dysfunction, cardiomyocyte apoptosis and autophagy, platelet aggregation, and microembolization [69–74]. However, the molecular mechanisms underlying MIRI are not completely understood.

Circulating exosomes can be markedly altered after MIRI and may serve as extracellular messengers through endocytosis, membrane fusion, and cell-receptor interaction to facilitate cell-cell communication [32]. Mounting evidence has shown that exosomes, especially stem cell-derived exosomes, have protective effects against MIRI [19, 28, 75, 76]. Mesenchymal stem cell-derived exosomes were demonstrated to promote cardiomyocyte viability and prevent adverse remodeling after MIRI, by enhancing the generation of ATP, reducing oxidative stress, and activating the PI3K/Akt pathway [28]. More interestingly, circulating exosomes isolated from healthy human and rats were also proved to be able to transmit signals to the heart and provide protective effects against MIRI [57]. The exosomes packed with HSP70, could activate Toll-like receptor 4 (TLR4) signaling and induce ERK1/2 and p38MAPK activation and subsequent HSP27 phosphorylation in cardiac myocytes (Fig. 17.2) [57]. Increasing evidence suggests that the activation of ERK1/2 and/or PI3K/AKT signaling pathways are crucial for the cardioprotective effects [77, 78]. HSP70, a member of small HSP family, can be loaded in exosomes [33] and is present in the circulation of normal individuals [79]. Moreover, the HSPs, especially HSP27 which is abundant in the myocardium, can be generated upon adverse stresses (e.g. heat) thus offering protective effects for the heart [80]. These studies highly suggest that circulating exosomes may provide a promosing non-cellular approach for the treatment of MIRI.

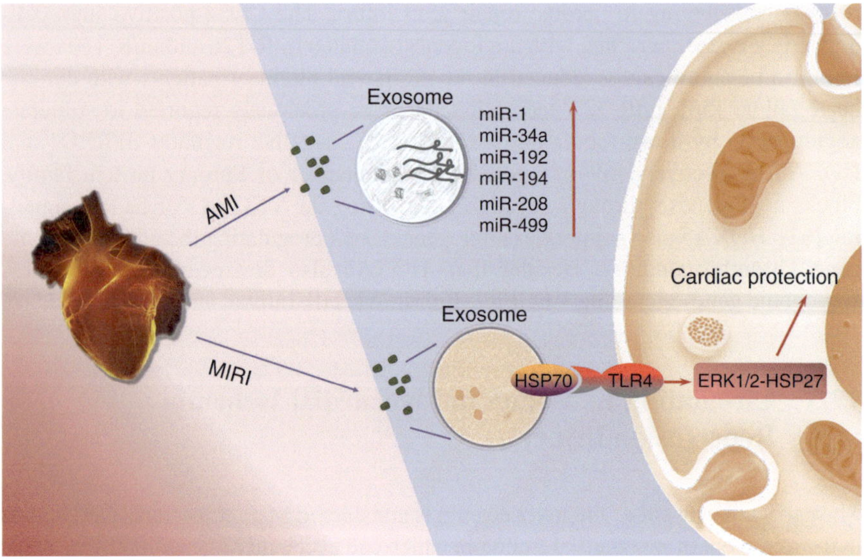

Fig. 17.2 Circulating exosomes contribute to the pathogenesis of myocardial infarction (MI) and myocardial ischemia reperfusion injury (MIRI)

17.4 Circulating Exosomes in Myocardial Infarction

Myocardial infarction (MI) is occurred when the flow of oxygen-rich blood is blocked in a section of myocardium, which is frequently caused by atherosclerosis-related coronary artery luminal occlusion and plaque rupture [81]. Simultaneously, MI is usually associated with a dramatic decrease of myocardial contractility and reduction of cardiac output [82]. In addition, MI may cause arrhythmia, cardiogenic shock, and heart failure. In pathophysiological aspects, cardiomyocyte apoptosis and necrosis are the essential causes of cardiomyocyte damage and loss in MI [83]. In the late stage, severe MI will ultimately progress to adverse cardiac remodeling and heart failure [84]. In these cases, controlling excessive inflammatory response, inhibiting cardiomyocyte death, preventing ventricular fibrosis, and facilitating angiogenesis are considered as potential therapeutic strategies for improving the prognosis of MI patients.

It has been reported that exosomes are highly involved in the pathophysiological processes of MI [20, 29]. Some exosomes derived from stem cells such as embryonic stem cells (ESCs), mesenchymal stem cells (MSCs), and cardiac progenitor cells (CPCs) were proved to improve cardiac function after MI, likely by reducing cardiomyocyte apoptosis, inhibiting myocardial fibrosis, and promoting angiogenesis [75, 85, 86]. However, some exosomes may exacerbate myocardial injury after MI and also be associated with vascular damage and cardiovascular risk [87, 88].

For example, exosomes containing HSP60, released from highly differentiated adult cardiomyocytes in an anoxic condition, are detrimental to cardiomyocytes during acute MI [34, 89]. Extracellular HSP60 was shown to cause cardiomyocyte apoptosis through activating TLR4 [90]. Nonetheless, HSP20 contained in circulating exosomes derived from cardiomyocytes was identified as a novel cardiokine which may promote myocardial neovascularization via activating vascular endothelial growth factor receptor 2 (VEGFR2) after MI [91].

Intriguingly, circulating miRNAs that are changed upon MI could also be packaged in the exosomes (Fig. 17.2). It was found that miR-1 and miR-208 which might be contained in exosomes were significantly increased in the serum of rats with AMI and in the urine of AMI patients [92]. Equally, the cardiac muscle-specific miRNAs including miR-208b and miR-499 were shown to be increased in the circulation of MI patients [93, 94]. As well, circulating p53-responsive miR-192, miR-194, and miR-34a, particularly enriched in exosomes, were significantly increased in the early stage of MI [95]. Notably, the miR-194 and miR-34a levels were correlated with left ventricle end-diastolic dimension 1 year after MI, indicating that circulating miR-194 and miR-34a might serve as predictors for heart failure development in MI patients [95].

17.5 Circulating Exosomes in Other Cardiovascular Diseases

17.5.1 Atherosclerosis

Atherosclerosis, the primary cause of MI, is a chronic inflammatory-immune disease of vasculature [96]. Atherosclerosis is associated with the thickening of vessel walls and the formation and deposition of lipid plaques in the cerebral, aortic, and peripheral arteries, which can be regulated by multiple cellular and molecular mechanisms. It was previously reported that high shear-stress or the shear-responsive transcription factor Krüppel-like factor 2 (KLF2) can induce vascular endothelial cells to secret exosomes enriched with miR-143 and miR-145 and subsequently regulate the target genes such as CAMK2d and ELK1 in smooth muscle cells [97], thus may regulate proliferation and de-differentiation of smooth muscle cells [98]. In addition, extracellular vesicles derived from KLF2-expressing endothelial cells can attenuate atherosclerosis formation *in vivo* [97]. Equally important, macrophage-derived exosomes from both atherosclerotic plaques and the peripheral blood were demonstrated to participate in the development of atherosclerosis [99, 100]. The atherosclerotic patients have higher levels of leucocyte-derived extracellular vesicles in the circulation compared to healthy participants [101]. Furthermore, the circulating exosomes originated from macrophage foam cells were proved to promote smooth muscle cell adhesion and migration in atherosclerotic lesion through activating the ERK and AKT pathways [101].

17.5.2 Hypertension

The renin-angiotensin system (RAS), principally composed of renin, angiotensinogen, angiotensin-converting enzyme (ACE), angiotensin II (Ang II), and Ang II type 1 and type 2 receptors (AT1R and AT2R), plays key roles in the development of hypertension. It was previously reported that the AT1R-enriched exosomes were secreted from cardiomyocytes into the serum of mice undergoing cardiac pressure overload, thus regulating the blood pressure under hemodynamic stress [102]. Moreover, exogenously delivered AT1R-enriched exosomes were demonstrated to be uptaken by recipient cells such as smooth muscle cells and endothelial cells, which contributed to the regain of blood pressure response induced by Ang II in AT1R knockout mice [102]. Thus, the circulating exosomes containing AT1R, released from cardiomyocytes during pressure overload, may play important roles in regulating the blood pressure in detrimental conditions such as hypertension and heart failure.

17.5.3 Sepsis Cardiomyopathy

Sepsis cardiomyopathy is common in clinic and is predominantly caused by systemic bacterial infection. Although the pathogenesis of sepsis cardiomyopathy is quite complex, the out-of-control immuno-inflammatory response, oxidative stress, cardiomyocyte apoptosis, and mitochondrial dysfunction are recognized as critical mechanisms. The platelet-derived extracellular vesicles isolated from septic patients were previously shown to induce vascular cell apoptosis through the NADPH oxidase-dependent release of superoxide [103]. The nitric oxide (NO) and bacterial toxin were proved to be positive factors for the secretion of platelet-derived exosomes. The circulating exosomes may further induce endothelial cell apoptosis via generating the peroxinitrite radical and activating Caspase 3 [104]. Further studies will be needed to investigate the potential of circulating exosomes and exosomal cargos in the diagnosis and prognosis of sepsis cardiomyopathy.

17.6 Perspective and Future Directions

Cardiovascular diseases are one of the major threats to human health [105, 106]. To date, a detailed understanding is available for stem cell transplantation in the treatment of myocardial injury and heart failure, however, there are still many problems in stem cell therapy such as ethical issue, limited source, low viability in local damaged myocardium, and immune rejection [107–109]. Although the induced pluripotent stem cells (iPSCs) are more likely to survive in the damaged myocardium compared to mesenchymal stem cells (MSCs) [110], iPSCs-associated tumorigenesis remains a critical issue. Initially, it is thought that stem cells can differentiate into cardiomyocytes and promote cardiac regeneration and repair. Nevertheless, subsequent detection revealed few new cardiomyocytes derived from transplanted stem cells, suggesting that stem cells are likely to promote the process of myocardial regeneration and angiogenesis by other mechanisms [111]. Circulating exosomes enriched with various types of bioactive molecules can be changed not only in the number but also in the composite cargos upon cardiac injury, which may influence cardiomyocyte function and contribute to cardiac regeneration and repair [57, 112]. In particular, compared with stem cell therapy, exosome-based therapeutic strategy would also decrease the risk of disordered differentiation and tumorigenesis induced by stem cells [75, 112, 113].

Circulating exosomes can mediate local and distant cell communication through the horizontal transfer of their contents such as miRNAs and proteins or the activation of signaling pathways in the target cells [12, 36]. Notably, the exosomal contents can be selectively enriched or modified by bioengineering, thus providing desirable effects in the treatment of cardiovascular diseases [114]. Moreover, given the particular lipid bilayer structure, exosomes can be used as a new drug carrier though it remains to be solved whether and how the delivered exosomes would reach the specific target tissues and cells to exert their biological therapeutic effects [115–117]. Also importantly, exosomes are naturally secreted into the extracellular environments, which may faultlessly overcome immunogenicity compared with other developed delivery devices. Last but not least, more preclinical and clinical studies will be needed to investigate the potential of circulating exosomes as biomarkers for the diagnosis, risk stratification, treatment, and prognosis of cardiovascular diseases [118, 119].

Acknowledgements This work was supported by the grants from National Natural Science Foundation of China (81570362, 91639101 and 81200169 to JJ Xiao and 81400647 to Y Bei), and the development fund for Shanghai talents (to JJ Xiao).

Competing Financial Interests The authors declare no competing financial interests.

References

1. Choi DS, Kim DK, Kim YK, Gho YS (2015) Proteomics of extracellular vesicles: exosomes and ectosomes. Mass Spectrom Rev 34(4):474–490
2. Raposo G, Stoorvogel W (2013) Extracellular vesicles: exosomes, microvesicles, and friends. J Cell Biol 200(4):373–383
3. Ventimiglia LN, Alonso MA (2016) Biogenesis and function of T cell-derived exosomes. Front Cell Dev Biol 4:84
4. Rayner KJ, Hennessy EJ (2013) Extracellular communication via microRNA: lipid particles have a new message. J Lipid Res 54(5):1174–1181
5. Boon RA, Vickers KC (2013) Intercellular transport of microRNAs. Arterioscler Thromb Vasc Biol 33(2):186–192
6. Hu G, Drescher KM, Chen XM (2012) Exosomal miRNAs: biological properties and therapeutic potential. Front Genet 3:56

7. Villarroya-Beltri C, Gutierrez-Vazquez C, Sanchez-Madrid F, Mittelbrunn M (2013) Analysis of microRNA and protein transfer by exosomes during an immune synapse. Methods Mol Biol 1024:41–51
8. Corcoran C, Friel AM, Duffy MJ, Crown J, O'Driscoll L (2011) Intracellular and extracellular microRNAs in breast cancer. Clin Chem 57(1):18–32
9. Friel AM, Corcoran C, Crown J, O'Driscoll L (2010) Relevance of circulating tumor cells, extracellular nucleic acids, and exosomes in breast cancer. Breast Cancer Res Treat 123(3):613–625
10. Melo SA, Luecke LB, Kahlert C, Fernandez AF, Gammon ST, Kaye J, LeBleu VS, Mittendorf EA, Weitz J, Rahbari N, Reissfelder C, Pilarsky C, Fraga MF, Piwnica-Worms D, Kalluri R (2015) Glypican-1 identifies cancer exosomes and detects early pancreatic cancer. Nature 523(7559):177–182
11. Fleury A, Martinez MC, Le Lay S (2014) Extracellular vesicles as therapeutic tools in cardiovascular diseases. Front Immunol 5:370
12. Valadi H, Ekstrom K, Bossios A, Sjostrand M, Lee JJ, Lotvall JO (2007) Exosome-mediated transfer of mRNAs and microRNAs is a novel mechanism of genetic exchange between cells. Nat Cell Biol 9(6):654–659
13. Mittelbrunn M, Gutierrez-Vazquez C, Villarroya-Beltri C, Gonzalez S, Sanchez-Cabo F, Gonzalez MA, Bernad A, Sanchez-Madrid F (2011) Unidirectional transfer of microRNA-loaded exosomes from T cells to antigen-presenting cells. Nat Commun 2:282
14. Hessvik NP, Phuyal S, Brech A, Sandvig K, Llorente A (2012) Profiling of microRNAs in exosomes released from PC-3 prostate cancer cells. Biochim Biophys Acta 1819(11–12):1154–1163
15. Bellingham SA, Coleman BM, Hill AF (2012) Small RNA deep sequencing reveals a distinct miRNA signature released in exosomes from prion-infected neuronal cells. Nucleic Acids Res 40(21):10937–10949
16. Dong H, Lei J, Ding L, Wen Y, Ju H, Zhang X (2013) MicroRNA: function, detection, and bioanalysis. Chem Rev 113(8):6207–6233
17. Ahlin F, Arfvidsson J, Vargas KG, Stojkovic S, Huber K, Wojta J (2016) MicroRNAs as circulating biomarkers in acute coronary syndromes: a review. Vasc Pharmacol 81:15–21
18. Emanueli C, Shearn AI, Angelini GD, Sahoo S (2015) Exosomes and exosomal miRNAs in cardiovascular protection and repair. Vasc Pharmacol 71:24–30
19. Zhang H, Xiang M, Meng D, Sun N, Chen S (2016) Inhibition of myocardial ischemia/reperfusion injury by exosomes secreted from mesenchymal stem cells. Stem Cells Int 2016:4328362
20. Kang K, Ma R, Cai W, Huang W, Paul C, Liang J, Wang Y, Zhao T, Kim HW, Xu M, Millard RW, Wen Z, Wang Y (2015) Exosomes secreted from CXCR4 overexpressing mesenchymal stem cells promote cardioprotection via Akt signaling pathway following myocardial infarction. Stem Cells Int 2015:659890
21. Davis ME (2016) Exosomes: what do we love so much about them? Circ Res 119(12):1280–1282
22. Emanueli C, Shearn AI, Laftah A, Fiorentino F, Reeves BC, Beltrami C, Mumford A, Clayton A, Gurney M, Shantikumar S, Angelini GD (2016) Coronary artery-bypass-graft surgery increases the plasma concentration of exosomes carrying a cargo of cardiac microRNAs: an example of exosome trafficking out of the human heart with potential for cardiac biomarker discovery. PLoS One 11(4):e0154274
23. Lawson C, Vicencio JM, Yellon DM, Davidson SM (2016) Microvesicles and exosomes: new players in metabolic and cardiovascular disease. J Endocrinol 228(2):R57–R71
24. Evans S, Mann DL (2013) Circulating p53-responsive microRNAs as predictive biomarkers in heart failure after acute myocardial infarction: the long and arduous road from scientific discovery to clinical utility. Circ Res 113(3):242–244
25. Sinning JM, Losch J, Walenta K, Bohm M, Nickenig G, Werner N (2011) Circulating CD31+/Annexin V+ microparticles correlate with cardiovascular outcomes. Eur Heart J 32(16):2034–2041
26. Aliotta JM, Pereira M, Wen S, Dooner MS, Del Tatto M, Papa E, Goldberg LR, Baird GL, Ventetuolo CE, Quesenberry PJ, Klinger JR (2016) Exosomes induce and reverse monocrotaline-induced pulmonary hypertension in mice. Cardiovasc Res 110(3):319–330

27. Abdelwahid E, Kalvelyte A, Stulpinas A, de Carvalho KA, Guarita-Souza LC, Foldes G (2016) Stem cell death and survival in heart regeneration and repair. Apoptosis 21(3):252–268
28. Arslan F, Lai RC, Smeets MB, Akeroyd L, Choo A, Aguor EN, Timmers L, van Rijen HV, Doevendans PA, Pasterkamp G, Lim SK, de Kleijn DP (2013) Mesenchymal stem cell-derived exosomes increase ATP levels, decrease oxidative stress and activate PI3K/Akt pathway to enhance myocardial viability and prevent adverse remodeling after myocardial ischemia/reperfusion injury. Stem Cell Res 10(3):301–312
29. Barile L, Lionetti V, Cervio E, Matteucci M, Gherghiceanu M, Popescu LM, Torre T, Siclari F, Moccetti T, Vassalli G (2014) Extracellular vesicles from human cardiac progenitor cells inhibit cardiomyocyte apoptosis and improve cardiac function after myocardial infarction. Cardiovasc Res 103(4):530–541
30. Chen L, Wang Y, Pan Y, Zhang L, Shen C, Qin G, Ashraf M, Weintraub N, Ma G, Tang Y (2013) Cardiac progenitor-derived exosomes protect ischemic myocardium from acute ischemia/reperfusion injury. Biochem Biophys Res Commun 431(3):566–571
31. Cosme J, Liu PP, Gramolini AO (2013) The cardiovascular exosome: current perspectives and potential. Proteomics 13(10–11):1654–1659
32. Ibrahim A, Marban E (2016) Exosomes: fundamental biology and roles in cardiovascular physiology. Annu Rev Physiol 78:67–83
33. Zhan R, Leng X, Liu X, Wang X, Gong J, Yan L, Wang L, Wang Y, Wang X, Qian LJ (2009) Heat shock protein 70 is secreted from endothelial cells by a non-classical pathway involving exosomes. Biochem Biophys Res Commun 387(2):229–233
34. Gupta S, Knowlton AA (2007) HSP60 trafficking in adult cardiac myocytes: role of the exosomal pathway. Am J Phys Heart Circ Phys 292(6):H3052–H3056
35. Waldenstrom A, Genneback N, Hellman U, Ronquist G (2012) Cardiomyocyte microvesicles contain DNA/RNA and convey biological messages to target cells. PLoS One 7(4):e34653
36. Skog J, Wurdinger T, van Rijn S, Meijer DH, Gainche L, Sena-Esteves M, Curry WT Jr, Carter BS, Krichevsky AM, Breakefield XO (2008) Glioblastoma microvesicles transport RNA and proteins that promote tumour growth and provide diagnostic biomarkers. Nat Cell Biol 10(12):1470–1476
37. Raimondo F, Morosi L, Chinello C, Magni F, Pitto M (2011) Advances in membranous vesicle and exosome proteomics improving biological understanding and biomarker discovery. Proteomics 11(4):709–720
38. Pant S, Hilton H, Burczynski ME (2012) The multifaceted exosome: biogenesis, role in normal and aberrant cellular function, and frontiers for pharmacological and biomarker opportunities. Biochem Pharmacol 83(11):1484–1494
39. Yamashita T, Kamada H, Kanasaki S, Maeda Y, Nagano K, Abe Y, Inoue M, Yoshioka Y, Tsutsumi Y, Katayama S, Inoue M, Tsunoda S (2013) Epidermal growth factor receptor localized to exosome membranes as a possible biomarker for lung cancer diagnosis. Pharmazie 68(12):969–973
40. Molkentin JD, Dorn GW 2nd (2001) Cytoplasmic signaling pathways that regulate cardiac hypertrophy. Annu Rev Physiol 63:391–426
41. Sahoo S, Losordo DW (2014) Exosomes and cardiac repair after myocardial infarction. Circ Res 114(2):333–344
42. Kuwabara Y, Ono K, Horie T, Nishi H, Nagao K, Kinoshita M, Watanabe S, Baba O, Kojima Y, Shizuta S, Imai M, Tamura T, Kita T, Kimura T (2011) Increased microRNA-1 and microRNA-133a levels in serum of patients with cardiovascular disease indicate myocardial damage. Circ Cardiovasc Genet 4(4):446–454
43. Zhao Y, Samal E, Srivastava D (2005) Serum response factor regulates a muscle-specific microRNA that targets Hand2 during cardiogenesis. Nature 436(7048):214–220
44. Care A, Catalucci D, Felicetti F, Bonci D, Addario A, Gallo P, Bang ML, Segnalini P, Gu Y, Dalton ND, Elia L, Latronico MV, Hoydal M, Autore C, Russo MA, Dorn GW 2nd, Ellingsen O, Ruiz-Lozano P, Peterson KL, Croce CM, Peschle C, Condorelli G (2007) MicroRNA-133 controls cardiac hypertrophy. Nat Med 13(5):613–618

45. Sayed D, Hong C, Chen IY, Lypowy J, Abdellatif M (2007) MicroRNAs play an essential role in the development of cardiac hypertrophy. Circ Res 100(3):416–424
46. Bagnall RD, Tsoutsman T, Shephard RE, Ritchie W, Semsarian C (2012) Global microRNA profiling of the mouse ventricles during development of severe hypertrophic cardiomyopathy and heart failure. PLoS One 7(9):e44744
47. Bang C, Batkai S, Dangwal S, Gupta SK, Foinquinos A, Holzmann A, Just A, Remke J, Zimmer K, Zeug A, Ponimaskin E, Schmiedl A, Yin X, Mayr M, Halder R, Fischer A, Engelhardt S, Wei Y, Schober A, Fiedler J, Thum T (2014) Cardiac fibroblast-derived microRNA passenger strand-enriched exosomes mediate cardiomyocyte hypertrophy. J Clin Investig 124(5):2136–2146
48. Pan W, Zhong Y, Cheng C, Liu B, Wang L, Li A, Xiong L, Liu S (2013) MiR-30-regulated autophagy mediates angiotensin II-induced myocardial hypertrophy. PLoS One 8(1):e53950
49. Buja LM (2005) Myocardial ischemia and reperfusion injury. Cardiovasc Pathol 14(4):170–175
50. Arumugam S, Mito S, Thandavarayan RA, Giridharan VV, Pitchaimani V, Karuppagounder V, Harima M, Nomoto M, Suzuki K, Watanabe K (2013) Mulberry leaf diet protects against progression of experimental autoimmune myocarditis to dilated cardiomyopathy via modulation of oxidative stress and MAPK-mediated apoptosis. Cardiovasc Ther 31(6):352–362
51. Kwak HB (2013) Effects of aging and exercise training on apoptosis in the heart. J Exerc Rehabil 9(2):212–219
52. Buja LM (1998) Modulation of the myocardial response to ischemia. Lab Investig 78(11):1345–1373
53. Reimer KA, Ideker RE (1987) Myocardial ischemia and infarction: anatomic and biochemical substrates for ischemic cell death and ventricular arrhythmias. Hum Pathol 18(5):462–475
54. Buja LM, Entman ML (1998) Modes of myocardial cell injury and cell death in ischemic heart disease. Circulation 98(14):1355–1357
55. Nadal-Ginard B, Kajstura J, Leri A, Anversa P (2003) Myocyte death, growth, and regeneration in cardiac hypertrophy and failure. Circ Res 92(2):139–150
56. Wang H, Bei Y, Huang P, Zhou Q, Shi J, Sun Q, Zhong J, Li X, Kong X, Xiao J (2016) Inhibition of miR-155 protects against LPS-induced cardiac dysfunction and apoptosis in mice. Mol Ther Nucleic Acids 5(10):e374
57. Vicencio JM, Yellon DM, Sivaraman V, Das D, Boi-Doku C, Arjun S, Zheng Y, Riquelme JA, Kearney J, Sharma V, Multhoff G, Hall AR, Davidson SM (2015) Plasma exosomes protect the myocardium from ischemia-reperfusion injury. J Am Coll Cardiol 65(15):1525–1536
58. Jakob P, Doerries C, Briand S, Mocharla P, Krankel N, Besler C, Mueller M, Manes C, Templin C, Baltes C, Rudin M, Adams H, Wolfrum M, Noll G, Ruschitzka F, Luscher TF, Landmesser U (2012) Loss of angiomiR-126 and 130a in angiogenic early outgrowth cells from patients with chronic heart failure: role for impaired in vivo neovascularization and cardiac repair capacity. Circulation 126(25):2962–2975
59. Fish JE, Santoro MM, Morton SU, Yu S, Yeh RF, Wythe JD, Ivey KN, Bruneau BG, Stainier DY, Srivastava D (2008) miR-126 regulates angiogenic signaling and vascular integrity. Dev Cell 15(2):272–284
60. Wang S, Aurora AB, Johnson BA, Qi X, McAnally J, Hill JA, Richardson JA, Bassel-Duby R, Olson EN (2008) The endothelial-specific microRNA miR-126 governs vascular integrity and angiogenesis. Dev Cell 15(2):261–271
61. Chen Y, Gorski DH (2008) Regulation of angiogenesis through a microRNA (miR-130a) that down-regulates antiangiogenic homeobox genes GAX and HOXA5. Blood 111(3):1217–1226
62. Taniguchi K, Kohno R, Ayada T, Kato R, Ichiyama K, Morisada T, Oike Y, Yonemitsu Y, Maehara Y, Yoshimura A (2007) Spreds are essential for embryonic lymphangiogenesis by regulating vascular endothelial growth factor receptor 3 signaling. Mol Cell Biol 27(12):4541–4550
63. Rhoads K, Arderiu G, Charboneau A, Hansen SL, Hoffman W, Boudreau N (2005) A role for Hox A5 in regulating angiogenesis and vascular patterning. Lymphat Res Biol 3(4):240–252

64. Zhao ZQ, Corvera JS, Halkos ME, Kerendi F, Wang NP, Guyton RA, Vinten-Johansen J (2003) Inhibition of myocardial injury by ischemic postconditioning during reperfusion: comparison with ischemic preconditioning. Am J Phys Heart Circ Phys 285(2):H579–H588
65. Agrawal V, Gupta JK, Qureshi SS, Vishwakarma VK (2016) Role of cardiac renin angiotensin system in ischemia reperfusion injury and preconditioning of heart. Indian Heart J 68(6):856–861
66. Evans CW, Iyer KS, Hool LC (2016) The potential for nanotechnology to improve delivery of therapy to the acute ischemic heart. Nanomedicine (Lond) 11(7):817–832
67. Zhao W, Zheng XL, Zhao SP (2015) Exosome and its roles in cardiovascular diseases. Heart Fail Rev 20(3):337–348
68. Buja LM, Vela D (2008) Cardiomyocyte death and renewal in the normal and diseased heart. Cardiovasc Pathol 17(6):349–374
69. Radomski MW, Palmer RM, Moncada S (1987) Endogenous nitric oxide inhibits human platelet adhesion to vascular endothelium. Lancet 2(8567):1057–1058
70. Beckman JS, Beckman TW, Chen J, Marshall PA, Freeman BA (1990) Apparent hydroxyl radical production by peroxynitrite: implications for endothelial injury from nitric oxide and superoxide. Proc Natl Acad Sci U S A 87(4):1620–1624
71. Xia Y, Zweier JL (1995) Substrate control of free radical generation from xanthine oxidase in the postischemic heart. J Biol Chem 270(32):18797–18803
72. Loke KE, McConnell PI, Tuzman JM, Shesely EG, Smith CJ, Stackpole CJ, Thompson CI, Kaley G, Wolin MS, Hintze TH (1999) Endogenous endothelial nitric oxide synthase-derived nitric oxide is a physiological regulator of myocardial oxygen consumption. Circ Res 84(7):840–845
73. Matsui Y, Takagi H, Qu X, Abdellatif M, Sakoda H, Asano T, Levine B, Sadoshima J (2007) Distinct roles of autophagy in the heart during ischemia and reperfusion: roles of AMP-activated protein kinase and Beclin 1 in mediating autophagy. Circ Res 100(6):914–922
74. Wang ZV, Rothermel BA, Hill JA (2010) Autophagy in hypertensive heart disease. J Biol Chem 285(12):8509–8514
75. Khan M, Nickoloff E, Abramova T, Johnson J, Verma SK, Krishnamurthy P, Mackie AR, Vaughan E, Garikipati VN, Benedict C, Ramirez V, Lambers E, Ito A, Gao E, Misener S, Luongo T, Elrod J, Qin G, Houser SR, Koch WJ, Kishore R (2015) Embryonic stem cell-derived exosomes promote endogenous repair mechanisms and enhance cardiac function following myocardial infarction. Circ Res 117(1):52–64
76. Barile L, Moccetti T, Marbán E, Vassalli G (2017) Roles of exosomes in cardioprotection. Eur Heart J 38(18):1372–1379
77. Rana A, Goyal N, Ahlawat A, Jamwal S, Reddy BV, Sharma S (2015) Mechanisms involved in attenuated cardio-protective role of ischemic preconditioning in metabolic disorders. Perfusion 30(2):94–105
78. Chen Q, Chen X, Han C, Wang Y, Huang T, Du Y, Dong Z (2016) FGF-2 transcriptionally down-regulates the expression of BNIP3L via PI3K/Akt/FoxO3a signaling and inhibits necrosis and mitochondrial dysfunction induced by high concentrations of hydrogen peroxide in H9c2 cells. Cell Physiol Biochem 40(6):1678–1691
79. Pockley AG, Shepherd J, Corton JM (1998) Detection of heat shock protein 70 (Hsp70) and anti-Hsp70 antibodies in the serum of normal individuals. Immunol Investig 27(6):367–377
80. Efthymiou CA, Mocanu MM, de Belleroche J, Wells DJ, Latchmann DS, Yellon DM (2004) Heat shock protein 27 protects the heart against myocardial infarction. Basic Res Cardiol 99(6):392–394
81. Orogo AM, Gustafsson AB (2013) Cell death in the myocardium: my heart won't go on. IUBMB Life 65(8):651–656
82. Goldstein JA (1998) Right heart ischemia: pathophysiology, natural history, and clinical management. Prog Cardiovasc Dis 40(4):325–341
83. Akodad M, Lattuca B, Nagot N, Georgescu V, Buisson M, Cristol JP, Leclercq F, Macia JC, Gervasoni R, Cung TT, Cade S, Cransac F, Labour J, Dupuy AM, Roubille F (2017) COLIN

trial: value of colchicine in the treatment of patients with acute myocardial infarction and inflammatory response. Arch Cardiovasc Dis. doi:10.1016/j.acvd.2016.10.004
84. Xanthopoulos A, Giamouzis G, Tryposkiadis K, Paraskevopoulou E, Paraskevopoulou P, Karagiannis G, Patsilinakos S, Parissis J, Farmakis D, Butler J, Skoularigis J, Triposkiadis F (2016) A simple score for early risk stratification in acute heart failure. Int J Cardiol 230:248–254
85. Teng X, Chen L, Chen W, Yang J, Yang Z, Shen Z (2015) Mesenchymal stem cell-derived exosomes improve the microenvironment of infarcted myocardium contributing to angiogenesis and anti-inflammation. Cell Physiol Biochem 37(6):2415–2424
86. Linke A, Muller P, Nurzynska D, Casarsa C, Torella D, Nascimbene A, Castaldo C, Cascapera S, Bohm M, Quaini F, Urbanek K, Leri A, Hintze TH, Kajstura J, Anversa P (2005) Stem cells in the dog heart are self-renewing, clonogenic, and multipotent and regenerate infarcted myocardium, improving cardiac function. Proc Natl Acad Sci U S A 102(25):8966–8971
87. Zampetaki A, Willeit P, Tilling L, Drozdov I, Prokopi M, Renard JM, Mayr A, Weger S, Schett G, Shah A, Boulanger CM, Willeit J, Chowienczyk PJ, Kiechl S, Mayr M (2012) Prospective study on circulating MicroRNAs and risk of myocardial infarction. J Am Coll Cardiol 60(4):290–299
88. Diehl P, Fricke A, Sander L, Stamm J, Bassler N, Htun N, Ziemann M, Helbing T, El-Osta A, Jowett JB, Peter K (2012) Microparticles: major transport vehicles for distinct microRNAs in circulation. Cardiovasc Res 93(4):633–644
89. Rizzo M, Macario AJ, de Macario EC, Gouni-Berthold I, Berthold HK, Rini GB, Zummo G, Cappello F (2011) Heat shock protein-60 and risk for cardiovascular disease. Curr Pharm Des 17(33):3662–3668
90. Kim SC, Stice JP, Chen L, Jung JS, Gupta S, Wang Y, Baumgarten G, Trial J, Knowlton AA (2009) Extracellular heat shock protein 60, cardiac myocytes, and apoptosis. Circ Res 105(12):1186–1195
91. Zhang X, Wang X, Zhu H, Kranias EG, Tang Y, Peng T, Chang J, Fan GC (2012) Hsp20 functions as a novel cardiokine in promoting angiogenesis via activation of VEGFR2. PLoS One 7(3):e32765
92. Cheng Y, Wang X, Yang J, Duan X, Yao Y, Shi X, Chen Z, Fan Z, Liu X, Qin S, Tang X, Zhang C (2012) A translational study of urine miRNAs in acute myocardial infarction. J Mol Cell Cardiol 53(5):668–676
93. van Rooij E, Quiat D, Johnson BA, Sutherland LB, Qi X, Richardson JA, Kelm RJ Jr, Olson EN (2009) A family of microRNAs encoded by myosin genes governs myosin expression and muscle performance. Dev Cell 17(5):662–673
94. Miyata S, Minobe W, Bristow MR, Leinwand LA (2000) Myosin heavy chain isoform expression in the failing and nonfailing human heart. Circ Res 86(4):386–390
95. Matsumoto S, Sakata Y, Suna S, Nakatani D, Usami M, Hara M, Kitamura T, Hamasaki T, Nanto S, Kawahara Y, Komuro I (2013) Circulating p53-responsive microRNAs are predictive indicators of heart failure after acute myocardial infarction. Circ Res 113(3):322–326
96. Yun TJ, Lee JS, Shim D, Choi JH, Cheong C (2017) Isolation and characterization of aortic dendritic cells and lymphocytes in atherosclerosis. Methods Mol Biol 1559:419–437
97. Cordes KR, Sheehy NT, White MP, Berry EC, Morton SU, Muth AN, Lee TH, Miano JM, Ivey KN, Srivastava D (2009) miR-145 and miR-143 regulate smooth muscle cell fate and plasticity. Nature 460(7256):705–710
98. Hergenreider E, Heydt S, Treguer K, Boettger T, Horrevoets AJ, Zeiher AM, Scheffer MP, Frangakis AS, Yin X, Mayr M, Braun T, Urbich C, Boon RA, Dimmeler S (2012) Atheroprotective communication between endothelial cells and smooth muscle cells through miRNAs. Nat Cell Biol 14(3):249–256
99. Leroyer AS, Isobe H, Leseche G, Castier Y, Wassef M, Mallat Z, Binder BR, Tedgui A, Boulanger CM (2007) Cellular origins and thrombogenic activity of microparticles isolated from human atherosclerotic plaques. J Am Coll Cardiol 49(7):772–777
100. Rautou PE, Leroyer AS, Ramkhelawon B, Devue C, Duflaut D, Vion AC, Nalbone G, Castier Y, Leseche G, Lehoux S, Tedgui A, Boulanger CM (2011) Microparticles from human atherosclerotic plaques promote endothelial ICAM-1-dependent monocyte adhesion and transendothelial migration. Circ Res 108(3):335–343

101. Niu C, Wang X, Zhao M, Cai T, Liu P, Li J, Willard B, Zu L, Zhou E, Li Y, Pan B, Yang F, Zheng L (2016) Macrophage foam cell-derived extracellular vesicles promote vascular smooth muscle cell migration and adhesion. J Am Heart Assoc 5(10). pii: e004099
102. Pironti G, Strachan RT, Abraham D, Mon-Wei Yu S, Chen M, Chen W, Hanada K, Mao L, Watson LJ, Rockman HA (2015) Circulating exosomes induced by cardiac pressure overload contain functional angiotensin II type 1 receptors. Circulation 131(24):2120–2130
103. Janiszewski M, Do Carmo AO, Pedro MA, Silva E, Knobel E, Laurindo FR (2004) Platelet-derived exosomes of septic individuals possess proapoptotic NAD(P)H oxidase activity: a novel vascular redox pathway. Crit Care Med 32(3):818–825
104. Gambim MH, do Carmo Ade O, Marti L, Verissimo-Filho S, Lopes LR, Janiszewski M (2007) Platelet-derived exosomes induce endothelial cell apoptosis through peroxynitrite generation: experimental evidence for a novel mechanism of septic vascular dysfunction. Crit Care 11(5):R107
105. Wang Z, Ge J (2014) Managing hypercholesterolemia and preventing cardiovascular events in elderly and younger Chinese adults: focus on rosuvastatin. Clin Interv Aging 9:1–8
106. Moran AE, Forouzanfar MH, Roth GA, Mensah GA, Ezzati M, Murray CJ, Naghavi M (2014) Temporal trends in ischemic heart disease mortality in 21 world regions, 1980 to 2010: the global burden of disease 2010 study. Circulation 129(14):1483–1492
107. Laflamme MA, Murry CE (2011) Heart regeneration. Nature 473(7347):326–335
108. Li TS, Cheng K, Malliaras K, Smith RR, Zhang Y, Sun B, Matsushita N, Blusztajn A, Terrovitis J, Kusuoka H, Marban L, Marban E (2012) Direct comparison of different stem cell types and subpopulations reveals superior paracrine potency and myocardial repair efficacy with cardiosphere-derived cells. J Am Coll Cardiol 59(10):942–953
109. Matar AA, Chong JJ (2014) Stem cell therapy for cardiac dysfunction. Spring 3:440
110. Shiba Y, Gomibuchi T, Seto T, Wada Y, Ichimura H, Tanaka Y, Ogasawara T, Okada K, Shiba N, Sakamoto K, Ido D, Shiina T, Ohkura M, Nakai J, Uno N, Kazuki Y, Oshimura M, Minami I, Ikeda U (2016) Allogeneic transplantation of iPS cell-derived cardiomyocytes regenerates primate hearts. Nature 538(7625):388–391
111. Nguyen BK, Maltais S, Perrault LP, Tanguay JF, Tardif JC, Stevens LM, Borie M, Harel F, Mansour S, Noiseux N (2010) Improved function and myocardial repair of infarcted heart by intracoronary injection of mesenchymal stem cell-derived growth factors. J Cardiovasc Transl Res 3(5):547–558
112. Gaceb A, Martinez MC, Andriantsitohaina R (2014) Extracellular vesicles: new players in cardiovascular diseases. Int J Biochem Cell Biol 50:24–28
113. Tolar J, Nauta AJ, Osborn MJ, Panoskaltsis Mortari A, McElmurry RT, Bell S, Xia L, Zhou N, Riddle M, Schroeder TM, Westendorf JJ, McIvor RS, Hogendoorn PC, Szuhai K, Oseth L, Hirsch B, Yant SR, Kay MA, Peister A, Prockop DJ, Fibbe WE, Blazar BR (2007) Sarcoma derived from cultured mesenchymal stem cells. Stem Cells 25(2):371–379
114. Zhang D, Lee H, Zhu Z, Minhas JK, Jin Y (2016) Enrichment of selective miRNAs in exosomes and delivery of exosomal miRNAs in vitro and in vivo. Am J Physiol Lung Cell Mol Physiol 312(1):L110–L121
115. Lee Y, El Andaloussi S, Wood MJ (2012) Exosomes and microvesicles: extracellular vesicles for genetic information transfer and gene therapy. Hum Mol Genet 21(R1):R125–R134
116. El Andaloussi S, Lakhal S, Mager I, Wood MJ (2013) Exosomes for targeted siRNA delivery across biological barriers. Adv Drug Deliv Rev 65(3):391–397
117. Haney MJ, Klyachko NL, Zhao Y, Gupta R, Plotnikova EG, He Z, Patel T, Piroyan A, Sokolsky M, Kabanov AV, Batrakova EV (2015) Exosomes as drug delivery vehicles for Parkinson's disease therapy. J Control Release 207:18–30
118. Azevedo LC, Pedro MA, Laurindo FR (2007) Circulating microparticles as therapeutic targets in cardiovascular diseases. Recent Pat Cardiovasc Drug Discov 2(1):41–51
119. Alvarez-Llamas G, de la Cuesta F, Barderas ME, Darde V, Padial LR, Vivanco F (2008) Recent advances in atherosclerosis-based proteomics: new biomarkers and a future perspective. Expert Rev Proteomics 5(5):679–691

Chapter 18
Therapeutic Effects of Ischemic-Preconditioned Exosomes in Cardiovascular Diseases

Shengguang Ding, Zhiqing Fan, Che Lin, Qiying Dai, Jinzhe Zhou, Haitao Huang, Yiming Xu, and Chongjun Zhong

18.1 Introduction

Cardiovascular disease (CVD) has been an immense health and economic burdens globally for years. From 2003 to 2013, death rates attributable to CVD declined 28.8%. In the same 10-year period, the actual number of CVD deaths per year declined by 11.7%. Yet in 2013, CVD still accounted for 30.8% (800,937) of all 2,596,993 deaths, or ≈1 of every 3 deaths in the United States [1].

Acute myocardial infarction (MI) as the hallmark of CVD has been considered as the leading cause of mortality worldwide. For now, percutaneous intervention is the most effective strategy to save dying myocardium. However, the reperfusion of acute ischemic myocardium itself is able to cause cardiomyocyte death. The

---

Shengguang Ding, Zhiqing Fan and Che Lin contributed equally to this work.

S. Ding • H. Huang • Y. Xu • C. Zhong (✉)
Department of Thoracic and Cardiovascular Surgery, The Second Affiliated Hospital of Nantong University, Nantong 226001, China
e-mail: chongjunzhong@hotmail.com

Z. Fan
Department of Cardiology, Daqing Qilfield General Hospital, Daqing 163001, China

C. Lin • J. Zhou
Department of Cardiology, Tongji Hospital, Tongji University School of Medicine, Shanghai 200065, China

Q. Dai
Metrowest Medical Center, Framingham 01702, MA, USA

Department of Cardiology, First Affiliated Hospital of Nanjing Medical University, Nanjing 210029, China

J. Xiao, S. Cretoiu (eds.), *Exosomes in Cardiovascular Diseases*, Advances in Experimental Medicine and Biology 998, DOI 10.1007/978-981-10-4397-0_18

underlying oxidative stress, intracellular Ca^{2+} overloading, rapid change in PH, inflammatory reaction and mitochondrial dysfunction all put the myocardium in danger [2]. Ischemic preconditioning (IPC) is a phenomenon that produce resistance to loss of blood supply by creating intermittent short episodes of ischemia. Having been reviewed detailed, IPC might be a potential treatment for ischemic/reperfusion (IR) injury [3–5]. A new concept that extracellular vesicles encompassing exosomes participates in the ischemic preconditioning has been brought out [6]. In this chapter, we summarize the protective effects of IPC exosomes in CVD and the most relevant discoveries in basic science.

18.2 Ischemic Preconditioning

As early as 1990s, it was hypothesized that episodes of brief ischemia would precondition the myocardium for the following ischemia. It is a strategy of creating brief short cycles of non-lethal ischemia-reperfusion stimulus and followed by persistent ischemia. It is expected that IPC would initiate a cardioprotective phenotype and render the myocardium resistant to a subsequent more severe sustained ischemic insult. The principle is to increase the myocardium tolerance to ischemia in various pathways.

To prove this, an ischemic model on the dogs was created. One group was treated with four 5 min circumflex occlusions, each separated by 5 min of reperfusion, followed by a sustained 40 min occlusion. The other group got a single 40 min occlusion. Results shown that preconditioned group had a limited infarct size to 25% of that seen in the non-preconditioned group [7]. Encouraging by this, another similar study was carried out to testify whether this protective effect also works in the remote virgin myocardium. Conclusions agree with the hypothesis and imply that preconditioning may be mediated by factors transported throughout the heart during brief ischemia/reperfusion [8]. Later, several studies found that short periods of ischemia and reperfusion of a tissue can protect a distant tissue against subsequent ischemia [9–13]. Furthermore, a reduction in the coronary resistance and subsequent increase in coronary artery flow was observed in a model exposed to intermittent ischemic conditioning [14]. Similar results were also obtained in human study [15]. With these evidence, remote ischemic preconditioning (RIPC) has been increasingly accepted as an effective method to improve cardiac function after IR injury. Some studies suspected that it is opioid receptor dependent [9, 16], while others support that the activity of a vagal pre-ganglionic neurons is essential in the remote ischemic preconditioning [17]. With all the evidence, the role of RIPC in IR injury is strongly supported [18]. Apart from that, myocardial postconditioning has been shown to benefit in reducing myocardial infarct size [19]. In spite of this, disagreement still exist. Researchers have been arguing that RIPC does not decrease ischemia-associated mortality, nor it reduce major adverse cardiovascular events [20].

18.3 Mechanism of IPC

Several systems have been proven to participate in this process, including ATP-sensitive potassium channels, reactive oxygen species, nitric oxide and various protein kinases [21]. In an ischemic rat model, remote ischemic preconditioning (RIPC) group was treated with four cycles of 5 min of limb ischemia. Followed by 5 min of reperfusion and subjected to 45 min of sustained ischemia by occluding the left coronary artery. Controlled group were treated just with 45 min of sustained artery occlusion. Mitochondrial ATP-sensitive K(+) (K(ATP)) channels were identified as an effector mechanisms for remote preconditioning [22]. Comparable conclusion was also made in a study for modulation of K(ATP) channels in endothelial IPC in human [23]. Another well explored mechanism is the regulation of inflammatory response during IPC. It has been proven that RIPC stimulus modifies human inflammatory gene expression, leading to cardioprotective effect due to affecting the inflammatory process [24]. Circulating cytokines and hypoxia induced factor-1α were found to be influenced as well [25]. Other factors, including oxygen radicals [26], neurotransmission [27–29], cannabinoids [30], nitric oxide synthase [31], connexin 43 phosphorylation [32], mitogen-activated protein kinases (MAPKs) [33], miR-144 [34] and phosphatidylinositol-3-kinase system [35] are all testified. However, little is known about the role of exosomes in IPC. Exosomes has recently been gaining attention with regards to its inter-cellular communication during IR injury. It contains nucleic acid and other important messenger factors. Understanding the underlying mechanism will help us understand how the heart respond to injury and stress at a deeper level.

18.4 Exosomes

Exosomes are small microvesicles (EV) that are released from late endosomal compartments of cells [36]. They are 40–199-nm vesicles released during reticulocyte differentiation as a consequence of multivesicular endosome fusion with the plasma membrane. They have been isolated from diverse body fluids, including semen, saliva, breast milk, amniotic fluid, ascites fluid, cerebrospinal fluid, and bile. EVs can be secreted and specifically taken up by other cells, mediating intercellular signal exchange [37]. Similarly to cytokines that constitute a network of communication, EVs may also exert their functions in a network, affecting distal organs [38]. In a study, rat's heart was exposed to 3 × 5–5 min global ischemia and reperfusion or 30 min aerobic perfusion. The presence or absence of EVs was confirmed by dynamic light scattering, the EV marker HSP60 based on Western blot, and electron microscopy. It was found that IPC markedly increased EV release from the heart, indicating that EV is necessary for cardioprotection by RIPC [37]. mRNA intended for both small and large ribosomal subunits as well as mRNA coding for proteins involved in mitochondrial energy generation are found in the cardiomyocyte-derived EVs, which implies EVs might participate in some protein production in the targeted cells.

These EVs, proven to belong to the exosome family, could be denoted "cardiosomes". Microscopic findings suggested its role in metabolism of microenvironment [39]. Furthermore, by introducing the exosomes from the newts to the rat's heart tissue, new proliferation of the rat cardiomyocyte and improvement in its function were observed [40]. These evidence confirm that exosome is closely associated with cardiac restoration.

18.5 Exosomes and IPC

Ischemic preconditioning effects can be transferred to nonpreconditioned animals via whole blood transfusion [41] or directly cell implantation [40]. These findings suggested a humoral mechanism for preconditioning at a distance. Exosomes contain many unique features like surface proteins/receptors, lipids, mRNAs, microRNAs, transcription factors and other proteins [41]. Stimulated by RIPC, exosomes acutely activate pro-survival kinases that rapidly prepare the heart against ischemia-reperfusion injury [42]. For now, it has been well established that exosome play an essential role in tumor and infection disease, but increasing studies proposed that it is also crucial for cardioprotection during IR injury.

Cells from different organ systems are able to produce exosomes and working actively in RIPC.

18.6 Cardiogenic Exosomes

It was discovered that human cardiomyocytes can produce exosomes-like vesicles via multivesicular endosome (MVE)-dependent pathway [43]. Released from the injured tissue, it carries signaling molecules to activate tissue repair. Isolated cardiac progenitor cells (CPC)-exosomes are found to express cardiac transcription factor, GATA4 and could be recognized by cardiac cells by H9C2. *In vivo* study demonstrated that exosomes from CPC could inhibit apoptosis induced by IR injury [44]. Emerging evidence demonstrated that exosomes participate in RIPC-induced cardioprotection. Coronary perfusates from preconditioned hearts contained more EVs than perfusates isolated from control. Correlating to the result that preconditioned group had smaller infarct size than the control group, it is concluded that the release of EVs from the heart after preconditioning stimuli is increased and that EVs are responsible for the transmission of remote conditioning signals for cardioprotection [37].

18.6.1 Mesenchymal Stem Cell (MSC) Derived Exosome

MSC is one type of adult stem cell that have great plasticity and has shown great potential for the replacement of damaged tissues such as bone, cartilage, tendon, and ligament [45]. In skeletal muscle, it has been proven that hypoxic preconditioned murine MSC enhanced muscle regeneration [46]. MSC are emerging as an extremely promising therapeutic agent for tissue regeneration and repair, proven by animal models [47, 48]. Exosomes have been recognized as part of MSC's paracrine system that potentiates its cardioprotective effect. These exosomes carry various of miRNA and humoral factors to the target cells [49]. miR-22, miR-210, miR-21 and HIF-1α are found in exosomes isolated from preconditioned MSCs. miR-22, previously known as a critical regulator of cardiomyocyte hypertrophy and cardiac remodeling [50], was shown to protect ischemic hearts by targeting Mecp2 [6]. Preceding study established miR-210 exerts cytoprotective effects in cardiomyocytes [51]. It is elucidated that miR-210 as a potent negative regulator of stem cell apoptosis during ischemic preconditioning downstream of HIF-1α. During ischemic injury, MSC acts as a major source to deliver miR-210 to protect heart tissue [52]. More researches looking into the clinical therapeutic effect of MSC derived exosomes suggested the potential for using human embryonic stem-cell derived vascular cells on rescuing peri-scar border zone in myocardial infarction [53]. On the other hand, in an in *vitro* cardiac injury model, insulin-like growth factor 1 (IGF1) is proven to be part of the signal pathway in exosome-mediated cardiac repair [54].

18.7 Endothelial Cell Derived Exosome

Cardiac endothelial cells could also communicate and regulate myocardium by producing exosomes. Similarly, these exosomes are testified to have nearly twofold increase after preconditioning, and have more potent in reducing cardiac cell death [55]. What's more, endothelial derived exosomes are found to overexpress hypoxia-inducible factor-1 (HIF1) and higher contents of microRNAs. These factors increase tolerance of cardiac progenitor cells under hypoxic stress [56].

18.8 IPC and Proteasome

Proteasome protects against ischemic injury by removing damaged proteins. It is a major intracellular proteolytic system which degrades oxidized and ubiquitinated forms of protein intracellularly. One of important mechanism of cardiac injury during IR injury is the decrease in its function by oxidative modification and inhibition of fluorogenic peptide hydrolysis [57]. In recent studies, it has been proven that MSC derived exosomes ameliorates IR injury through proteomic complementation [58].

The combination of proteasome, ubiquitin, the ubiquitination machinery and the deubiquitinases, is called ubiquitin proteasome system (UPS). The major function of UPS is to prevent accumulation of non-functional, potentially toxic proteins. It contains one 20S subunit and two 19S regulatory cap units. The 20S proteasome is the central proteolytic structure which consists of two pairs of rings each contains seven subunits while the 19S subunit contains multiple ATPase active sites and ubiquitin binding sites. It confers selectivity for ubiquitin-conjugated proteins. Dysfunction of UPS was observed during IR injury, which could be one of the important factor contributing to the heart injury. Recent studies also revealed that IPC protect against ischemic injury by preserving UPS function [59]. IPC protects

UPS by diminishing oxidative damage to 19S regulatory subunits [60] and increasing the degradation of δPKC [61]. A way to quantify the cardioprotective effect from IPC could be to measure the postischemic levels of oxidized and/or ubiquitinated proteins. These levels could predict eventual cardiac function [62].

The 20S subunit of the UPS is found to be attached to the cell plasma membrane and certain observations are interpreted as to suggest that they may be released into the extracellular medium [63]. Once released, they are recognized as circulating proteasomes. Study comparing the features of circulating proteasomes with those of proteasomes isolated from major blood cells found out that the subtype patterns of the circulating ones are clearly different [64]. Circulating proteasome is related to cell damage. Increased serum level is seen in various autoimmune disease [65].

18.9 Exosome and Proteasome

Several studies have been done to explore the correlation between exosome and proteasome. Profiled by mass spectrometry and antibody array, proteasomes of exosomes have been found to contain 857 unique gene products. A predominant feature of MSC exosome proteome is the presence of α and β chain of the 20S proteasome. Further work was done to explore the proteomic profiling of exosome. In *vivo* mouse myocardial infarction model was created by temporarily ligation of the LCA. Exosomes were injected in the treatment group before reperfusion. Proteins in the LCA ligated area was extracted, using a cell extraction buffer. Then sequenced protein analysis confirm the hypothesis that 20S proteasome exists in exosomes and could contribute to the cardioprotective activity. The presence of 20S proteasome in MSC exosomes further suggested that cells extruded 20S proteasome through exosomes [66]. Using the exosome as a carrier, we could deliver therapeutic proteasome specifically to different part of the organ systems.

18.10 Therapeutic Effect of Exosomes and Undetermined Questions

Exosomes have great impact on recipient cells. The distinct transmembrane proteins of exosomes directly interact with the receptors from the target cells [67]. This protein-receptor relationship makes exosomes as ideal carriers to deliver treatment or miRNA to specific cells. What's more, exosomes are non-immunogenic in nature, and have no accumulation in the liver. Based on this, exosomes are hypothesized as a promising medication-carrier [68]. Also, many aspects of exosome suggested itself as a novel means to identify cancer biomarkers for early detection and therapeutic targets, and therapeutic devices to ameliorate the progression of the disease [69].

Great interest has been raising on the remedial role of exosomes in coronary artery disease. Both in *vitro* and in *vivo* study have proven that MSC preserve cardiac function by paracrine system in which exosomes is fundamental. Since there is concern for potential tumor formation in *vivo* in stem cell therapy [70], the paracrine theory provide an alternative method for using MSC in treatment of coronary artery disease. In addition, exosomes could accumulate in human atherosclerotic plaques, where they affect the physiologic balance [71]. The emergence of exosomes as alternative to cell therapy, opens a new insight into the treatment of cardiac disease. However, our knowledge of the transport, target cell biologic reaction and the complexity of interaction of pathways in exosomes remains immature. It is highly urgent to determine if exosomes from plasma after IPC would be more cardioprotective.

18.11 Summary

CVD has been considered as the number one killer worldwide. Tons of researches have been done to look into the ischemic process and mechanisms during ischemia. IPC is cardioprotective. Various factors, such as inflammatory factors, miRNA and proteins have been proved to play a role in mediating the cardioprotective effects of IPC. Increasing evidence suggested that exosome, a well-known messenger in cell-to-cell communication, is associated with IPC-related cardioprotection. Encouraged by this, exosomes are testified to apply to the injured heart tissue and was found that it improves cardiac function. These finding brings up a possible new treatment for CVD.

Traditional management for ischemia is timely effective restoration of blood flow. Besides that, cell and targeted molecular therapy have gained increasing interest as potential therapies. Large amount of cardiomyocytes dead during ischemia. Although emerging evidence support that human heart has the capacity to regenerate itself [72, 73], the endogenous proliferation rate of cardiomyocytes is too low to compensate for the loss of cardiomyocytes [54]. Stem cell based therapy solved this problem but still has its limitations, since it may also has the tumorigenic potential [70]. Based on this, the idea of exosome-centered therapy was developed and testified [74, 75]. However, more clinical studies need to be done to confirm the therapeutic effect of exosomes.

Acknowledgements This work was supported by the grants from National Natural Science Foundation of China (81472158), and Natural Science Foundation of Shanghai (14ZR1438300).

Competing Financial Interests The authors declare no competing financial interests.

References

1. Writing Group M, Mozaffarian D, Benjamin EJ, Go AS, Arnett DK, Blaha MJ, Cushman M, Das SR, de Ferranti S, Despres JP, Fullerton HJ, Howard VJ, Huffman MD, Isasi CR, Jimenez MC, Judd SE, Kissela BM, Lichtman JH, Lisabeth LD, Liu S, Mackey RH, Magid DJ, DK MG, Mohler ER 3rd, Moy CS, Muntner P, Mussolino ME, Nasir K, Neumar RW, Nichol G, Palaniappan L, Pandey DK, Reeves MJ, Rodriguez CJ, Rosamond W, Sorlie PD, Stein J, Towfighi A, Turan TN, Virani SS, Woo D, Yeh RW, Turner MB, American Heart Association Statistics C, Stroke Statistics S (2016) Heart disease and stroke statistics-2016 update: a report from the American Heart Association. Circulation 133(4):e38–360
2. Hausenloy DJ, Yellon DM (2013) Myocardial ischemia-reperfusion injury: a neglected therapeutic target. J Clin Invest 123(1):92–100
3. Przyklenk K, Whittaker P (2011) Remote ischemic preconditioning: current knowledge, unresolved questions, and future priorities. J Cardiovasc Pharmacol Ther 16(3–4):255–259

4. Przyklenk K (2013) Reduction of myocardial infarct size with ischemic "conditioning": physiologic and technical considerations. Anesth Analg 117(4):891–901
5. Przyklenk K, Whittaker P (2013) Genesis of remote conditioning: action at a distance—'hypotheses non fingo'? J Cardiovasc Med 14(3):180–186
6. Feng Y, Huang W, Wani M, Yu X, Ashraf M (2014) Ischemic preconditioning potentiates the protective effect of stem cells through secretion of exosomes by targeting Mecp2 via miR-22. PLoS One 9(2):e88685
7. Murry CE, Jennings RB, Reimer KA (1986) Preconditioning with ischemia: a delay of lethal cell injury in ischemic myocardium. Circulation 74(5):1124–1136
8. Przyklenk K, Bauer B, Ovize M, Kloner RA, Whittaker P (1993) Regional ischemic 'preconditioning' protects remote virgin myocardium from subsequent sustained coronary occlusion. Circulation 87(3):893–899
9. Shimizu M, Tropak M, Diaz RJ, Suto F, Surendra H, Kuzmin E, Li J, Gross G, Wilson GJ, Callahan J, Redington AN (2009) Transient limb ischaemia remotely preconditions through a humoral mechanism acting directly on the myocardium: evidence suggesting cross-species protection. Clin Sci (Lond) 117(5):191–200
10. Kharbanda RK, Mortensen UM, White PA, Kristiansen SB, Schmidt MR, Hoschtitzky JA, Vogel M, Sorensen K, Redington AN, MacAllister R (2002) Transient limb ischemia induces remote ischemic preconditioning in vivo. Circulation 106(23):2881–2883
11. Gho BC, Schoemaker RG, van den Doel MA, Duncker DJ, Verdouw PD (1996) Myocardial protection by brief ischemia in noncardiac tissue. Circulation 94(9):2193–2200
12. Pell TJ, Baxter GF, Yellon DM, Drew GM (1998) Renal ischemia preconditions myocardium: role of adenosine receptors and ATP-sensitive potassium channels. Am J Phys 275(5 Pt 2):H1542–H1547
13. Birnbaum Y, Hale SL, Kloner RA (1997) Ischemic preconditioning at a distance: reduction of myocardial infarct size by partial reduction of blood supply combined with rapid stimulation of the gastrocnemius muscle in the rabbit. Circulation 96(5):1641–1646
14. Shimizu M, Konstantinov IE, Kharbanda RK, Cheung MH, Redington AN (2007) Effects of intermittent lower limb ischaemia on coronary blood flow and coronary resistance in pigs. Acta Physiol 190(2):103–109
15. Zhou K, Yang B, Zhou XM, Tan CM, Zhao Y, Huang C, Liao XB, Xiao HB (2007) Effects of remote ischemic preconditioning on the flow pattern of the left anterior descending coronary artery in normal subjects. Int J Cardiol 122(3):250–251
16. Dickson EW, Blehar DJ, Carraway RE, Heard SO, Steinberg G, Przyklenk K (2001) Naloxone blocks transferred preconditioning in isolated rabbit hearts. J Mol Cell Cardiol 33(9):1751–1756
17. Mastitskaya S, Marina N, Gourine A, Gilbey MP, Spyer KM, Teschemacher AG, Kasparov S, Trapp S, Ackland GL, Gourine AV (2012) Cardioprotection evoked by remote ischaemic preconditioning is critically dependent on the activity of vagal pre-ganglionic neurones. Cardiovasc Res 95(4):487–494
18. Tapuria N, Kumar Y, Habib MM, Abu Amara M, Seifalian AM, Davidson BR (2008) Remote ischemic preconditioning: a novel protective method from ischemia reperfusion injury—a review. J Surg Res 150(2):304–330
19. Hansen PR, Thibault H, Abdulla J (2010) Postconditioning during primary percutaneous coronary intervention: a review and meta-analysis. Int J Cardiol 144(1):22–25
20. Brevoord D, Kranke P, Kuijpers M, Weber N, Hollmann M, Preckel B (2012) Remote ischemic conditioning to protect against ischemia-reperfusion injury: a systematic review and meta-analysis. PLoS One 7(7):e42179
21. Kolar F, Ostadal B (2004) Molecular mechanisms of cardiac protection by adaptation to chronic hypoxia. Physiol Res 53(Suppl 1):S3–13
22. Kristiansen SB, Henning O, Kharbanda RK, Nielsen-Kudsk JE, Schmidt MR, Redington AN, Nielsen TT, Botker HE (2005) Remote preconditioning reduces ischemic injury in the explanted heart by a KATP channel-dependent mechanism. Am J Phys Heart Circ Phys 288(3):H1252–H1256

23. Broadhead MW, Kharbanda RK, Peters MJ, MacAllister RJ (2004) KATP channel activation induces ischemic preconditioning of the endothelium in humans in vivo. Circulation 110(15):2077–2082
24. Konstantinov IE, Arab S, Kharbanda RK, Li J, Cheung MM, Cherepanov V, Downey GP, Liu PP, Cukerman E, Coles JG, Redington AN (2004) The remote ischemic preconditioning stimulus modifies inflammatory gene expression in humans. Physiol Genomics 19(1):143–150
25. Albrecht M, Zitta K, Bein B, Wennemuth G, Broch O, Renner J, Schuett T, Lauer F, Maahs D, Hummitzsch L, Cremer J, Zacharowski K, Meybohm P (2013) Remote ischemic preconditioning regulates HIF-1alpha levels, apoptosis and inflammation in heart tissue of cardiosurgical patients: a pilot experimental study. Basic Res Cardiol 108(1):314
26. Baines CP, Goto M, Downey JM (1997) Oxygen radicals released during ischemic preconditioning contribute to cardioprotection in the rabbit myocardium. J Mol Cell Cardiol 29(1):207–216
27. Lim SY, Yellon DM, Hausenloy DJ (2010) The neural and humoral pathways in remote limb ischemic preconditioning. Basic Res Cardiol 105(5):651–655
28. Oxman T, Arad M, Klein R, Avazov N, Rabinowitz B (1997) Limb ischemia preconditions the heart against reperfusion tachyarrhythmia. Am J Phys 273(4):H1707–H1712
29. Ding YF, Zhang MM, He RR (2001) Role of renal nerve in cardioprotection provided by renal ischemic preconditioning in anesthetized rabbits. Sheng Li Xue Bao 53(1):7–12
30. Hajrasouliha AR, Tavakoli S, Ghasemi M, Jabehdar-Maralani P, Sadeghipour H, Ebrahimi F, Dehpour AR (2008) Endogenous cannabinoids contribute to remote ischemic preconditioning via cannabinoid CB2 receptors in the rat heart. Eur J Pharmacol 579(1):246–252
31. Li G, Labruto F, Sirsjo A, Chen F, Vaage J, Valen G (2004) Myocardial protection by remote preconditioning: the role of nuclear factor kappa-B p105 and inducible nitric oxide synthase. Eur J Cardiothorac Surg 26(5):968–973
32. Brandenburger T, Huhn R, Galas A, Pannen BH, Keitel V, Barthel F, Bauer I, Heinen A (2014) Remote ischemic preconditioning preserves Connexin 43 phosphorylation in the rat heart in vivo. J Transl Med 12(1):228
33. Heidbreder M, Naumann A, Tempel K, Dominiak P, Dendorfer A (2008) Remote vs. ischaemic preconditioning: the differential role of mitogen-activated protein kinase pathways. Cardiovasc Res 78(1):108–115
34. Li J, Rohailla S, Gelber N, Rutka J, Sabah N, Gladstone RA, Wei C, Hu P, Kharbanda RK, Redington AN (2014) MicroRNA-144 is a circulating effector of remote ischemic preconditioning. Basic Res Cardiol 109(5):423
35. Tong H, Chen W, Steenbergen C, Murphy E (2000) Ischemic preconditioning activates phosphatidylinositol-3-kinase upstream of protein kinase C. Circ Res 87(4):309–315
36. Keller S, Sanderson MP, Stoeck A, Altevogt P (2006) Exosomes: from biogenesis and secretion to biological function. Immunol Lett 107(2):102–108
37. Giricz Z, Varga ZV, Baranyai T, Sipos P, Paloczi K, Kittel A, Buzas EI, Ferdinandy P (2014) Cardioprotection by remote ischemic preconditioning of the rat heart is mediated by extracellular vesicles. J Mol Cell Cardiol 68:75–78
38. Gyorgy B, Szabo TG, Pasztoi M, Pal Z, Misjak P, Aradi B, Laszlo V, Pallinger E, Pap E, Kittel A, Nagy G, Falus A, Buzas EI (2011) Membrane vesicles, current state-of-the-art: emerging role of extracellular vesicles. Cell Mol Life Sci 68(16):2667–2688
39. Waldenstrom A, Genneback N, Hellman U, Ronquist G (2012) Cardiomyocyte microvesicles contain DNA/RNA and convey biological messages to target cells. PLoS One 7(4):e34653
40. Tseliou E, Weixin L, Valle J, Sun B, Mirotsou M, Marbán E (2015) Abstract 15925: newt exosomes are bioactive on mammalian heart, enhancing proliferation of rat cardiomyocytes and improving recovery after myocardial infarction. Circulation 132(Suppl 3):A15925–A15925
41. Ailawadi S, Wang X, Gu H, Fan GC (2015) Pathologic function and therapeutic potential of exosomes in cardiovascular disease. Biochim Biophys Acta (BBA) Mol Basis Dis 1852(1):1–11
42. Vicencio J, Boi-Doku C, Das D, Sivaraman V, Kearney J, Hall A, Arjun S, Zheng Y, Yellon D, Davidson S (2014) 24 protecting the heart at a distance: exosomes for nano-sized cardioprotection. Heart 100(Suppl 1):A9–A9

43. Sahoo S, Losordo DW (2014) Exosomes and cardiac repair after myocardial infarction. Circ Res 114(2):333–344
44. Chen L, Wang Y, Pan Y, Zhang L, Shen C, Qin G, Ashraf M, Weintraub N, Ma G, Tang Y (2013) Cardiac progenitor-derived exosomes protect ischemic myocardium from acute ischemia/reperfusion injury. Biochem Biophys Res Commun 431(3):566–571
45. Beyer Nardi N, da Silva ML (2006) Mesenchymal stem cells: isolation, in vitro expansion and characterization. Handb Exp Pharmacol 174:249–282
46. Leroux L, Descamps B, Tojais NF, Seguy B, Oses P, Moreau C, Daret D, Ivanovic Z, Boiron JM, Lamaziere JD, Dufourcq P, Couffinhal T, Duplaa C (2010) Hypoxia preconditioned mesenchymal stem cells improve vascular and skeletal muscle fiber regeneration after ischemia through a Wnt4-dependent pathway. Mol Ther 18(8):1545–1552
47. Minguell JJ, Erices A (2006) Mesenchymal stem cells and the treatment of cardiac disease. Exp Biol Med 231(1):39–49
48. Gnecchi M, Danieli P, Cervio E (2012) Mesenchymal stem cell therapy for heart disease. Vasc Pharmacol 57(1):48–55
49. Chen TS, Lai RC, Lee MM, Choo AB, Lee CN, Lim SK (2010) Mesenchymal stem cell secretes microparticles enriched in pre-microRNAs. Nucleic Acids Res 38(1):215–224
50. Huang ZP, Chen J, Seok HY, Zhang Z, Kataoka M, Hu X, Wang DZ (2013) MicroRNA-22 regulates cardiac hypertrophy and remodeling in response to stress. Circ Res 112(9):1234–1243
51. Mutharasan RK, Nagpal V, Ichikawa Y, Ardehali H (2011) microRNA-210 is upregulated in hypoxic cardiomyocytes through Akt- and p53-dependent pathways and exerts cytoprotective effects. Am J Phys Heart Circ Phys 301(4):H1519–H1530
52. Kim HW, Haider HK, Jiang S, Ashraf M (2009) Ischemic preconditioning augments survival of stem cells via miR-210 expression by targeting caspase-8-associated protein 2. J Biol Chem 284(48):33161–33168
53. Xiong Q, Ye L, Zhang P, Lepley M, Swingen C, Zhang L, Kaufman DS, Zhang J (2012) Bioenergetic and functional consequences of cellular therapy: activation of endogenous cardiovascular progenitor cells. Circ Res 111(4):455–468
54. Li Y, L-b S, Zhang Y, Z-w Z, Wang J-j, B-b L, L-l Z, Zhu B, C-s H, Liang C, Pan X, Y-j G, Shen Z, Yu X-y (2016) Abstract 17615: MSC exosome mediated myocardial repair is inhibited by knockdown of Igf-1 signaling pathway through CRISPR/Cas9 genome editing. Circulation 134(Suppl 1):A17615–A17615
55. Zheng Y, Vicencio JM, Yellon DM, Davidson SM (2014) 27 exosomes released from endothelial cells are cardioprotective. Heart 100(Suppl 1):A10–A10
56. Ong SG, Lee WH, Huang M, Dey D, Kodo K, Sanchez-Freire V, Gold JD, Wu JC (2014) Cross talk of combined gene and cell therapy in ischemic heart disease: role of exosomal microRNA transfer. Circulation 130(11 Suppl 1):S60–S69
57. Bulteau AL, Lundberg KC, Humphries KM, Sadek HA, Szweda PA, Friguet B, Szweda LI (2001) Oxidative modification and inactivation of the proteasome during coronary occlusion/reperfusion. J Biol Chem 276(32):30057–30063
58. Lai RC, Yeo RW, Tan KH, Lim SK (2013) Mesenchymal stem cell exosome ameliorates reperfusion injury through proteomic complementation. Regen Med 8(2):197–209
59. Powell SR, Divald A (2010) The ubiquitin-proteasome system in myocardial ischaemia and preconditioning. Cardiovasc Res 85(2):303–311
60. Divald A, Kivity S, Wang P, Hochhauser E, Roberts B, Teichberg S, Gomes AV, Powell SR (2010) Myocardial ischemic preconditioning preserves postischemic function of the 26S proteasome through diminished oxidative damage to 19S regulatory particle subunits. Circ Res 106(12):1829–1838
61. Churchill EN, Ferreira JC, Brum PC, Szweda LI, Mochly-Rosen D (2010) Ischaemic preconditioning improves proteasomal activity and increases the degradation of deltaPKC during reperfusion. Cardiovasc Res 85(2):385–394
62. Powell SR, Wang P, Katzeff H, Shringarpure R, Teoh C, Khaliulin I, Das DK, Davies KJ, Schwalb H (2005) Oxidized and ubiquitinated proteins may predict recovery of postischemic cardiac function: essential role of the proteasome. Antioxid Redox Signal 7(5–6):538–546

63. Sixt SU, Dahlmann B (2008) Extracellular, circulating proteasomes and ubiquitin—incidence and relevance. Biochim Biophys Acta (BBA) Mol Basis Dis 1782(12):817–823
64. Zoeger A, Blau M, Egerer K, Feist E, Dahlmann B (2006) Circulating proteasomes are functional and have a subtype pattern distinct from 20S proteasomes in major blood cells. Clin Chem 52(11):2079–2086
65. Egerer K, Kuckelkorn U, Rudolph PE, Ruckert JC, Dorner T, Burmester GR, Kloetzel PM, Feist E (2002) Circulating proteasomes are markers of cell damage and immunologic activity in autoimmune diseases. J Rheumatol 29(10):2045–2052
66. Lai RC, Tan SS, Teh BJ, Sze SK, Arslan F, de Kleijn DP, Choo A, Lim SK (2012) Proteolytic potential of the MSC exosome proteome: implications for an exosome-mediated delivery of therapeutic proteasome. Int J Proteomics 2012:971907
67. Zhang J, Li S, Li L, Li M, Guo C, Yao J, Mi S (2015) Exosome and exosomal microRNA: trafficking, sorting, and function. Genomics Proteomics Bioinformatics 13(1):17–24
68. Ha D, Yang N, Nadithe V (2016) Exosomes as therapeutic drug carriers and delivery vehicles across biological membranes: current perspectives and future challenges. Acta Pharm Sin B 6(4):287–296
69. Munson P, Shukla A (2015) Exosomes: potential in cancer diagnosis and therapy. Medicine 2(4):310–327
70. Lee AS, Tang C, Rao MS, Weissman IL, Wu JC (2013) Tumorigenicity as a clinical hurdle for pluripotent stem cell therapies. Nat Med 19(8):998–1004
71. Boulanger CM, Loyer X, Rautou PE, Amabile N (2017) Extracellular vesicles in coronary artery disease. Nat Rev Cardiol. doi:10.1038/nrcardio.2017.7
72. Bergmann O, Bhardwaj RD, Bernard S, Zdunek S, Barnabe-Heider F, Walsh S, Zupicich J, Alkass K, Buchholz BA, Druid H, Jovinge S, Frisen J (2009) Evidence for cardiomyocyte renewal in humans. Science 324(5923):98–102
73. Graham E, Bergmann O (2017) Dating the heart: exploring cardiomyocyte renewal in humans. Physiology 32(1):33–41
74. Lai RC, Chen TS, Lim SK (2011) Mesenchymal stem cell exosome: a novel stem cell-based therapy for cardiovascular disease. Regen Med 6(4):481–492
75. Sahoo S, Mathiyalagan P, Hajjar RJ (2017) Pericardial fluid exosomes: a new material to treat cardiovascular disease. Mol Ther 25(3):568–569

Part VI
Future Prospects

Chapter 19
Exosomes: Outlook for Future Cell-Free Cardiovascular Disease Therapy

Bahar Barani, Sheeja Rajasingh, and Johnson Rajasingh

Abbreviations

AAA	Abdominal aortic aneurysm
AdMSCs	Adipose-derived MSCs
ASCs	Adipose stem cells
BMMSCs	Bone marrow-derived MSCs
CVDs	Cardiovascular diseases
EnMSCs	Endometrium-derived MSCs
ESCs	Embryonic stem cells
EVs	Extracellular vesicles
FIZZ1	Found in inflammatory zone 1
HIMF	Hypoxia-induced mitogenic factor
HSP	Heat shock protein
IL	Interleukin
iPSCs	Induced pluripotent stem cells
MCP1	Monocyte chemotactic protein 1
MI	Myocardial infarction
miRNA	Micro RNA

B. Barani • S. Rajasingh
Division of Cardiovascular Diseases, Department of Internal Medicine, Cardiovascular Research Institute, University of Kansas Medical Center, Kansas City, KS, USA

J. Rajasingh, Ph.D. (✉)
Division of Cardiovascular Diseases, Department of Internal Medicine, Cardiovascular Research Institute, University of Kansas Medical Center, Kansas City, KS, USA

Department of Biochemistry and Molecular Biology, University of Kansas Medical Center, Kansas City, KS 66160, USA
e-mail: rjohnson9@kumc.edu

J. Xiao, S. Cretoiu (eds.), *Exosomes in Cardiovascular Diseases*, Advances in Experimental Medicine and Biology 998, DOI 10.1007/978-981-10-4397-0_19

mRNA	Messenger RNA
MSC-CM	MSC-derived condition medium
MSCs	Mesenchymal stem cells
MVB	Microvesicular body
nCPC	Neonatal cardiac progenitor cell
nTCM	Neonatal total condition medium
siRNAs	Small interfering RNAs
VEGF	Vascular endothelial growth factor

19.1 Introduction

According to the World Health Organization, 17.3 million deaths were caused by various heart diseases in 2008 and heart disease is the number one reason for death globally. The large number of deaths are due to cardiovascular diseases (CVDs) which include atherosclerotic coronary artery disease and myocardial ischemia. The prevalence of CVDs is projected to rise to about 40.5% of the USA population by 2030 [1]. The prognosis of these diseases is poor despite presently available therapies. Therefore, intense CVDs research continues to grow with the quest of developing more targeted treatment options that promote the avenues of personalized medicine. Furthermore, early detections of functional or damage-related cardiac biomarkers are needed for proper diagnosis to reduce the death rates in humans. Even though there have been significant developments in the field of CVDs, stem cell-based therapies are one of the most promising approaches for improving cardiac regeneration and function. Recently, a number of studies reported that innovative stem cell therapies in patients with CVDs have proven effective in improving ventricular remodeling and quality of life. This increase in the number of beneficial effects further confirmed the therapeutic utility of stem cells in CVDs [2–5]. Although stem cell therapy is showing significant therapeutic benefits by forming new capillaries and cardiomyocytes around the infarcted zone by the transplanted cells, the major regeneration potential is resulted by the release of paracrine factors [6, 7]. Recent studies have revealed that the paracrine secretions from stem cells contained in membrane-bound nanovesicles called exosomes, when treated to an ischemic mouse heart, were therapeutic, and mimicked the beneficial effects of the parent stem cells [8–12].

Exosomes are endocytic vesicles that have been found to facilitate communication between cells [13]. They are membrane-bound nanovesicles and range between 30–150 nm in size [14]. The electron microscopic image analysis showes many different sizes of mouse bone marrow mesenchymal stem cell (MSC)-derived nanovesicles, including exosomes (Fig. 19.1). These exosomes bud within the endosomal compartments internally and form intraluminal vesicles. When these vesicles fuse with the plasma membrane they release their contents into the extracellular fluid, but they are not released by plasma membrane shedding [15, 16]. Another view points out that exosomes are endocytic vesicles which have common proteins. These proteins seem to originate from the cell cytosol and endosomal compartments, but not

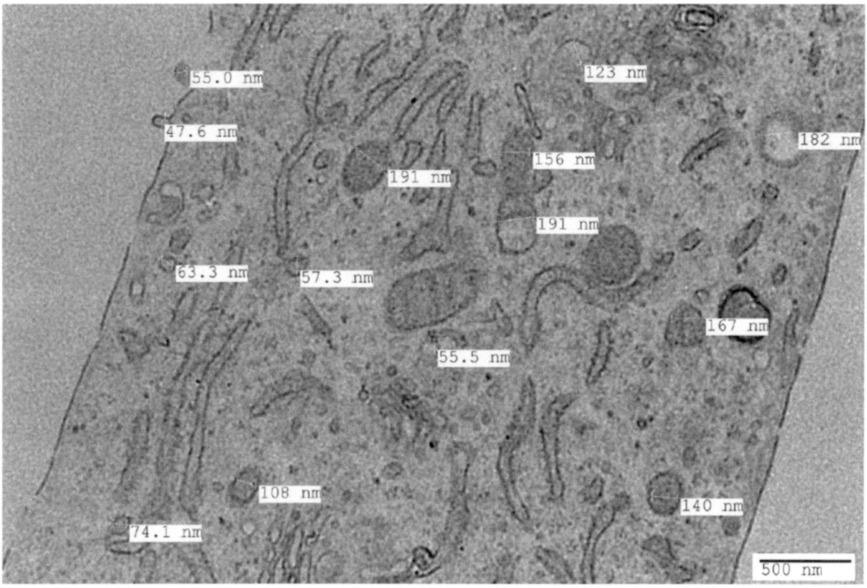

Fig. 19.1 Electron microscopic image of exosomes and nanovesicles derived from MSCs. The image shows many nanovesicles, including exosomes in LPS-induced mouse bone marrow derived MSCs Exosomes are less than 120 μm size and indicated in white arrows, magnification 10,000 times

from the endoplasmic reticulum, Golgi apparatus, mitochondria, or nucleus [17]. It is also thought that exosomes are derived from a conserved evolutionary process to allow for intercellular signaling processes [18]. More recently it has come to light that the cargos originally localized to the rough endoplasmic reticulum, and theoretically bound for the conventional secretory pathway, may possibly be diverted to unconventional pathways to generate exosomes and other extracellular vesicles [19, 20]. A greater understanding about the basic pathways that are being followed and in what situations cells utilize nonconventional pathways may provide insight into how secretory proteins are released in the context of myocardial repair and regeneration. Also, exosomes have been demonstrated to contribute a minor amount of adiponectin secretion from cells and were demonstrated to be released more when stimulated with docosahexaenoic acid [21]. Understanding the specific signaling events that cause exosome release is critical to study the role of exosomes in the body.

Exosomes are shown to be involved in cellular communication and the antigen presentation process, suggesting its role in fundamental immunological processes [22, 23]. Exosome may also serves as a vector for helping to transfer genetic information in the form of messenger RNAs (mRNAs) and microRNAs [24]. Furthermore, exosomes play a role in shuttling proteins from cell to cell [25]. While their function is not well understood, recent research indicates that exosomes play a critical role in modulating events such as inflammatory processes. Due to their cargos and carrying capacity to tissues, exosomes may have vast application in therapy. The therapeutic use of stem cells raises concerns regarding undesirable proliferative cells that are not relevant for the disease. In this context, stem cell-derived exosomes are considered as a major emerging tool for curing many diseases including CVDs. There are several avenues that have been explored related to the role of exosomes in cardiovascular health. This report gives a brief overview of topics that have been studied in relation to exosomes in cardiovascular research, and then to more broadly discuss the future of exosomes in cardiovascular therapy.

19.2 Exosomes and Their Composition

The inner contents of exosomes are varying among different cell types. Exosomes have been found to carry a variety of important biological molecules, including lipids, proteins, mRNAs, non-coding RNAs, and rarely DNA out of cells and into other areas [26, 27]. The important proteins are cytoskeletal proteins, metabolic enzymes, proteins involved in microvesicular body (MVB) formation, synuclein, flotillin and heat shock proteins (HSP). Apart from important lipids such as sphingomyelin, phosphatidylcholine, phosphatidylserine and phosphatidylinositol, they also contain mRNAs, microRNAs (miRNAs) and other noncoding RNAs [15]. The non-coding RNAs are involved in post-transcriptional regulation [28], and some miRNAs are carried within the exosomes [29]. Exosomes have been shown to have roles in both paracrine and endocrine mechanisms [30]. Recent studies have demonstrated that circulating miRNAs may be utilized as biomarkers for early detection of many diseases, including CVDs, cancers, diabetes and kidney diseases [18, 27]. Moreover, these miRNAs are often protected from degradation by being packaged inside the membrane bound microvesicles [26]. Studies have proven that miRNAs are found abundantly in micro vesicles and they are released by different types of cells [31, 32]. miRNAs containing exosomes are described as great biomarkers for diseases. Recent studies have shown that the miRNAs are considered as a key modulator of lung and cardiovascular diseases [33]. miRNAs specific knockdown studies have shown its involvement in various diseases. In many conditions, miRNAs are aberrantly expressed, demonstrating either up regulation or down regulation depending on the physiological state of the cells. Understanding the roles of these molecular modulators and the signaling pathways with their target genes is a great challenge in future research.

19.3 Exosomes Derived from Different Types of Cells

Comparing the utility of different types of stem cells will be useful in future research. It has been shown that the physicochemical and pharmacokinetic, characteristics of exosomes may be different in different cell types [34]. There is little known about the utility and benefits of exosomes from different MSC sources. There has been evidence that the source and culture conditions of the exosomes derived from stem cells can modulate the regenerative potential. Studies have shown that bioactive paracrine factors secreted by the transplanted cells may potentially influence inflammation, cytoprotection, neovascularization, fibrosis, contractility and regeneration of new cell formation, thereby improving cardiac function after injury [35, 36]. These outcomes may vary depending upon the cell types used for the treatment.

19.3.1 Exosomes from Induced Pluripotent Stem Cells (iPSCs)

Like embryonic stem cells (ESCs), iPSCs have the ability to proliferate and differentiate into any type of cell in our body regardless of the parent cells. Studies have shown that exosomes from ESCs promote cardiac repair and enhance heart function after acute myocardial ischemia [37]. Similarly, iPSC-derived exosomes and their content have recently been investigated. It has been shown that iPSC-derived exosomal contents are beneficial for wound repair and reduce fibrosis [38], provide protective effects on the injured myocardium [39] and attenuate limb ischemia by promoting angiogenesis [40]. In addition, promising results from iPSC derived microvesicles suggest that they are useful for differentiating cardiac derived MSC into cardiac and endothelial cells [41] and also in maintaining iPSC pluripotency [42]. These studies suggest that iPSC-derived exosomes are good for treating degenerative diseases and also useful for differentiating purpose.

19.3.2 Exosomes from Adult Cardiomyocytes and iPSC-Derived Cardiomyocytes

Discovery of the heart as a self-renewing organ capable of supporting limited cell replacement under pathological insult has altered the prevalent view of cardiac medicine. Studies have shown that iPSC-derived cardiomyocytes attenuate cardiac injury and regain cardiac function in acute myocardial infarction (MI) models [43–46]. Another study also has shown that the transplantation of cardiac progenitors derived from epigenetically modified bone marrow derived progenitor cells into infarcted mouse hearts significantly improved left ventricular function that was coupled to differentiation of the injected cells into cardiomyocytes and endothelial cells at sites of transplantation [47]. Recently, cell-free therapeutic approaches using stem cell-derived exosomes have been shown to successfully augment cardiac function by multiple means, such as reducing fibrosis, increasing angiogenesis and augmenting endogenous repair processes [8, 37, 48]. Moreover, transplantation of induced cardiomyocytes generated from human placenta amniotic MSCs have been shown to improve cardiac function and repair the injured cardiac tissues after MI. Similar kinds of benefits also have been noticed when human placenta amniotic MSCs induced cardiomyocytes-derived exosomes were used for post MI therapy [49, 50]. A study noted that when human iPSC-derived cardiomyocytes and neonatal rat ventricular myocytes were treated with serum from pediatric patients with dilated cardiomyopathy, there was an induction of pathological change in the normal cardiomyocytes exposed to patient serum but not the control cells. More specifically, exosomes from the serum of dilated cardiomyopathy patients induced pathological changes in the normal cells [51]. Additionally, cardiomyocyte-derived exosomes have been shown to be a part of glucose transport in the form of glucose transporters and glycolytic enzymes, making these exosomes a portion of the cardio-endothelial communication system [52]. Thus, exosomes could be purified from a variety of cardiovascular disease states and examined to see their role in modulating gene expression of healthy cells to understand more about the disease pathogenesis and target for therapy. Overall, the exosomes derived from iPSC-derived cardiomyocytes has better therapeutic potential for cardiac related diseases when compared to the exosomes from other cell types. The beneficial effects of this novel cell-free approach in the cardiac regenerative medicine field remains to be further determined.

In addition to the MSCs discussed, there are also tissue-specific stem cells such as cardiac progenitor cells and other fetal stem cells. With the growth of the stem cell field, there is room now to ask how exosomes related to different kinds of mesenchymal stem cells may be used either therapeutically or as biomarkers. The question of how exosomes from other branches play a role in the future of cardiovascular therapy. Human cardiospheres have been put forward as having multipotent stem and progenitor cell potential [53]. These have been demonstrated to have primitive cells and committed progenitors for cardiomyocytes, endothelial cells, and smooth muscle cells. These cardiosphere-derived cells have been demonstrated to secrete a variety of components, including exosomes, in the case of myocardial infarction [54]. This was accomplished via the modulation of macrophage polarization. While the study focused on the role of YRNA in the modulation of interleukin-10 (IL-10) protein secretion, the study highlights the fact that better understanding the way exosomes from different types of stem cells interact with other organelles, may provide novel ways to manipulate pathogenesis such as that related to myocardial infarction [54]. Transplantation of exosomes are involved in protection against cancer, autoimmune, vascular damage, regenerative therapy for brain, heart, liver, kidney and lungs ischemic injury and infectious diseases (Fig. 19.2).

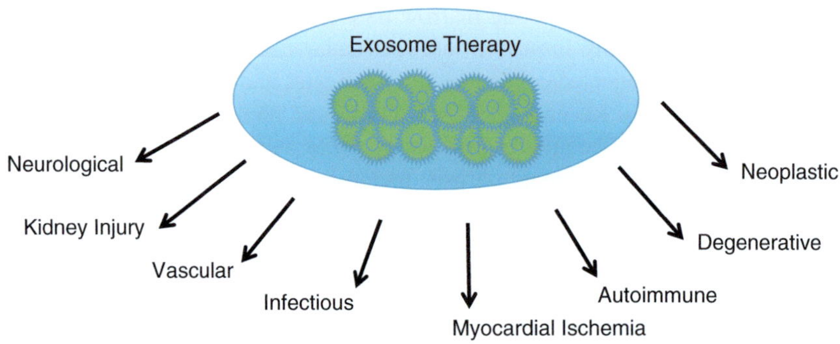

Fig. 19.2 Therapeutic potential of exosomes on different types of diseases. Transplantation of exosomes are involved in protection against cancer, autoimmune, infectious diseases, vascular damages, regenerative therapy for brain, heart, liver, kidney and lung ischemic injuries

19.3.3 Exosomes from MSCs

MSCs are adult multipotent cells that have been identified in various adult tissues, including adipose tissue, in dental pulp, placenta, umbilical cord blood, amniotic fluid, umbilical cord Wharton's jelly and the brain [14, 25, 55–59]. MSCs are most widely being used stem cells in clinical trials for cardiac diseases. It has been shown that intravenous and intracoronary treatment with MSC-derived condition medium (MSC-CM) showed significant reduction in infarct size in porcine myocardial ischemia and reperfusion model [60]. Moreover, the administration of purified exosomes from the MSC-CM revealed cardioprotection in the mouse MI injury model [61]. Therapeutic potential of MSCs has been studied both animals and human patients indicating its therapeutic efficiency in a variety of diseases [62, 63]. Among these MSCs, the use of umbilical cord derived MSCs is promising due to its noninvasive isolation and proliferative and immunomodulatory properties [48, 64]. Studies have provides and analyzed the therapeutic benefit of transplantation of exosomes from endometrium-derived MSCs (EnMSCs), human bone marrow-derived MSCs (BMMSCs) or adipose-derived MSCs (AdMSCs) on the rat myocardium after myocardial infarction [59]. The results indicated a greater protection of myocardial tissue by the exosomes derived from EnMSCs compared to BMMSCs or AdMSCs. These beneficial effects may be due to miR-21 from secreted exosomes from EnMSCs compared to the other types of MSCs [59]. This is one of few studies that compare different origins of MSC-derived exosomes and provide better understanding of future therapeutic efforts. Another study focused specifically on adipose stem-cell (ASC) derived nanovesicles used to study the possible therapeutic options in emphysema [57].

Thus, there is room not only to further compare the different stem cells sources in regenerative attempts compared to exosomes from those stem cell sources, but also to do even more detailed studies about the subtypes of these additional stem cell sources. For example, one study found that MSC-derived exosomes provided superior effectiveness in repairing infarcted heart models when compared to MSCs alone by modulating specific miRNA expression profiles [65]. This type of comparative analysis would be beneficial with other types of stem cell sources as well. Another group demonstrated the importance of the role of exosomes from cardiomyocyte progenitor cells and mesenchymal stem cells in vessel formation and angiogenesis and how this is largely via their modulation of extracellular matrix metalloproteinase inducer [66].

19.4 Exosomes in Drug Delivery System

Drug delivery systems such as liposomes and polymeric nanoparticles have been used to deliver drugs like anti-cancer drugs, anti-fungal drugs and analgesics [67]. However, these systems have their own limitations such as immunogenicity, stability, toxicity, biocompatibility and safety [68, 69]. Exosomes are considered as a useful drug carrier because of overcoming the limitations observed by the liposomes and polymeric delivery systems. Research has been done on the therapeutic advantages of exosomes being used as vehicles for drug delivery. It has been shown that exosomes are employed to deliver small molecules such as curcumin, doxyrubicin for the anti-inflammatory and anti-cancer effects, respectively [70, 71].

Studies have shown that exosomes are being used for nucleic acid delivery such as small interfering RNAs (siRNAs) and miRNAs [72, 73]. There have been varying attempts to use the exosomes as delivery agents of the siRNAs. siRNAs have been promoted as potential cancer therapeutics, however, pose the issue of being degraded rapidly [74]. While research has been conducted on the therapeutic advantages of exosomes in areas such as cancer research [75], little research has been done on using exosomes as drug delivery vehicles for cardiovascular disease. Relating to cardiovascular health, it has been noted that using stents in combination with delivering DNA, siRNA or miRNA, instead of the conventional use of anti-inflammatory and anti-thrombotic drugs may be beneficial to preventing plaque formation at the site of the stent [76]. Thus, there should be greater exploration of the utility of exosomes as a drug delivery system because of size and nature's own biological molecules. There are many hurdles and challenges existing for the use of exosomes as a safe, therapeutic drug delivery system in terms of drug loading and assembly. There are concerns about the exosomes from donor origin eliciting unwanted immune reactions. These immune reactions can be overcome by using autologous exosomes.

19.5 Exosomes and Gene Therapy

Studies have shown that exosomes have been used as a therapeutic delivery system of genetic materials that improve gene expression in certain diseases. Unlike other vectors, exosomes have been shown to have several advantages for gene delivery. These exosomes are a cell-free natural system capable of ferrying mRNA and miRNAs between cells. The membrane-bound inner contents of exosomes are protected from cellular digestion and easily taken up by the target cells. Studies have shown that use of exosomes as a delivery vehicle for exogenous siRNAs into the recipient cells has been successful. The combination of exosomes with siRNAs technology is a reliable and promising strategy to control tumor growth [77–79]. This technology has been used for modulating the target genes for therapy [77, 80]. Recently, it has been shown that exosome-associated adeno-associated virus can serve as an effective vector to deliver transgenes and rescue hearing in mouse models with hereditary deafness [81]. Thus, there is potential to use exosomes as vectors in battling hereditary diseases.

Recently, with the notion of extracellular vesicles as carriers for biological therapeutics, exosome mimetics has been created with the help of natural exosomes and liposomes through the biological engineering technology [82, 83]. It is more likely that exosomes can be targeted to specific tissues, such as cardiac tissues and other tissues. One study found that tetherin along with glycosylphosphatidylinositol-anchor was found to be a critical component of the exosomes that was responsible for appropriate interactions with the target tissues [84]. A greater understanding of the specific molecules on the surface of exosomes that promote interactions in a tissue specific way is critical for the potential use of exosomes as a future drug-delivery vehicle. Cardiac amyloidosis has been studied as a model to pursue gene therapy using siRNAs and antisense oligonucleotides to treat cardiac related pathologies [85]. These studies suggest that exosomes and exosome mimetics are the important biological tool for treating various types of diseases.

19.6 Exosomes as Biomarkers

Exosomes also may play a role in the future as a biomarker for cardiovascular injury. Biomarkers are useful in evaluating normal biological and pathogenic processes, and also to evaluate in response to therapeutic intervention [86]. Studies have indicated that different types of diabetes demonstrate different exosomal biomolecules. This was discussed in context of how blood-based exosomal markers could be used to measure the progression of various types of diabetes [86]. Studies pointed to the fact that miRNAs that often undergo degradation can be considered as potential markers to examine endothelial regeneration or circulate in fibroblast-derived exosomes to target cardiac cells [87]. Related to myocardial infarction, it has been noted that after a myocardial infarction, exosomes are the key communicator for cardiac repair and monitoring these exosomes may serve as diagnostic and prognostic indicators [88]. Furthermore, there has been an association described between ficolin-3 isolated from microparticles obtained from activated platelets and the abdominal aortic aneurysm (AAA). AAA is often deadly because they are often asymptomatic for long periods of time. Increased ficolin-3 levels are associated with the presence of AAA and progression [89]. More recently, the roles of exosomal miRNAs have been reported and considered as diagnostic biomarkers for lung cancer, ovarian cancer, and cardiovascular diseases [56, 90]. The exosomes containing miRNA-21 and miRNA-146a, are up regulated in cervical cancers, and are considered as potential novel biomarkers for cervical cancer diagnosis [91]. miRNA-208a is identified and determined as a potential marker for the early stage of myocardial infarction. Dysfunctional endothelial cells released microvesicles and it is a kind of biomarker for endothelial damage [92]. A recent study has shown that the heart can have damage directly to the myocardium by myocardial infarction, and also face damage due to cardiotoxicity from chemotherapeutic agents [93]. There may be differences in the exosome expression based on the etiology underlying the cardiovascular disease and research. These studies suggest that the exosome cargo varies depending on the health status of the cell of origin. Thus, exosomes can serve as a biomarker for health status. It would be beneficial for future studies to focus on comparing these differences in etiology and targeting the important molecules that are abundantly present in the exosomes.

19.7 Exosomes in Modulating Immunity

Studies have shown the role of exosomes in inducing an antigen specific response restricted to major histocompatibility complex of T-cells [94]. Dendritic cells are key antigen presenting cells involved in immune and inflammatory function. They work to capture antigens and present these to lymphoid organs, utilizing cytokines as the stimulatory molecules to differentiate T cells [95]. Studies have demonstrated that dendritic cell-derived exosomes can stimulate an immune response by activating naïve T lymphocytes both in vivo and in vitro [17]. Many facets of cardiovascular disease such as atherosclerosis and cardiac ischemia are modulated by the immune response [96, 97]. Furthermore, modulation of immunity via exosomes has also been shown to play a role in tumor growth [98]. Understanding the link between exosomes, tumor growth, inflammation and immunity may reveal the exact mechanisms by which exosomes can be protective or pathogenic. One study noted that exosomes derived from human amniotic epithelial cells have positive effects on scarless wound healing [99]. This study demonstrates the critical role that exosomes play in modulating the inflammatory response that leads to scars. Thus, the modulation of the immune response with exosomes provides an avenue for further exploration in finding therapeutic options in CVDs and many other diseases.

19.8 Exosome Involvement in Systemic and Pulmonary Hypertension

Hypertension is a critical risk factor related to cardiovascular disease. Exosomes from macrophages were found to induce inflammatory factors in endothelial cells on hypertensive rat models. Under hypertension induced conditions in rats via continuous Angiotensin II infusion, exosomes increased expression of intracellular adhesion molecule-1 and plasminogen activator inhibitor-1 in human coronary artery endothelial cells. The results demonstrated that inflammatory processes related to hypertension are in some ways modulated by macrophage-derived exosomes. This area of research provides potential for modulation in the control of hypertension [100]. It has been shown that MSC-derived exosomes mediate the cryoprotective beneficial effect on hypoxia-induced pulmonary hypertension. In this study, they identified that miRNA-17 family of miRNAs are the key effector for MSC paracrine function [101].

Pulmonary hypertension is an affliction of high blood pressure within the transport of blood to and from the lung and is associated with various types of cardiac malfunctions [102]. MSC-derived exosomes contributed to the prevention of hypoxic signaling. Monocyte chemoattractant protein-1 (MCP-1) and hypoxia-induced mitogenic factor (HIMF)/found in inflammatory zone-1 (FIZZ1), along with other pro-inflammatory mediators, are upregulated in hypoxia and have been shown to contribute to pulmonary hypertension development in animal models and in humans. MCP-1 contributed to be thickening of arterioles within the pulmonary system, and HIMF/FIZZ1 was a marker for activated macrophages. It was demonstrated that the administration of MSC-derived exosomes modulated the MCP-1 and HIMF/FIZZ1 responses. This was associated with a deterrence of pulmonary hypertension. Furthermore, IL6 which generally activates STAT3 was also reduced [25]. This study demonstrates that exosome research can propel forward therapeutic options related to pulmonary hypertension.

19.9 Exosomes in Cardiovascular Diseases

Early research has been conducted studying the role of exosomes in varying cardiovascular disease processes, some of which are mentioned above. However, there is much left to be discovered. Every facet discussed can be analyzed for etiologies of cardiovascular disease rooted in vascular, infectious, neoplastic, degenerative, iatrogenic, congenital, autoimmune, traumatic, and endocrine or metabolic origins. Studies have shown that the survivors of acute MI eventually develop chronic heart failure, and there are an estimated over five million cases in the United States alone [103]. Stem cell therapy has emerged as a promising therapeutic tool for the treatment of acute or chronic MI. However, an ideal stem cell source remains elusive, and they have drawbacks such as limited engraftments and differentiation potential, ethical issues, immunologic incompatibility, or teratoma formation [104–107]. Several studies suggested MSC-derived exosomes, which promote angiogenesis activity, but therapeutic mechanism of MSC-exosomes on an ischemic heart is unclear. It has been recently reported that the role of human bone marrow MSCs derived extracellular vesicles promoted angiogenesis in a rat myocardial infarction model [108]. This study suggests that extracellular vesicles contain some of the important bioactive factors to induce protection against myocardial infarction and promote angiogenesis.

Growing evidence indicates that the protective and regenerative angiogenic effects of several stem cell therapies have been attributed to the paracrine actions in many CVDs [109]. miRNA containing exosomes have been reported in a variety of cell types, including vascular cells, cardiomyocytes, cardiac fibroblasts and cardiac progenitors [110]. One study analyzed patients with aortic stenosis who also had left ventricular hypertrophy undergoing aortic valve replacement. Upon examination, the study found that pericardial fluid-derived exosomes and their miRNAs protected myocardial tissue in times of ischemia by promoting angiogenesis both in vitro and in vivo. Overall, these data suggest that perivascular fluid derived exosomes orchestrate aortic vascular repair [110]. Vascular endothelial growth factor (VEGF) induces angiogenesis and has been found to be useful for stem cell based treatments. In an in vivo study, an inducible VEGF secreting MSCs that controls the expression of VEGF seeding onto a cardiac patch significantly improved the left ventricular ejection fraction and fractional shortening in a rat MI model upon VEGF activation. Moreover, the controlled usage of VEGF has been found to improve stem cell based therapeutic efficacy in myocardial infarction in animal models [111]. This study demonstrates that the VEGF-secreting stem cell system is an effective therapeutic approach for the myocardial ischemia. In addition, when exosomes derived from Akt-overexpressing MSCs were placed into an acute myocardial infarction model, improved cardiac function was noted with an enhanced endothelial cell proliferation, migration, and tube like structure formation. Furthermore, platelet derived growth factor-D was found to be upregulated which is responsible for AKT mediated angiogenesis [112]. Another study which implicated the interaction of MiR-15b-5p and AKT3 in describing potential pathogenesis of collateral artery formation, found that an injection of Chol-AKT3-siRNA induced an AKT3 deficiency which helped to promote blood recovery in an ischemic heart model [113]. Another risk factor for heart disease is hyperlipidemia and miRNAs involved with exosomes may be implicated in hyperlipidemia and the progression of atherosclerosis [114]. Understanding the role of exosomes more clearly in terms of risk factors is critical. These types of studies indicate exosomes may provide novel targets in therapeutic treatment of risk factors for cardiovascular diseases. These studies documented that exosomes are the potential therapeutic biomolecules that can be used for regulating the pathological conditions of ischemic injury.

Pu et al., have attempted to understand how ASCs and their derivatives can affect ischemia/reperfusion related to flap necrosis [58]. Cardiovascular research can be propelled forward by more fully investigating how the content of exosomes from various cardiovascular tissues in normal states and in different disease states differ from one another. This brings about important questions of whether endocardium, myocardium, and pericardium may interact with different exosomes. Evidence demonstrates that epicardial progenitor cells, and more specifically the pericardial fluid, interacted with exosomes in a way that was potentially protective in the case of infarction [115]. This suggests that specific parts of the heart should be analyzed both individually and collectively when discussing the modulation of phenotypes based on exosomes (Fig. 19.3).

There is a lack of consensus on the most dependable route for exosome isolation from any given source. Thus, another avenue that needs exploration is to establish a

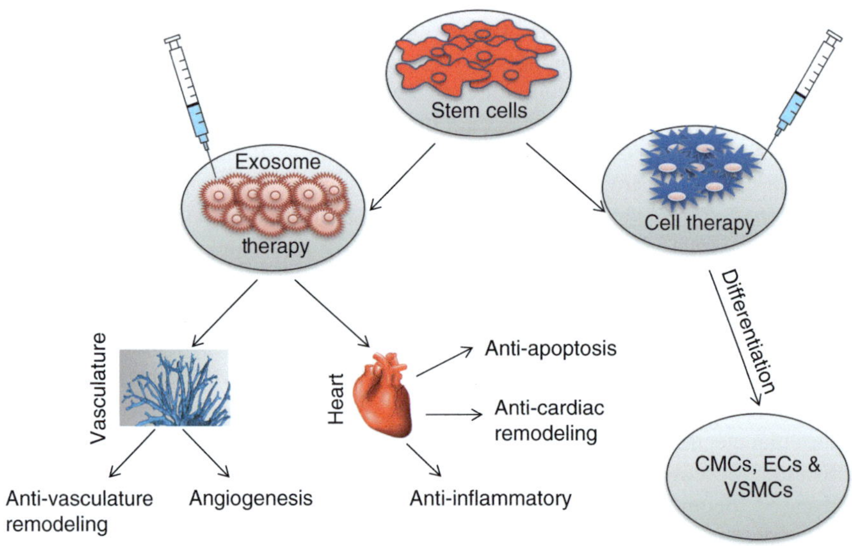

Fig. 19.3 Exosomes for cardiovascular Therapy. Stem cells are multipotent stem cells and are capable of differentiating into cardiomyocytes (CMCs), endothelial cells (ECs) and vascular smooth muscle cells (VSMCs). Stem cells also predominantly secrete paracrine factors in the form of exosomes that are essential for protecting injuries in the heart

reliable universal protocol for exosome purification and analysis [116]. There are studies that suggest that using specific types of isolation techniques can promote or inhibit certain factors. One study demonstrated that secreted extracellular vesicles (EVs) derived from umbilical cord MSCs when isolated by size-exclusion chromatography (SEC) had higher immunosuppressive effect when compared to non-EV fraction or when less purified EV fractions were utilized [117]. In contrast, a recent study has shown that treatment of post MI with neonatal total condition medium (nTCM) showed significantly higher functional benefits when compared with neonatal cardiac progenitor cell (nCPC) derived exosomes [118]. The superior beneficial effect observed in nTCM is because of harboring multiple biological factors such as proteins, mRNAs, extracellular vesicles and exosomes. Further studies are needed to delineate the importance of purified exosomes and total conditioned medium. Thus, understanding the immunosuppressive factors that are abundantly present in exosomes based on the cell types and isolation protocol for the therapeutic exosome usage needs to be uncovered. Another study found that when comparing exosome preparations from plasma and urine of healthy people and oncological disease patients, the blood was found to contain the highest level of non-vesicle contaminants which were similar in size to the exosomes [119]. To overcome these drawbacks from natural exosomes, a recent study has shown that artificial nanovesicles generated from ASC has similar beneficial effects to the natural nanovesicles derived from ASCs [57]. Additional work may also be completed in classifying specific types of exosomes and their contents based on their biological function within the body. Exosomal miRNAs from adipose have recently been classified as a form of adipokine [120]. Extracellular vesicles can play a role in cell-cell communication between macrophage foam cells and vascular smooth muscle cells (VSMCs) in atherosclerotic lesion. Moreover, authors also found that foam cell-derived EVs from macrophage could enter the VSMC and transfer integrins to the surface of these cells [121]. This study suggests that one can consider the fact that artificial nanovesicles (exosome mimics) may have economic and other purification advantages for the future therapeutic purpose.

It is also of vital importance to improve purification techniques if there ever reaches a time where there are further attempts to use exosomes as drug delivery vessels. One novel mechanism to detect exosomes is via digital detection by Single Particle Interferometric Reflectance Imaging Sensor [122]. These kinds of advancements are of high importance if exosomes are going to serve as a novel mechanism to understand and treat cardiovascular disease. There is vast room for growth in finding markers that determine prognosis in a broad range of cardiac dysfunctions. Work should be conducted to further improve the sensitivity and specificity of these types of modalities.

19.10 Exosomes in Other Diseases

A greater understanding of exosome interactions from different organelles may provide a greater overall understanding of the mechanisms and identify new targets for future therapy. This also means that the research needs to be conducted in an interdisciplinary manner, focusing, on exosomes at different disease states. While analyzing exosomes from an interdisciplinary standpoint, a recent finding suggested that urinary exosomal activating transcriptional factor-3 may serve as a diagnostic marker for acute kidney injury in the case of sepsis [123]. Similarly, there is little information on the progression of sepsis-induced Takotsubo cardiomyopathy [124]. This disease could be studied further to see any specific markers exist to identify this disease.

One example is the use of exosomal miRNAs in determining prognosis of multiple myelomas [125]. Another study showed that the exosomes derived from Alzheimer's patients have increased amount of amyloid beta 42 and the exosomes from prion disease patients have the abundant amount of prion proteins being transported [12]. Research related to traumatic cardiac damage such as that seen in cases of motor vehicle collisions or gunshot wounds may also benefit from exploring the utility of exosomes. It has been demonstrated that when traumatic brain injury rat models were given intravenous MSC-derived exosomes, there was an improvement of neuroplasticity and neurological recovery [126]. In a similar way, perhaps traumatic cardiac pathologies could benefit from examining the utility of exosomes to improve patient outcomes. Not only this, but exosomes provide a novel way to understand unknown mechanisms underlying a huge variety of cardiovascular injury. One study depicted how this could be done by using exosomes and HL-60 cells to demonstrate how miRNA-mRNA networks underlie the toxicity of toluene, a substance toxic to many systems of the body, including the cardiovascular system [127]. These different types of materials available in the exosomes serve as a cargo to communicate the status of the disease type as well as to treat the diseased condition. Using exosomes to learn mechanisms of disease should be further explored.

Recent work has suggested that the exosomes from specific tumor cells have certain functional miRNA that may promote their growth [13]. Studies have shown that the miRNAs present in the serum exosome is considered as a promising potential biomarker for cancer [128–130]. Tumor derived exosomes which are found in the plasma could be a good biomarker for cancers [131].

It has been reported that the placenta releases exosomes and that the total number of exosomes can be higher when there are complications during pregnancy. Furthermore, maternal body mass index was reported to influence the variability of exosomes from the placenta during pregnancy [132]. Another study pointed to the fact that placenta-derived exosomes may serve as a marker for gestational diabetes mellitus in early pregnancy [133]. It has been noted that gestational diabetes may cause harm to the child prior to its birth with fetal autonomic nervous system potentially altered by maternal gestational diabetes [134]. It was also found that maternal exosomal miRNA from mice with gestational diabetes were different than those without gestational diabetes, and also found that these exosomes could cross the placenta barrier and infiltrate cardiac tissues [55]. Thus, an area of cardiovascular research should focus on how exosomes may play roles in impacting fetuses' future cardiovascular health. There is room to study how the release of exosomes from the mother may be different during pregnancy, what factors may reach the fetus, how the exosomes may impact the fetus, and the type of exosomes the fetus may release in various conditions including in normal and in times of stress. Furthermore, CVDs can look to the way other fields have utilized exosomes as prognostic indicators.

19.11 Summary

Beneficial effects of exosomes derived from different stem cell types in cardiac regeneration have recently unfolded (Fig. 19.4). Growing evidence suggests that exosomes are intercellular communicators and carry signals to distant places during disease progression and prevention. Exosomes, unlike cells, need not to be viable to be functional. Exosomes possess many membrane adhesion proteins for efficient binding and retention in the target tissues. Furthermore, they have a resistant membrane that helps for long-term storage and high self-life period which will be an advantage for multiple transplantation options for cardiovascular disease patients. The cell-free exosome size and nature's own biological molecules have the capacity to evade the macrophagic degradation and circulate all over the body to perform its function. However, the utility of exosomes in the field of regenerative medicine is currently at infancy due to an inadequate understanding of the nature of exosomes, making it unpredictable for long-term therapeutic safety. Currently, the exosome field faces many problems beginning from lack of optimal isolation procedure to at the end of transplantation stage. The low quantities of exosomes from the isolation method and the clinical drug approval is expensive. One main concern that remains in the field is due to heterogeneous content present. Another concern is minimal human leukocyte antigen present in the membrane may show immunogenicity effects to the recipients. In the future, one may consider personalized exosome mimetics that have the ability to overcome the unwanted immune reaction.

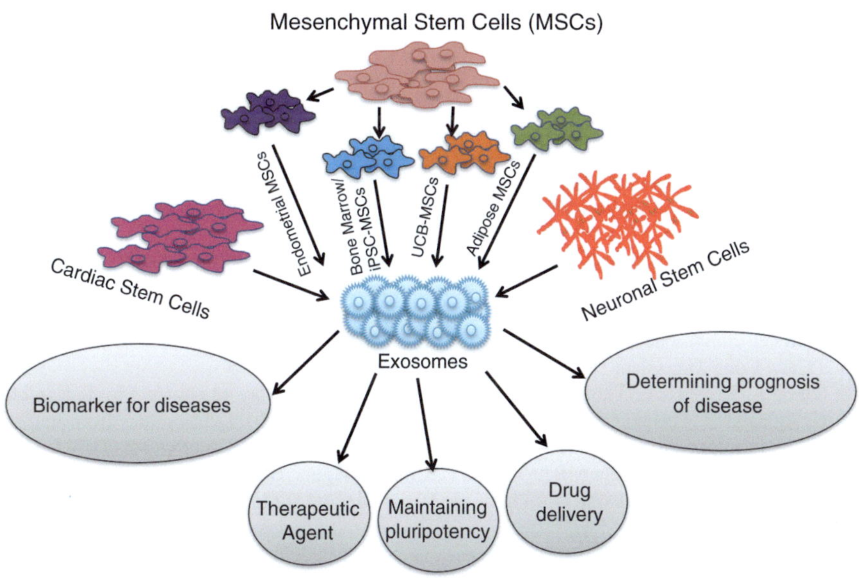

Fig. 19.4 MSC-derived exosomes for cardiovascular therapy. Schematic illustration is showing the perspective of stem cell derived exosomes for various functions

19.12 Conclusion

Although, there are a few drawbacks that exist with stem cell-derived exosomes, they have shown great promises in cell-free application in regenerative medicine. Many avenues are open for exploration towards developing novel methods of manufacturing pure populations of exosomes. Once these problems are resolved, the exosomes will have the potential to change the sphere of cardiovascular research and eventually therapeutic interventions in cardiovascular disease.

Acknowledgements This work was supported, in part, by American Heart Association Grant-in-Aid 16GRNT30950010 and National Institutes of Health COBRE grant P20GM104936 (to JR).

References

1. Heidenreich PA, Trogdon JG, Khavjou OA, Butler J, Dracup K, Ezekowitz MD, Finkelstein EA, Hong Y, Johnston SC, Khera A, Lloyd-Jones DM, Nelson SA, Nichol G, Orenstein D, Wilson PW, Woo YJ (2011) Forecasting the future of cardiovascular disease in the United States: a policy statement from the American Heart Association. Circulation 123(8):933–944
2. Lee N, Thorne T, Losordo DW, Yoon YS (2005) Repair of ischemic heart disease with novel bone marrow-derived multipotent stem cells. Cell Cycle 4(7):861–864
3. Losordo DW, Henry TD, Davidson C, Sup Lee J, Costa MA, Bass T, Mendelsohn F, Fortuin FD, Pepine CJ, Traverse JH, Amrani D, Ewenstein BM, Riedel N, Story K, Barker K, Povsic TJ, Harrington RA, Schatz RA (2011) Intramyocardial, autologous CD34+ cell therapy for refractory angina. Circ Res 109(4):428–436
4. Williams AR, Trachtenberg B, Velazquez DL, McNiece I, Altman P, Rouy D, Mendizabal AM, Pattany PM, Lopera GA, Fishman J, Zambrano JP, Heldman AW, Hare JM (2011) Intramyocardial stem cell injection in patients with ischemic cardiomyopathy: functional recovery and reverse remodeling. Circ Res 108(7):792–796
5. Bolli R, Chugh AR, D'Amario D, Loughran JH, Stoddard MF, Ikram S, Beache GM, Wagner SG, Leri A, Hosoda T, Sanada F, Elmore JB, Goichberg P, Cappetta D, Solankhi NK, Fahsah I, Rokosh DG, Slaughter MS, Kajstura J, Anversa P (2011) Cardiac stem cells in patients with ischaemic cardiomyopathy (SCIPIO): initial results of a randomised phase 1 trial. Lancet 378(9806):1847–1857
6. Uemura R, Xu M, Ahmad N, Ashraf M (2006) Bone marrow stem cells prevent left ventricular remodeling of ischemic heart through paracrine signaling. Circ Res 98(11):1414–1421
7. Angoulvant D, Ivanes F, Ferrera R, Matthews PG, Nataf S, Ovize M (2011) Mesenchymal stem cell conditioned media attenuates in vitro and ex vivo myocardial reperfusion injury. J Heart Lung Transplant 30(1):95–102
8. Sahoo S, Klychko E, Thorne T, Misener S, Schultz KM, Millay M, Ito A, Liu T, Kamide C, Agrawal H, Perlman H, Qin G, Kishore R, Losordo DW (2011) Exosomes from human CD34(+) stem cells mediate their proangiogenic paracrine activity. Circ Res 109(7):724–728
9. Arslan F, Lai RC, Smeets MB, Akeroyd L, Choo A, Aguor EN, Timmers L, van Rijen HV, Doevendans PA, Pasterkamp G, Lim SK, de Kleijn DP (2013) Mesenchymal stem cell-derived exosomes increase ATP levels, decrease oxidative stress and activate PI3K/Akt pathway to enhance myocardial viability and prevent adverse remodeling after myocardial ischemia/reperfusion injury. Stem Cell Res 10(3):301–312
10. Braccioli L, van Velthoven C, Heijnen CJ (2014) Exosomes: a new weapon to treat the central nervous system. Mol Neurobiol 49(1):113–119

11. Dorronsoro A, Robbins PD (2013) Regenerating the injured kidney with human umbilical cord mesenchymal stem cell-derived exosomes. Stem Cell Res Ther 4(2):39
12. Kalani A, Tyagi N (2015) Exosomes in neurological disease, neuroprotection, repair and therapeutics: problems and perspectives. Neural Regen Res 10(10):1565–1567
13. Valadi H, Ekstrom K, Bossios A, Sjostrand M, Lee JJ, Lotvall JO (2007) Exosome-mediated transfer of mRNAs and microRNAs is a novel mechanism of genetic exchange between cells. Nat Cell Biol 9(6):654–659
14. Marote A, Teixeira FG, Mendes-Pinheiro B, Salgado AJ (2016) MSCs-derived exosomes: cell-secreted nanovesicles with regenerative potential. Front Pharmacol 7:231
15. Balasubramanian S, Rajasingh S, Thangavel J, Dawn B, Rajasingh J (2015) Chapter 3—Exosome function in miRNA-mediated paracrine effects. In: Mesenchymal stem cell derived exosomes. Academic, Boston, pp 37–62
16. Yellon DM, Davidson SM (2014) Exosomes: nanoparticles involved in cardioprotection? Circ Res 114(2):325–332
17. Thery C, Duban L, Segura E, Veron P, Lantz O, Amigorena S (2002) Indirect activation of naive CD4+ T cells by dendritic cell-derived exosomes. Nat Immunol 3(12):1156–1162
18. Konstantinell A, Bruun JA, Olsen R, Aspar A, Skalko-Basnet N, Sveinbjornsson B, Moens U (2016) Secretomic analysis of extracellular vesicles originating from polyomavirus-negative and polyomavirus-positive Merkel cell carcinoma cell lines. Proteomics 16(19):2587–2591
19. Glembotski CC (2017) Expanding the paracrine hypothesis of stem cell-mediated repair in the heart: when the unconventional becomes conventional. Circ Res 120(5):772–774
20. Zhang M, Schekman R (2013) Cell biology. Unconventional secretion, unconventional solutions. Science 340(6132):559–561
21. DeClercq V, d'Eon B, McLeod RS (2015) Fatty acids increase adiponectin secretion through both classical and exosome pathways. Biochim Biophys Acta 1851(9):1123–1133
22. Thery C, Zitvogel L, Amigorena S (2002) Exosomes: composition, biogenesis and function. Nat Rev Immunol 2(8):569–579
23. Bang C, Thum T (2012) Exosomes: new players in cell-cell communication. Int J Biochem Cell Biol 44(11):2060–2064
24. Hu G, Drescher KM, Chen XM (2012) Exosomal miRNAs: biological properties and therapeutic potential. Front Genet 3:56
25. Lee C, Mitsialis SA, Aslam M, Vitali SH, Vergadi E, Konstantinou G, Sdrimas K, Fernandez-Gonzalez A, Kourembanas S (2012) Exosomes mediate the cytoprotective action of mesenchymal stromal cells on hypoxia-induced pulmonary hypertension. Circulation 126(22):2601–2611
26. Rashed MH, Bayraktar E, Helal GK, Abd-Ellah MF, Amero P, Chavez-Reyes A, Rodriguez-Aguayo C (2017) Exosomes: from garbage bins to promising therapeutic targets. Int J Mol Sci 18(3)
27. Samanta S, Balasubramanian S, Rajasingh S, Patel U, Dhanasekaran A, Dawn B, Rajasingh J (2016) MicroRNA: a new therapeutic strategy for cardiovascular diseases. Trends Cardiovasc Med 26(5):407–419
28. Min PK, Chan SY (2015) The biology of circulating microRNAs in cardiovascular disease. Eur J Clin Investig 45(8):860–874
29. Kumar S, Vijayan M, Bhatti JS, Reddy PH (2017) MicroRNAs as peripheral biomarkers in aging and age-related diseases. Prog Mol Biol Transl Sci 146:47–94
30. Soria FN, Pampliega O, Bourdenx M, Meissner WG, Bezard E, Dehay B (2017) Exosomes, an unmasked culprit in neurodegenerative diseases. Front Neurosci 11:26
31. Camussi G, Deregibus MC, Bruno S, Grange C, Fonsato V, Tetta C (2011) Exosome/microvesicle-mediated epigenetic reprogramming of cells. Am J Cancer Res 1(1):98–110
32. Yang M, Chen J, Su F, Yu B, Su F, Lin L, Liu Y, Huang JD, Song E (2011) Microvesicles secreted by macrophages shuttle invasion-potentiating microRNAs into breast cancer cells. Mol Cancer 10:117
33. van Rooij E, Purcell AL, Levin AA (2012) Developing microRNA therapeutics. Circ Res 110(3):496–507

34. Charoenviriyakul C, Takahashi Y, Morishita M, Matsumoto A, Nishikawa M, Takakura Y (2017) Cell type-specific and common characteristics of exosomes derived from mouse cell lines: yield, physicochemical properties, and pharmacokinetics. Eur J Pharm Sci 96:316–322
35. Gnecchi M, Zhang Z, Ni A, Dzau VJ (2008) Paracrine mechanisms in adult stem cell signaling and therapy. Circ Res 103(11):1204–1219
36. Duran JM, Makarewich CA, Sharp TE, Starosta T, Zhu F, Hoffman NE, Chiba Y, Madesh M, Berretta RM, Kubo H, Houser SR (2013) Bone-derived stem cells repair the heart after myocardial infarction through transdifferentiation and paracrine signaling mechanisms. Circ Res 113(5):539–552
37. Khan M, Nickoloff E, Abramova T, Johnson J, Verma SK, Krishnamurthy P, Mackie AR, Vaughan E, Garikipati VN, Benedict C, Ramirez V, Lambers E, Ito A, Gao E, Misener S, Luongo T, Elrod J, Qin G, Houser SR, Koch WJ, Kishore R (2015) Embryonic stem cell-derived exosomes promote endogenous repair mechanisms and enhance cardiac function following myocardial infarction. Circ Res 117(1):52–64
38. Gazdhar A, Grad I, Tamo L, Gugger M, Feki A, Geiser T (2014) The secretome of induced pluripotent stem cells reduces lung fibrosis in part by hepatocyte growth factor. Stem Cell Res Ther 5(6):123
39. Wang Y, Zhang L, Li Y, Chen L, Wang X, Guo W, Zhang X, Qin G, He SH, Zimmerman A, Liu Y, Kim IM, Weintraub NL, Tang Y (2015) Exosomes/microvesicles from induced pluripotent stem cells deliver cardioprotective miRNAs and prevent cardiomyocyte apoptosis in the ischemic myocardium. Int J Cardiol 192:61–69
40. Hu GW, Li Q, Niu X, Hu B, Liu J, Zhou SM, Guo SC, Lang HL, Zhang CQ, Wang Y, Deng ZF (2015) Exosomes secreted by human-induced pluripotent stem cell-derived mesenchymal stem cells attenuate limb ischemia by promoting angiogenesis in mice. Stem Cell Res Ther 6:10
41. Bobis-Wozowicz S, Kmiotek K, Sekula M, Kedracka-Krok S, Kamycka E, Adamiak M, Jankowska U, Madetko-Talowska A, Sarna M, Bik-Multanowski M, Kolcz J, Boruczkowski D, Madeja Z, Dawn B, Zuba-Surma EK (2015) Human induced pluripotent stem cell-derived microvesicles transmit RNAs and proteins to recipient mature heart cells modulating cell fate and behavior. Stem Cells 33(9):2748–2761
42. Zhou J, Ghoroghi S, Benito-Martin A, Wu H, Unachukwu UJ, Einbond LS, Guariglia S, Peinado H, Redenti S (2016) Characterization of induced pluripotent stem cell microvesicle genesis, morphology and pluripotent content. Sci Rep 6:19743
43. Rajasingh J, Lambers E, Hamada H, Bord E, Thorne T, Goukassian I, Krishnamurthy P, Rosen KM, Ahluwalia D, Zhu Y, Qin G, Losordo DW, Kishore R (2008) Cell-free embryonic stem cell extract-mediated derivation of multipotent stem cells from NIH3T3 fibroblasts for functional and anatomical ischemic tissue repair. Circ Res 102(11):e107–e117
44. Chong JJ, Yang X, Don CW, Minami E, Liu YW, Weyers JJ, Mahoney WM, Van Biber B, Cook SM, Palpant NJ, Gantz JA, Fugate JA, Muskheli V, Gough GM, Vogel KW, Astley CA, Hotchkiss CE, Baldessari A, Pabon L, Reinecke H, Gill EA, Nelson V, Kiem HP, Laflamme MA, Murry CE (2014) Human embryonic-stem-cell-derived cardiomyocytes regenerate non-human primate hearts. Nature 510(7504):273–277
45. Caspi O, Huber I, Kehat I, Habib M, Arbel G, Gepstein A, Yankelson L, Aronson D, Beyar R, Gepstein L (2007) Transplantation of human embryonic stem cell-derived cardiomyocytes improves myocardial performance in infarcted rat hearts. J Am Coll Cardiol 50(19):1884–1893
46. Rajasingh J, Bord E, Hamada H, Lambers E, Qin G, Losordo DW, Kishore R (2007) STAT3-dependent mouse embryonic stem cell differentiation into cardiomyocytes: analysis of molecular signaling and therapeutic efficacy of cardiomyocyte precommitted mES transplantation in a mouse model of myocardial infarction. Circ Res 101(9):910–918
47. Rajasingh J, Thangavel J, Siddiqui MR, Gomes I, Gao XP, Kishore R, Malik AB (2011) Improvement of cardiac function in mouse myocardial infarction after transplantation of epigenetically-modified bone marrow progenitor cells. PLoS One 6(7):e22550
48. Bobis-Wozowicz S, Kmiotek K, Kania K, Karnas E, Labedz-Maslowska A, Sekula M, Kedracka-Krok S, Kolcz J, Boruczkowski D, Madeja Z, Zuba-Surma EK (2017) Diverse impact of xeno-free conditions on biological and regenerative properties of hUC-MSCs and their extracellular vesicles. J Mol Med (Berlin, Germany) 95(2):205–220

49. Ge X, Wang IN, Toma I, Sebastiano V, Liu J, Butte MJ, Reijo Pera RA, Yang PC (2012) Human amniotic mesenchymal stem cell-derived induced pluripotent stem cells may generate a universal source of cardiac cells. Stem Cells Dev 21(15):2798–2808
50. Kim PJ, Mahmoudi M, Ge X, Matsuura Y, Toma I, Metzler S, Kooreman NG, Ramunas J, Holbrook C, McConnell MV, Blau H, Harnish P, Rulifson E, Yang PC (2015) Direct evaluation of myocardial viability and stem cell engraftment demonstrates salvage of the injured myocardium. Circ Res 116(7):e40–e50
51. Jiang X, Sucharov J, Stauffer BL, Miyamoto SD, Sucharov CC (2017) Exosomes from pediatric dilated cardiomyopathy patients modulate a pathological response in cardiomyocytes. Am J Phys Heart Circ Phys 312(4):H818–H826. doi:10.1152/ajpheart.00673.2016
52. Garcia NA, Moncayo-Arlandi J, Sepulveda P, Diez-Juan A (2016) Cardiomyocyte exosomes regulate glycolytic flux in endothelium by direct transfer of GLUT transporters and glycolytic enzymes. Cardiovasc Res 109(3):397–408
53. Chimenti I, Gaetani R, Barile L, Forte E, Ionta V, Angelini F, Frati G, Messina E, Giacomello A (2012) Isolation and expansion of adult cardiac stem/progenitor cells in the form of cardiospheres from human cardiac biopsies and murine hearts. Methods Mol Biol 879: 327–338
54. Cambier L, de Couto G, Ibrahim A, Echavez AK, Valle J, Liu W, Kreke M, Smith RR, Marban L, Marban E (2017) Y RNA fragment in extracellular vesicles confers cardioprotection via modulation of IL-10 expression and secretion. EMBO Mol Med 9(3):337–352
55. Shi R, Zhao L, Cai W, Wei M, Zhou X, Yang G, Yuan L (2017) Maternal exosomes in diabetes contribute to the cardiac development deficiency. Biochem Biophys Res Commun 483(1):602–608
56. Gallo A, Tandon M, Alevizos I, Illei GG (2012) The majority of microRNAs detectable in serum and saliva is concentrated in exosomes. PLoS One 7(3):e30679
57. Kim YS, Kim JY, Cho R, Shin DM, Lee SW, Oh YM (2017) Adipose stem cell-derived nanovesicles inhibit emphysema primarily via an FGF2-dependent pathway. Exp Mol Med 49(1):e284
58. Pu CM, Liu CW, Liang CJ, Yen YH, Chen SH, Jiang-Shieh YF, Chien CL, Chen YC, Chen YL (2017) Adipose-derived stem cells protect skin flaps against ischemia/reperfusion injury via interleukin-6 expression. J Investig Dermatol. doi:10.1016/j.jid.2016.12.030
59. Wang K, Jiang Z, Webster KA, Chen J, Hu H, Zhou Y, Zhao J, Wang L, Wang Y, Zhong Z, Ni C, Li Q, Xiang C, Zhang L, Wu R, Zhu W, Yu H, Hu X, Wang J (2016) Enhanced Cardioprotection by human endometrium mesenchymal stem cells driven by exosomal microRNA-21. Stem Cells Transl Med. doi:10.5966/sctm.2015-0386
60. Timmers L, Lim SK, Arslan F, Armstrong JS, Hoefer IE, Doevendans PA, Piek JJ, El Oakley RM, Choo A, Lee CN, Pasterkamp G, de Kleijn DP (2007) Reduction of myocardial infarct size by human mesenchymal stem cell conditioned medium. Stem Cell Res 1(2):129–137
61. Lai RC, Arslan F, Lee MM, Sze NS, Choo A, Chen TS, Salto-Tellez M, Timmers L, Lee CN, El Oakley RM, Pasterkamp G, de Kleijn DP, Lim SK (2010) Exosome secreted by MSC reduces myocardial ischemia/reperfusion injury. Stem Cell Res 4(3):214–222
62. Sharma RR, Pollock K, Hubel A, McKenna D (2014) Mesenchymal stem or stromal cells: a review of clinical applications and manufacturing practices. Transfusion 54(5):1418–1437
63. Trounson A, McDonald C (2015) Stem cell therapies in clinical trials: progress and challenges. Cell Stem Cell 17(1):11–22
64. Hass R, Kasper C, Bohm S, Jacobs R (2011) Different populations and sources of human mesenchymal stem cells (MSC): a comparison of adult and neonatal tissue-derived MSC. Cell Commun Signal 9:12
65. Shao L, Zhang Y, Lan B, Wang J, Zhang Z, Zhang L, Xiao P, Meng Q, Geng YJ, Yu XY, Li Y (2017) MiRNA-sequence indicates that mesenchymal stem cells and exosomes have similar mechanism to enhance cardiac repair. Biomed Res Int 2017:4150705
66. Vrijsen KR, Maring JA, Chamuleau SA, Verhage V, Mol EA, Deddens JC, Metz CH, Lodder K, van Eeuwijk EC, van Dommelen SM, Doevendans PA, Smits AM, Goumans MJ, Sluijter JP (2016) Exosomes from cardiomyocyte progenitor cells and mesenchymal stem cells stimulate angiogenesis via EMMPRIN. Adv Healthc Mater 5(19):2555–2565

67. Ha D, Yang N, Nadithe V (2016) Exosomes as therapeutic drug carriers and delivery vehicles across biological membranes: current perspectives and future challenges. Acta Pharm Sin B 6(4):287–296
68. Raemdonck K, Braeckmans K, Demeester J, De Smedt SC (2014) Merging the best of both worlds: hybrid lipid-enveloped matrix nanocomposites in drug delivery. Chem Soc Rev 43(1):444–472
69. Li C, Zhang J, Zu YJ, Nie SF, Cao J, Wang Q, Nie SP, Deng ZY, Xie MY, Wang S (2015) Biocompatible and biodegradable nanoparticles for enhancement of anti-cancer activities of phytochemicals. Chin J Nat Med 13(9):641–652
70. Sun D, Zhuang X, Xiang X, Liu Y, Zhang S, Liu C, Barnes S, Grizzle W, Miller D, Zhang HG (2010) A novel nanoparticle drug delivery system: the anti-inflammatory activity of curcumin is enhanced when encapsulated in exosomes. Mol Ther 18(9):1606–1614
71. Tian Y, Li S, Song J, Ji T, Zhu M, Anderson GJ, Wei J, Nie G (2014) A doxorubicin delivery platform using engineered natural membrane vesicle exosomes for targeted tumor therapy. Biomaterials 35(7):2383–2390
72. Banizs AB, Huang T, Dryden K, Berr SS, Stone JR, Nakamoto RK, Shi W, He J (2014) In vitro evaluation of endothelial exosomes as carriers for small interfering ribonucleic acid delivery. Int J Nanomedicine 9:4223–4230
73. Ohno SI, Takanashi M, Sudo K, Ueda S, Ishikawa A, Matsuyama N, Fujita K, Mizutani T, Ohgi T, Ochiya T, Gotoh N, Kuroda M (2013) Systemically injected exosomes targeted to EGFR deliver antitumor microRNA to breast cancer cells. Mol Ther 21(1):185–191
74. Arnold AE, Czupiel P, Shoichet M (2017) Engineered polymeric nanoparticles to guide the cellular internalization and trafficking of small interfering ribonucleic acids. J Control Release. doi:10.1016/j.jconrel.2017.02.019
75. Hadla M, Palazzolo S, Corona G, Caligiuri I, Canzonieri V, Toffoli G, Rizzolio F (2016) Exosomes increase the therapeutic index of doxorubicin in breast and ovarian cancer mouse models. Nanomedicine (Lond) 11(18):2431–2441
76. Lekshmi KM, Che HL, Cho CS, Park IK (2017) Drug- and gene-eluting stents for preventing coronary restenosis. Chonnam Med J 53(1):14–27
77. Zhang Y, Li L, Yu J, Zhu D, Zhang Y, Li X, Gu H, Zhang CY, Zen K (2014) Microvesicle-mediated delivery of transforming growth factor beta1 siRNA for the suppression of tumor growth in mice. Biomaterials 35(14):4390–4400
78. Kumar L, Verma S, Vaidya B, Gupta V (2015) Exosomes: natural carriers for siRNA delivery. Curr Pharm Des 21(31):4556–4565
79. Lasser C (2015) Exosomes in diagnostic and therapeutic applications: biomarker, vaccine and RNA interference delivery vehicle. Expert Opin Biol Ther 15(1):103–117
80. Alvarez-Erviti L, Seow Y, Yin H, Betts C, Lakhal S, Wood MJ (2011) Delivery of siRNA to the mouse brain by systemic injection of targeted exosomes. Nat Biotechnol 29(4):341–345
81. Gyorgy B, Sage C, Indzhykulian AA, Scheffer DI, Brisson AR, Tan S, Wu X, Volak A, Mu D, Tamvakologos PI, Li Y, Fitzpatrick Z, Ericsson M, Breakefield XO, Corey DP, Maguire CA (2017) Rescue of hearing by gene delivery to inner-ear hair cells using exosome-associated AAV. Mol Ther 25(2):379–391
82. Barile L, Vassalli G (2017) Exosomes: therapy delivery tools and biomarkers of diseases. Pharmacol Ther. doi:10.1016/j.pharmthera.2017.02.020
83. Kooijmans SA, Vader P, van Dommelen SM, van Solinge WW, Schiffelers RM (2012) Exosome mimetics: a novel class of drug delivery systems. Int J Nanomedicine 7:1525–1541
84. Edgar JR, Manna PT, Nishimura S, Banting G, Robinson MS (2016) Tetherin is an exosomal tether. elife 22:5
85. Sperry BW, Tang WH (2017) Amyloid heart disease: genetics translated into disease-modifying therapy. Heart. doi:10.1136/heartjnl-2016-309914
86. Lee MJ, Park DH, Kang JH (2016) Exosomes as the source of biomarkers of metabolic diseases. Ann Pediatr Endocrinol Metab 21(3):119–125
87. Hoefer IE, Steffens S, Ala-Korpela M, Back M, Badimon L, Bochaton-Piallat ML, Boulanger CM, Caligiuri G, Dimmeler S, Egido J, Evans PC, Guzik T, Kwak BR, Landmesser U, Mayr

M, Monaco C, Pasterkamp G, Tunon J, Weber C (2015) Novel methodologies for biomarker discovery in atherosclerosis. Eur Heart J 36(39):2635–2642

88. Yuan MJ, Maghsoudi T, Wang T (2016) Exosomes mediate the intercellular communication after myocardial infarction. Int J Med Sci 13(2):113–116
89. Fernandez-Garcia CE, Burillo E, Lindholt JS, Martinez-Lopez D, Pilely K, Mazzeo C, Michel JB, Egido J, Garred P, Blanco-Colio LM, Martin-Ventura JL (2017) Association of ficolin-3 with abdominal aortic aneurysm presence and progression. J Thromb Haemost 15(3):575–585
90. Kuwabara Y, Ono K, Horie T, Nishi H, Nagao K, Kinoshita M, Watanabe S, Baba O, Kojima Y, Shizuta S, Imai M, Tamura T, Kita T, Kimura T (2011) Increased microRNA-1 and microRNA-133a levels in serum of patients with cardiovascular disease indicate myocardial damage. Circ Cardiovasc Genet 4(4):446–454
91. Liu J, Sun H, Wang X, Yu Q, Li S, Yu X, Gong W (2014) Increased exosomal microRNA-21 and microRNA-146a levels in the cervicovaginal lavage specimens of patients with cervical cancer. Int J Mol Sci 15(1):758–773
92. Horstman LL, Jy W, Jimenez JJ, Ahn YS (2004) Endothelial microparticles as markers of endothelial dysfunction. Front Biosci 9:1118–1135
93. Garrido V, Mendoza-Torres E, Riquelme JA, Diaz A, Pizarro M, Bustamante M, Chavez MN, Ocaranza MP, Mellado R, Corbalan R, Allende ML, Lavandero S (2017) Novel therapies targeting cardioprotection and regeneration. Curr Pharm Des
94. Utsugi-Kobukai S, Fujimaki H, Hotta C, Nakazawa M, Minami M (2003) MHC class I-mediated exogenous antigen presentation by exosomes secreted from immature and mature bone marrow derived dendritic cells. Immunol Lett 89(2–3):125–131
95. Banchereau J, Steinman RM (1998) Dendritic cells and the control of immunity. Nature 392(6673):245–252
96. Kollgaard T, Enevold C, Bendtzen K, Hansen PR, Givskov M, Holmstrup P, Nielsen CH (2017) Cholesterol crystals enhance TLR2- and TLR4-mediated pro-inflammatory cytokine responses of monocytes to the proatherogenic oral bacterium Porphyromonas gingivalis. PLoS One 12(2):e0172773
97. Wang C, Zhang C, Liu L, A X, Chen B, Li Y, Du J (2017) Macrophage-derived mir-155-containing exosomes suppress fibroblast proliferation and promote fibroblast inflammation during cardiac injury. Mol Ther 25(1):192–204
98. Sadallah S, Eken C, Schifferli JA (2011) Ectosomes as modulators of inflammation and immunity. Clin Exp Immunol 163(1):26–32
99. Zhao B, Zhang Y, Han S, Zhang W, Zhou Q, Guan H, Liu J, Shi J, Su L, Hu D (2017) Exosomes derived from human amniotic epithelial cells accelerate wound healing and inhibit scar formation. J Mol Histol 48(2):121–132
100. Osada-Oka M, Shiota M, Izumi Y, Nishiyama M, Tanaka M, Yamaguchi T, Sakurai E, Miura K, Iwao H (2016) Macrophage-derived exosomes induce inflammatory factors in endothelial cells under hypertensive conditions. Hypertens Res 40:353–360
101. Zhu YG, Hao Q, Monsel A, Feng XM, Lee JW (2013) Adult stem cells for acute lung injury: remaining questions and concerns. Respirology 18(5):744–756
102. Walther T, Tschope C, Sterner-Kock A, Westermann D, Heringer-Walther S, Riad A, Nikolic A, Wang Y, Ebermann L, Siems WE, Bader M, Shakibaei M, Schultheiss HP, Dorner A (2007) Accelerated mitochondrial adenosine diphosphate/adenosine triphosphate transport improves hypertension-induced heart disease. Circulation 115(3):333–344
103. Roger VL, Go AS, Lloyd-Jones DM, Benjamin EJ, Berry JD, Borden WB, Bravata DM, Dai S, Ford ES, Fox CS, Fullerton HJ, Gillespie C, Hailpern SM, Heit JA, Howard VJ, Kissela BM, Kittner SJ, Lackland DT, Lichtman JH, Lisabeth LD, Makuc DM, Marcus GM, Marelli A, Matchar DB, Moy CS, Mozaffarian D, Mussolino ME, Nichol G, Paynter NP, Soliman EZ, Sorlie PD, Sotoodehnia N, Turan TN, Virani SS, Wong ND, Woo D, Turner MB (2012) Heart disease and stroke statistics—2012 update: a report from the American Heart Association. Circulation 125(1):e2–e220
104. Blum B, Benvenisty N (2008) The tumorigenicity of human embryonic stem cells. Adv Cancer Res 100:133–158

105. Ben-David U, Benvenisty N (2011) The tumorigenicity of human embryonic and induced pluripotent stem cells. Nat Rev Cancer 11(4):268–277
106. Nussbaum J, Minami E, Laflamme MA, Virag JA, Ware CB, Masino A, Muskheli V, Pabon L, Reinecke H, Murry CE (2007) Transplantation of undifferentiated murine embryonic stem cells in the heart: teratoma formation and immune response. FASEB J 21(7):1345–1357
107. Lee AS, Tang C, Cao F, Xie X, van der Bogt K, Hwang A, Connolly AJ, Robbins RC, Wu JC (2009) Effects of cell number on teratoma formation by human embryonic stem cells. Cell Cycle 8(16):2608–2612
108. Bian S, Zhang L, Duan L, Wang X, Min Y, Yu H (2013) Extracellular vesicles derived from human bone marrow mesenchymal stem cells promote angiogenesis in a rat myocardial infarction model. J Mol Med (Berlin, Germany). doi:10.1007/s00109-013-1110-5
109. Burchfield JS, Dimmeler S (2008) Role of paracrine factors in stem and progenitor cell mediated cardiac repair and tissue fibrosis. Fibrogenesis Tissue Repair 1(1):4
110. Beltrami C, Besnier M, Shantikumar S, Shearn AI, Rajakaruna C, Laftah A, Sessa F, Spinetti G, Petretto E, Angelini GD, Emanueli C (2017) Human pericardial fluid contains exosomes enriched with cardiovascular-expressed microRNAs and promotes therapeutic angiogenesis. Mol Ther 25(3):679–693
111. Cho HM, Kim PH, Chang HK, Shen YM, Bonsra K, Kang BJ, Yum SY, Kim JH, Lee SY, Choi MC, Kim HH, Jang G, Cho JY (2017) Targeted genome engineering to control VEGF expression in human umbilical cord blood-derived mesenchymal stem cells: potential implications for the treatment of myocardial infarction. Stem Cells Transl Med 6(3):1040–1051
112. Ma J, Zhao Y, Sun L, Sun X, Zhao X, Sun X, Qian H, Xu W, Zhu W (2017) Exosomes derived from Akt-modified human umbilical cord mesenchymal stem cells improve cardiac regeneration and promote angiogenesis via activating platelet-derived growth factor D. Stem Cells Transl Med 6(1):51–59
113. Zhu LP, Zhou JP, Zhang JX, Wang JY, Wang ZY, Pan M, Li LF, Li CC, Wang KK, Bai YP, Zhang GG (2017) MiR-15b-5p regulates collateral artery formation by targeting AKT3 (protein kinase B-3). Arterioscler Thromb Vasc Biol 116:308905
114. Novak J, Bienertova-Vasku J, Kara T, Novak M (2014) MicroRNAs involved in the lipid metabolism and their possible implications for atherosclerosis development and treatment. Mediat Inflamm 2014:275867
115. Foglio E, Puddighinu G, Fasanaro P, D'Arcangelo D, Perrone GA, Mocini D, Campanella C, Coppola L, Logozzi M, Azzarito T, Marzoli F, Fais S, Pieroni L, Marzano V, Germani A, Capogrossi MC, Russo MA, Limana F (2015) Exosomal clusterin, identified in the pericardial fluid, improves myocardial performance following MI through epicardial activation, enhanced arteriogenesis and reduced apoptosis. Int J Cardiol 197:333–347
116. Malik ZA, Liu TT, Knowlton AA (2016) Cardiac myocyte exosome isolation. Methods Mol Biol 1448:237–248
117. Monguio-Tortajada M, Roura S, Galvez-Monton C, Pujal JM, Aran G, Sanjurjo L, Franquesa M, Sarrias MR, Bayes-Genis A, Borras FE (2017) Nanosized UCMSC-derived extracellular vesicles but not conditioned medium exclusively inhibit the inflammatory response of stimulated T cells: implications for nanomedicine. Theranostics 7(2):270–284
118. Sharma S, Mishra R, Bigham GE, Wehman B, Khan MM, Xu H, Saha P, Goo YA, Datla SR, Chen L, Tulapurkar ME, Taylor BS, Yang P, Karathanasis S, Goodlett DR, Kaushal S (2017) A deep proteome analysis identifies the complete secretome as the functional unit of human cardiac progenitor cells. Circ Res 120(5):816–834
119. Grigor'eva AE, Dyrkheeva NS, Bryzgunova OE, Tamkovich SN, Chelobanov BP, Ryabchikova EI (2017) [Contamination of exosome preparations, isolated from biological fluids]. Biomed Khim 63(1):91–96
120. Thomou T, Mori MA, Dreyfuss JM, Konishi M, Sakaguchi M, Wolfrum C, Rao TN, Winnay JN, Garcia-Martin R, Grinspoon SK, Gorden P, Kahn CR (2017) Adipose-derived circulating miRNAs regulate gene expression in other tissues. Nature 542(7642):450–455
121. Niu C, Wang X, Zhao M, Cai T, Liu P, Li J, Willard B, Zu L, Zhou E, Li Y, Pan B, Yang F, Zheng L (2016) Macrophage foam cell-derived extracellular vesicles promote vascular smooth muscle cell migration and adhesion. J Am Heart Assoc 5(10)

122. Daaboul GG, Gagni P, Benussi L, Bettotti P, Ciani M, Cretich M, Freedman DS, Ghidoni R, Ozkumur AY, Piotto C, Prosperi D, Santini B, Unlu MS, Chiari M (2016) Digital detection of exosomes by interferometric imaging. Sci Rep 6:37246
123. Panich T, Chancharoenthana W, Somparn P, Issara-Amphorn J, Hirankarn N, Leelahavanichkul A (2017) Urinary exosomal activating transcriptional factor 3 as the early diagnostic biomarker for sepsis-induced acute kidney injury. BMC Nephrol 18(1):10
124. Hung MJ, Kao YC, Chen WS, Mao CT, Chen TH, Yang NI, Ko T, Liang CY (2016) Layer-specific quantification of myocardial deformation in sepsis-induced Takotsubo cardiomyopathy: three case reports of a serial 2-dimensional speckle-tracking echocardiographic study. Medicine (Baltimore) 95(44):e5250
125. Manier S, Liu CJ, Avet-Loiseau H, Park J, Shi J, Campigotto F, Salem KZ, Huynh D, Glavey SV, Rivotto B, Sacco A, Roccaro AM, Bouyssou J, Minvielle S, Moreau P, Facon T, Leleu X, Weller E, Trippa L, Ghobrial IM (2017) Prognostic role of circulating exosomal miRNAs in multiple myeloma. Blood. doi:10.1182/blood-2016-09-742296
126. Xiong Y, Mahmood A, Chopp M (2017) Emerging potential of exosomes for treatment of traumatic brain injury. Neural Regen Res 12(1):19–22
127. Lim JH, Song MK, Cho Y, Kim W, Han SO, Ryu JC (2017) Comparative analysis of microRNA and mRNA expression profiles in cells and exosomes under toluene exposure. Toxicology in Vitro 41:92–101
128. Lv LL, Cao YH, Ni HF, Xu M, Liu D, Liu H, Chen PS, Liu BC (2013) MicroRNA-29c in urinary exosome/microvesicle as a biomarker of renal fibrosis. Am J Physiol Renal Physiol 305(8):F1220–F1227
129. Kosaka N, Iguchi H, Ochiya T (2010) Circulating microRNA in body fluid: a new potential biomarker for cancer diagnosis and prognosis. Cancer Sci 101(10):2087–2092
130. Yang C, Chalasani G, Ng YH, Robbins PD (2012) Exosomes released from mycoplasma infected tumor cells activate inhibitory B cells. PLoS One 7(4):e36138
131. Taylor DD, Gercel-Taylor C (2008) MicroRNA signatures of tumor-derived exosomes as diagnostic biomarkers of ovarian cancer. Gynecol Oncol 110(1):13–21
132. Elfeky O, Longo S, Lai A, Rice GE, Salomon C (2017) Influence of maternal BMI on the exosomal profile during gestation and their role on maternal systemic inflammation. Placenta 50:60–69
133. Powe CE (2017) Early pregnancy biochemical predictors of gestational diabetes mellitus. Curr Diab Rep 17(2):12
134. Fehlert E, Willmann K, Fritsche L, Linder K, Mat-Husin H, Schleger F, Weiss M, Kiefer-Schmidt I, Brucker S, Haring HU, Preissl H, Fritsche A (2016) Gestational diabetes alters the fetal heart rate variability during an oral glucose tolerance test: a fetal magnetocardiography study. BJOG Int J Obstet Gynaecol. doi:10.1111/1471-0528.14474

MIX
Papier aus verantwortungsvollen Quellen
Paper from responsible sources
FSC® C105338

If you have any concerns about our products,
you can contact us on
ProductSafety@springernature.com

In case Publisher is established outside the EU,
the EU authorized representative is:
Springer Nature Customer Service Center GmbH
Europaplatz 3, 69115 Heidelberg, Germany

Printed by Libri Plureos GmbH
in Hamburg, Germany